Energiepolitik und Klimaschutz
Energy Policy and Climate Protection

Herausgegeben von
L. Mez, Berlin, Deutschland
A. Brunnengräber, Berlin, Deutschland

Weltweite Verteilungskämpfe um knappe Energieressourcen und der Klimawandel mit seinen Auswirkungen führen zu globalen, nationalen, regionalen und auch lokalen Herausforderungen, die Gegenstand dieser Publikationsreihe sind. Die Beiträge der Reihe sollen Chancen und Hemmnisse einer präventiv orientierten Energie- und Klimapolitik vor dem Hintergrund komplexer energiepolitischer und wirtschaftlicher Interessenlagen und Machtverhältnisse ausloten. Themenschwerpunkte sind die Analyse der europäischen und internationalen Liberalisierung der Energiesektoren und -branchen, die internationale Politik zum Schutz des Klimas, Anpassungsmaßnahmen an den Klimawandel in den Entwicklungs-, Schwellen und Industrieländern, die Produktion von biogenen Treibstoffen zur Substitution fossiler Energieträger oder die Probleme der Atomenergie und deren nuklearen Hinterlassenschaften.

Die Reihe bietet empirisch angeleiteten, quantitativen und international vergleichenden Arbeiten, Untersuchungen von grenzüberschreitenden Transformations- und Mehrebenenprozessen oder von nationalen „best practice"-Beispielen ebenso ein Forum wie theoriegeleiteten, qualitativen Untersuchungen, die sich mit den grundlegenden Fragen des gesellschaftlichen Wandels in der Energiepolitik und beim Klimaschutz beschäftigen.

Herausgegeben von
PD Dr. Lutz Mez
Freie Universität Berlin

PD Dr. Achim Brunnengräber
Freie Universität Berlin

Achim Brunnengräber
Maria Rosaria Di Nucci
Ana María Isidoro Losada
Lutz Mez · Miranda A. Schreurs (Eds.)

Nuclear Waste Governance

An International Comparison

Springer VS

Editors
Achim Brunnengräber
Maria Rosaria Di Nucci
Ana María Isidoro Losada
Lutz Mez
Miranda A. Schreurs

Freie Universität Berlin, Germany

This volume is a contribution of the Environmental Policy Research Centre (Forschungszentrum für Umweltpolitik – FFU) of the Freie Universität Berlin to the project "Multi-level Governance-Perspective on Management of Nuclear Waste Disposal. A Comparative Analysis: Actors, Instruments and Institutions" that is part of the interdisciplinary ENTRIA research platform funded by the German Federal Ministry of Education and Research (BMBF/02S9082B).

Energiepolitik und Klimaschutz
Energy Policy and Climate Protection
ISBN 978-3-658-08961-0 ISBN 978-3-658-08962-7 (eBook)
DOI 10.1007/978-3-658-08962-7

Library of Congress Control Number: 2015933513

Springer VS

Printed on acid-free paper

Springer VS is a brand of Springer Fachmedien Wiesbaden
Springer Fachmedien Wiesbaden is part of Springer Science+Business Media
(www.springer.com)

Contents

IV. Countries With Direct Disposal of Spent Nuclear Fuel (Including Cases with Additional Disposal of Radioactive Waste from Reprocessing)

V. Countries with Long-Term Surface Storage for High Level Radioactive Waste

Preface

This book was conceptualised as part of the research project, "Multi Level Governance-Perspective on Management of Nuclear Waste Disposal. A Comparative Analysis: Actors, Instruments and Institutions", located at the Environmental Policy Research Centre (Forschungszentrum für Umweltpolitik (FFU)), Freie Universität Berlin. This research project is part of the interdisciplinary research initiative, "ENTRIA: Disposal Options for Radioactive Residues: Interdisciplinary Analyses and Development of Evaluation Principles." The ENTRIA project brings together twelve German universities and one Swiss partner to examine nuclear waste management in relation to technical options, decision making processes, risk and safety as well as social challenges. Nuclear waste management represents a particularly pressing issue for politics, science and society. It is an issue that has been fraught with many challenges. In many countries, there have been thorny debates about feasible solutions and suitable sites for a central nuclear repository that have been ongoing for decades. Pressures to address the issue are growing. European Directive 2011/70/Euratom requires all member states to submit a report on the implementation of their national programmes for the safe management of spent fuel and radioactive waste by August 23, 2015.

This book examines the national plans that countries are developing to address high-level radioactive waste storage and disposal. Country reports on nuclear waste governance are introduced for ten Euratom member states: Belgium, the Czech Republic, Finland, France, Germany, Italy, the Netherlands, Spain, Sweden, the United Kingdom, as well as for Switzerland and the United States. A detailed comparative and multi-level social and political analysis sheds light on the efforts being made and the difficulties associated with trying to find socially, politically, economically, and technologically acceptable strategies for high-level radioactive nuclear waste storage and/or disposal. The progress these countries have made and the obstacles they face are discussed in detail in the case studies. This is done looking at their regulations, technology choices, safety criteria, monitoring systems, compensation schemes, institutional structures, and approaches to public involvement. The book shows that while some countries have opted for medium-term storage solutions (of about a century), others are looking at various approaches to deep geological disposal. Site selection is a common challenge, but only a small number of states, e.g. Finland and Sweden are in an advanced stage of implementation.

The chapters included in this book identify the primary stakeholders in the debate and their interests, the responsibilities and authority of different actors in relevant decision-making processes, and the value systems that are influencing the different national policy choices. The views and expectations of different communities regarding participatory decision making and compensation and the steps that have been or are being taken to promote dialogue and constructive problem-solving are also considered.

This book was developed in cooperation with a long-standing energy research initiative known as the REFORM Group, short for Restructuring Energy Systems For Optimal Resource Management. The REFORM Group is an international network of energy experts coming from research organisations, universities, companies and decision makers. Established in 1991 with the aim of fostering interaction, sharing knowledge among researchers, policy makers, administrators and industry; and understanding the transformation of energy systems towards low-carbon energy systems, the core group has grown to over 50 members from more than 20 countries. Nuclear power has been critically researched by the group since the first annual conference was held at Schloss Leopoldskron in Salzburg, Austria in 1995. Nuclear waste was first taken up as a topic in 2012. Experts have come together annually since then to examine nuclear waste governance in comparative perspective. This book grew out of these papers and discussions and has benefited from outside comments and critique.

The German Federal Ministry for the Environment, Nature Protection, Building and Nuclear Safety provided funding to support the meetings at Schloss Leopoldskron in Salzburg, Austria. The German Federal Ministry for Education and Research is providing the core funding for the ENTRIA project, including the work being conducted by the FFU. The editors are grateful for this support. We also wish to thank all participants at the workshops in Salzburg for their stimulating and insightful inputs. Special thanks go to ENTRIA spokesman, Professor Klaus-Jürgen Röhlig and deputy spokesman, Peter Hocke whose inputs were critical to the success of the workshops as well as to this book project. The texts were skilfully proofread by Jessica Wallach and David Buchanan. Special acknowledgement also goes to Vlasta Wallat who assisted in organising the REFORM Group. Finally, we would like to thank Britta Göhrisch-Radmacher at Springer VS for her constructive and skilful copy-editing and for her patience throughout the publication process. Any mistakes are the responsibility of the authors and editors.

Berlin, November 2014 Lutz Mez for the editorial team.

Abbreviations

AGR	Advanced Gas Reactor
AEEG	Authority for Electric Energy and Gas, Italy
AC	Autonomous Communities, regional governments, Spain
ACP	African, Caribbean and Pacific Group of States
AGR	Advanced Gas-cooled Reactor
AKA	Använt Kärnbränsle och radioativt avfall (Commission on Radioactive Waste), Sweden
AkEnd	Arbeitskreis Auswahlverfahren Endlagerstandorte (Committee on a Site Selection Procedure for Repository Sites), Germany
AKW	Atomkraftwerk (Nuclear Power Plant), Germany
ALARA	As Low as Reasonably Achievable
AMAC	Associación de Municipios en Áreas de Centrales Nucleares (Association of Municipalities in Areas of NPPs), Spain
ANI	American Nuclear Insurers, USA
APA	Administrative Procedures Act, USA
ARE	Swiss Federal Office for Spatial Development
AREVA	French multinational nuclear group
Arius	Association for Regional and International Underground Storage
ASN	Agenzia per la Sicurezza Nucleare (Nuclear Safety Agency), Italy (suppressed)
ASN	Autorité de Sûreté Nucléaire (National Agency for Nuclear Safety), France
ASND	Autorité de Sûreté Nucléaire de Défense (Defense National Agency for Nuclear Safety), France
ATC	Almacén Temporal Centralizado (Centralised Temporary Storage Facility), Spain
AtG	Gesetz über die friedliche Verwendung der Kernenergie und den Schutz gegen ihre Gefahren (Atomic Energy Act), Germany

ATI	Almacén Temporal Individualizado (Temporary Storage Facility), Spain
bn	Billion
BfKE	Bundesamt für kerntechnische Entsorgung (Federal Office for Nuclear Waste Management), Germany
BfS	Bundesamt für Strahlenschutz (Federal Office for Radiation Protection), Germany
BGR	Bundesanstalt für Geowissenschaft und Rohstoffe (Federal Institute for Geosciences and Natural Resources), Germany
BIS	Department for Business Innovation and Skills, United Kingdom
BMBF	Bundesministerium für Bildung und Forschung (Federal Ministry of Education and Research), Germany
BMUB	Bundesministerium für Umwelt, Naturschutz, Bau und Reaktorsicherheit (Federal Ministry for the Environment, Nature Conservation, Building and Nuclear Safety), Germany, (in former times BMU)
BMWi	Bundesministeriums für Wirtschaft und Energie (Federal Ministry for Economic Affairs and Energy), Germany
BNFL	British Nuclear Fuel Limited, United Kingdom
BP	Belgoprocess, manages and executes the industrial activities of NIRAS/ONDRAF, Belgium
BOE	Boletin Oficial del Estado (Official Gazette), Spain
BRC	Blue Ribbon Commission on America's Nuclear Future, USA
BSS	Basic Safety Standard
BUND	German organization linked to "Friends of the Earth"
BWR	Boiling Water Reactor
CANDU	CANada Deuterium Uranium, Canadian, pressurized heavy water reactor
CDU	Christlich Demokratische Union Deutschlands (Christian Democratic Union), Germany
CEEI	Central Europe Engineering & Investment, Czech Republic
CGI	Czech Geological Institute
CHN	High Level Committee, France

CIEMAT	Centro de Investigaciones Energéticas, Medioambientales y Tecnológicas, public research body, Spain
Cigéo	Centre Industriel de Stockage Géologique, geological repository, France
Clab	Centralt mellanlager för använt kärnbränsle (Centralized Intermediate Storage Site for Spent Nuclear Fuel), Sweden
CLIS	Commission Locale d'Information et de Suivi (Local Information and Oversight Committee), France
CNDP	Commission Nationale du Débat Public (National Commission for Public Debate), France
CNE	Commission Nationale d'Évaluation (National Assessment Board), France
CNEN	Comitato Nazionale per l'Energia Nucleare (former National Committee on Nuclear Energy), Italy
CNSE	Comisión Nacional del Sistema Eléctrico (National Commission of the Electric System), Spain
CoRWM	Committee on Radioactive Waste Management, United Kingdom
COVRA	Centrale Organisatie Voor Radioactief Afval, publicly owned company, for transportation and storage of NW and for research on longer term repository options, The Netherlands
COWAM	Communities Waste Management
CPDP	Commission Particulière du Débat Public, public commission set up by the French government, France
CSC	Coal and Steel Community
CSN	Consejo de Seguridad Nuclear (Nuclear Safety Council), Spain
CSU	Christlich Soziale Union (Christian Social Union), Germany
CTSF	Centralised Temporary Storage Facility, Spain
CZK	Czech Crown
DADA	Decide-Announce-Defend-Abandon
DBE	Deutsche Gesellschaft zum Bau und Betrieb von Endlagern für Abfallstoffe (German Service Company for the Construction and Operation of Waste Repositories

DDPS	Eidgenössisches Departement für Verteidigung, Bevölkerungsschutz und Sport (Federal Department of Defence, Civil Protection and Sport), Switzerland
DECC	Department of Energy and Climate Change, United Kingdom
Defra	Department of Environment Food and Rural Affairs, United Kingdom
DETEC	Eidgenössische Departement für Umwelt, Verkehr, Energie und Kommunikation (Federal Department of the Environment, Transport, Energy and Communications), Switzerland
DGD	Deep Geological Disposal
DGEC	Direction Générale de l'Énergie et du Climat (Directorate General for Energy and Climate), France
DISP	Direzione Sicurezza Nucleare e Protezione Sanitaria (Nuclear Safety and Health Protection Direction of ENEA), Italy
DOE	Department of Energy, USA
DSND	Délégué à la Sûreté Nucléaire et à la Radioprotection pour les Activités et Installations Intéressant de la Défense, France
EA	Environment Agency, United Kingdom
EC	European Commission
EdF	Electricité de France
EEC	European Economic Community
EGT	Expert Group on Nuclear Waste Disposal, Switzerland
EKRA	Expertengruppe Entsorgungsgruppe für radioaktive Abfälle (Expert Group on Nuclear Waste Disposal), Switzerland
EIA	United States Energy Information Agency
EIA	Environmental Impact Assessment
ENEA	National Agency for new Technologies, Energy and Sustainable Economic Development, Italy
ENEL	Ente Nazionale Energia Elettrica, Italian electric utility company, Italy
ENRESA	Empresa Nacional de Residuos Radioactivos, S.A. (National Company for Radioactive Waste Management), Spain
ENSI	Swiss Federal Nuclear Safety Inspectorate
ENSREG	European Nuclear Safety Regulators' Group

EPA	Environmental Protection Agency, USA
EPR	European Pressurized Reactor
EPWR	European Pressurized Water Reactor
ERAM	Endlager für radioactive Abfälle Morsleben (Morsleben Repository for Radioactive Waste), Germany
ERDO	European Repository Development Organisation
ESK	Entsorgungskommission (Nuclear Waste Management Commission), Germany
EU	European Union
EURIDICE	European Underground Research Infrastructure for Disposal of Nuclear Waste in Clay Environment, Belgium
FANC/AFCN	Federaal agentschap voor nucleaire controle (Federal Agency for Nuclear Control), Belgium
FDP	Freie Demokratische Partei (Free Democratic Union), Germany
FED	Forum Endlager-Dialog (Disposal Dialogue Forum), Germany
FOA	Swedish Defence Research Establishment
FOIA	Freedom of Information Act, USA
FFU	Forschungszentrum für Umweltpolitik (Environmental Policy Research Centre), Freie Universität Berlin
FLT	Fund for the Long-Term Management of all Radioactive Wastes, Belgium
FMT	Fund for the Middle-Term Management of all Radioactive Wastes, Belgium
FOEN	Swiss Federal Office for Environment
FOPH	Swiss Federal Office for Public Health
GDR	German Democratic Republic
GG	Grundgesetz (Basic Law), Germany
GHG	Greenhouse Gas
GIP	Groupement d'Intérêt Public (Public Interest Group), France
GMF	Group of European Municipalities with Nuclear Facilities
GRWP	General Radioactive Waste Plan, Spain
GWe	Gigawatt of Electricity

HABOG	Hoogradioactief Afval Behandelingsen Opslag Gebouw, surface repository at the NPP Borssele in the Netherlands
HADES	High Activity Disposal Experimental Site, Belgian Underground Research Laboratory
HCTISN	Haut Comité pour la Transparence et l' Information sur la Sécurité Nucléaire (High Commission for Transparency and Information on Nuclear Security), France
HLW	High-level Radioactive Waste
HM	Heavy Metals
IAEA	International Atomic Energy Agency
IEA	International Energy Agency
ICPR	International Commission on Radiological Protection
ILW	Intermediate Level Radioactive Waste
ILW-LL	Intermediate-level Long-lived Radioactive Waste
IPCC	International Panel on Climate Change
IPFM	International Panel on Fissile Materials
I-Power	Innovative, Passive, Optimised, Worldwide Economical Reactor
IRSN	Institut de Radioprotection et de Sûreté Nucléaire (Institute for Radiological Protection and Nuclear Safety), France
ISPRA	Istituto Superiore per la. Protezione e la Ricerca Ambientale (Institute for Environmental Protection and Research), Italy
ITAS	Insitute of Technology Assessment and Systems Analysis, Germany
IVO	Imatran Voima Oy, state-owned company, Finland (now Fortum Oyj)
KBS	A repository concept for high-level reprocessing waste or spent nuclear fuel, developed in Sweden.
KBS	King Baudouin Foundation, Belgium
KIT	Karlsruhe Institute of Technology, Germany
KNE	Kommission für Nukleare Entsorgung (Commission for Nuclear Waste Disposal), Switzerland
LILW	Low- and Intermediate-level, Short Lived Waste, Belgium
LLW	Low Level Nuclear Waste

LLW	Low or Intermediate-level Waste, Italy
m	Million
MEE	Ministry of Employment and the Economy, Finland
MINETUR	Ministerio de Industria, Energía y Turismo (Ministry of Industry, Energy and Tourism), Spain
MNA	Multinational Nuclear Approach Expert Group, IAEA
MoE	Ministry of Environment , Czech Republic
MoIT	Ministry of Industry and Trade, Czech Republic
MONA	Mol's Negotiation Platform on Nuclear Waste – Category A, Belgium
MOX	Mixed Oxide Fuel
MS	Member States
mSv	Millisievert
MW	Megawatt
MWh	Megawatt Hours
Nagra	Nationale Genossenschaft für die Lagerung radioaktiver Abfälle (National Cooperative for the Disposal of Radioactive Waste), Switzerland
NCW	Non-classified Waste
NDA	Nuclear Decommissioning Authority, United Kingdom
NEA	Nuclear Energy Agency
NEI	Nuclear Energy Institute, USA
NEPA	National Environmental Policy Act, USA
NEZ	Nukleares Entsorgungszentrum (Nuclear Waste Management Center), Germany
NGO	Non-governmental organizations
NIMBY	Not In My Backyard
NIRAS/ONDRAF (N/O)	Nationale Instelling voor Radioactief en verijkte Splijtstoffen, (National Agency for Management of Radioactive Waste and Enriched Fissile Material), Belgium
NLF	Nuclear Liabilities Fund, United Kingdom
NPP	Nuclear Power Plant

NPT	Non-Proliferation Treaty
NRC	Nuclear Regulatory Commission, USA
NRDC	Natural Resources Defense Council, USA
NSC	Swiss Federal Nuclear Safety Commission
Nuclenor	Centrales Nucleares del Norte, NPP operator, Spain
NWF	Nuclear Waste Fund, USA
NWPA	Nuclear Waste Policy Act, USA
NWPAA	Nuclear Waste Policy Act Amendment, USA
NWTRB	Nuclear Waste Technical Review Board, USA
OCRWM	Office of Civilian Radioactive Waste Management, USA
OECD	Organisation for Economic Cooperation and Development
ONR	Office for Nuclear Regulation, United Kingdom
Opecst	Parlamentary Office of Science and Technology, France
OPERA	Onderzoeks Programma Eindberging Radiactief Afval (Research Programme for the Geological Disposal of Radioactive Waste), The Netherlands
PBO	Parent Body Organisation, United Kingdom
PNGMDR	Plan National de Gestion des Matières et Déchets Radioactifs (National Management Plan for Radioactive Materials and Waste), France
PP	Partido Popular, conservative centre-right People's Party, Spain
PRIS	Power Reactor Information System
PRISM	Power Reactor Innovative Small Module
PSI	Paul Scherrrer Institute, Switzerland
PSOE	Partido Socialista Obrero Español, social-democratic centre-left Socialist Party, Spain
Pu	Plutonium
PV	Photovoltaic
PWR	Pressurised Water Reactor
RAWRA	Radioactive Waste Repository Authority, Czech Rep.
RD	Royal Decree
R&D	Research and Development

RD&D	Research, Development and Demonstration
RSK	Reaktorsicherheitskommission (Reactor Safety Commission), Germany
RW	Radioactive Waste
RWM	Radioactive Waste Management
RWMC	NEA Radioactive Waste Management Committee
SAPIERR	Strategic Action Plan for Implementation of European Regional Repositories
SCK•CEN	Belgian Nuclear Research Centre
SDP	Spatial Development Policy, Czech Republic
SEA	Strategic Environmental Impact Assessment
SEIA	Strategic Environmental Impact Assessment
SEPA	Scottish Environment Protection Agency
SF	Spent Fuel
SFL	Long-lived Intermediate Level Waste, Sweden
SFR	Swedish repository for short-lived operational radioactive waste
SFOE	Swiss Federal Office of Energy
SKB	Radioactive Waste Company, Sweden
SKBF	Swedish Nuclear Fuel Supply Company
SKI	Swedish Nuclear Power Inspectorate
SLC	Sites to Site Licence Company, United Kingdom
SNF	Spent Nuclear Fuel
SOGIN	Società Gestione Impianti Nucleari, nuclear waste decommissioning and disposal operator, Italy
SONS	State Office for Nuclear Safety, Czech Rep.
SPD	Sozialdemokratische Partei Deutschlands (German Social Democratic Party)
SRU	Sachverständigenrat für Umweltfragen, Germany
SSI	Swedish Radiation Protection Institute
SSK	Strahlenschutzkommission (Commission on Radiological Protection), Germany
SSM	Swedish Radiation Safety Authority

StandAG	Standortauswahlgesetz (Repository Site Selection Act), Germany
STJWG	IAEA and EC Working Group "Status and Trends"
STORA	Study and Consultative Group on Radioactive Waste, Dessel, Belgium
STUK	Säteilyturvakeskus (Radiation and Nuclear Safety Authority), Finland
swisstopo	Swiss Federal Office of Topography
Synatom	Societé Belge des Combustibles Nucléaires, Belgian nuclear fuel cycle company
TFS	Task Force Site, Italy
TFS	Technical Forum Safety, Switzerland
tHM	Metric Tonnes of Heavy Metal
TRU waste	Transuranic Waste
TVO	Teollisuuden Voima Oyj, energy company, Finland
TWh	TerraWatt Hour(s)
U	Uranium
UCS	Union of Concerned Scientists, USA
ÚJV	Nuclear Research Institute, Czech Republic
UK	United Kingdom
UKAEA	United Kingdom Atomic Energy Authority
UN	United Nations
UNFCCC	United Nations Framework Convention on Climate Change
UNSCEAR	United Nations Scientific Committee on the Effects of Atomic Radiation
U.S.NRC	United States Nuclear Regulatory Commission
URL	Underground Research Laboratory, France
VLLW	Very Low-level Waste
VTT	Technical Research Centre of Finland
WANO	World Association of Nuclear Operators
WNA	World Nuclear Association
WIPP	Waste Isolation Pilot Plant, USA

WWF	World Wide Fund for Nature
YMNWR	Yucca Mountain Nuclear Waste Repository, USA
ZWIBEZ	Zwischenlager des Kernkraftwerks Beznau, interim storage facility, Switzerland
ZWILAG	Zwischenlager Würenlager AG, interim storage facility, Switzerland

I. Introduction

Comparative Perspectives on Nuclear Waste Governance

Maria Rosaria Di Nucci, Achim Brunnengräber, Lutz Mez and Miranda Schreurs[1]

The contrasting arguments are well known: opponents of nuclear power argue that the nuclear industry should not continue to produce nuclear energy without having a deep geological disposal (DGD) repository for its radioactive high-level waste (HLW). The nuclear industry and parts of the scientific community claim that the necessary knowledge and technologies for radioactive waste management exist; they argue the main problem hindering a waste solution lies with missing societal acceptance and that the fault for this lies with politics.

From a political and social science perspective, managing and disposing spent fuel and radioactive waste represents a complex problem with technical, societal, and political dimensions and implications. For decades, governments and industry have promised HLW disposal solutions, but across the world there has been little progress in finding sites, let alone constructing facilities. There are no real models for other countries to follow and those approaches currently being developed will not be easy for other countries to imitate.

The nuclear waste disposal processes in the 12 countries discussed in this volume suggest just how multifaceted this problem is. Horst Rittel coined the term "wicked problem" to describe situations that are difficult or impossible to solve because of incomplete information; contradictory understandings or values; changing health, environmental, societal or other requirements; and uncertainty (Rittel and Webber 1973). This status describes the situation of HLW storage and disposal well. Nuclear waste governance (NWG) can be used as a heuristic tool to analyse the interdependency of actors, interests, conflicts and policies on different levels, but the complexity and challenges of nuclear waste management represent a "wicked problem", as the CARL report has argued (Bergmans et al. 2008).

---

1 This chapter is a contribution of the Environmental Policy Research Centre (Forschungszentrum für Umweltpolitik – FFU) of the Freie Universität Berlin to the project "Multi Level Governance-Perspective on Management of Nuclear Waste Disposal. A Comparative Analysis: Actors, Instruments and Institutions" that is part of the inter-disciplinary ENTRIA research platform funded by the German Federal Ministry of Education and Research (BMBF/02S9082B).

Nuclear waste governance requires sensitivity to the multiple dimensions and perspectives associated with nuclear waste challenges. In all countries and at all levels, nuclear waste is simultaneously a technical, political, and institutional matter. Various actors and factors, such as the nature of the political and legal systems, geographical and hydrological conditions, technical skills, the stock of available knowledge, formal and informal rules and procedures, political constraints, the degree of public rejection and acceptance, and a country's nuclear history can shape siting processes. Nuclear waste governance must also pay heed to processes at the local, national, and international levels.

How should we move forward with this wicked problem? Although nuclear power generation has been declining since 2006, there are 31 countries that actually possess civil nuclear energy programmes, and a number of others with experimental and research nuclear reactors. In August 2014, there were 436 operational commercial nuclear power plants (NPP) with a total capacity of 373.5 GW.[2] About one-quarter of spent nuclear fuel goes for reprocessing. In addition, there are about 270,000 tonnes of spent fuel in storage worldwide, much of it at reactor sites. About 90% of the world's used fuel is in storage ponds. Some of this has been there for decades. The remaining 10% is in dry storage.[3]

The World Nuclear Association (WNA) does not seem to consider this as a big challenge: "HLW is currently increasing by about 12,000 tonnes worldwide every year, which is the equivalent of a two-storey structure built on a basketball court or about 100 double-decker buses".[4] The WNA estimates "about 3 cubic metres per year of vitrified waste, or 25-30 tonnes of used fuel for a typical large nuclear reactor. The relatively small amount involved allows it to be effectively and economically isolated".[5]

Currently 50 countries have spent fuel stored in pools at reactor sites or in central interim sites, awaiting reprocessing or disposal. These interim sites are generally neither suitable for the long-term storage of HLW nor safe for long term disposal. There is still no country in the world with an operating repository for the final disposal of HLW.

The challenges to building a final disposal facility are manifold. The preferred option appears to be DGD, but securing societal acceptance has proven difficult. In the United States, the final repository planned for Yucca Mountain in

2 See http://www.iaea.org/pris/, last accessed 10 August 2014.

3 See http://www.world-nuclear.org/info/Nuclear-Fuel-Cycle/Nuclear-Wastes/Radioactive-Waste-Management/, last accessed 15 September 2014.

4 See http://www.world-nuclear.org/info/Nuclear-Fuel-Cycle/Nuclear-Wastes/Radioactive-Wastes-Myths-and-Realities/, last accessed 15 September 2014.

5 See http://www.world-nuclear.org/info/Nuclear-Fuel-Cycle/Nuclear-Wastes/Waste-Management-Overview/, last accessed 15 September 2014.

Nevada is presently at a standstill due to strong local opposition and a presidential decision to stop funding for the project (which has in turn been challenged by a court decision). In the European Union (EU), only Finland and Sweden are in advanced stages of implementing plans for the direct disposal of spent fuel and HLW. In both countries, sites have been selected and the construction has started. France is also in an advanced stage of planning.

Not all countries have the appropriate geological conditions for such a disposal. For countries with small nuclear programmes, the financial resources needed for the construction and operation of a disposal facility are excessive. Scientific studies on the subject stress the remaining technical uncertainties and gaps in knowledge, for example on corrosion, geological formations and chemical effects (Feiveson et al. 2011: 122ff).

1 The challenged paradigm of deep geological disposal (DGD)

Deep geological disposal has long been seen as the most adequate way of disposing of HLW.[6] A number of test facilities around the world have been studied in connection with different host rocks, varying from granite to clay to salt. A number of countries, including Belgium, Canada, Finland, France, Japan, Korea, Sweden, and Switzerland, have underground test facilities (NEA/ OECD 2008).

The paradigm of DGD as the sole solution started eroding in the last decade. Some advocates of nuclear energy have admitted that geological disposal may be a controversial but still reversible solution. The ENEF-Working Group "Risks" claimed that: "For final disposal of the types of wastes mentioned, the only available option that does not place continuing burdens on future generations is the implementation of geological repositories. This is a consensus opinion of the great majority of scientific and technical experts in the field and it is subscribed to by governments of most Member States. It is nevertheless recognized that there are diverging views in some groups and that there are remaining concerns in the public about geological repositories" (ENEF 2009: 3).[7] Although the

6 For a list of the arguments for DGD, see the collective statement, "Moving forward with Geological Disposal of Radioactive Waste" of the NEA/ OECD (2008), https://www.oecd-nea.org/rwm/documents/FSC_moving_flyer_A4.pdf (last accessed 30 September 2014). See also, The European Nuclear Safety Regulators Group (ENSREG), http://www.ensreg.eu/safe-management-spent-fuel-and-radioactive-waste/waste-management-routes-under-research, last accessed 12 September 2014.

7 Members of the ENEF Sub-Working Group "Waste Management", a sub-group of the Working Group "Risks" which prepared the cited documents included governmental

advocates of the permanent closure of radioactive waste in DGD consider this to be the option with the highest levels of security and safety for the long term, some countries are pursuing other options, essentially buying time to learn more about technological possibilities for addressing nuclear waste. Thus a number of countries have decided to store their HLW in interim facilities before making a decision on whether to pursue deep geological disposal. For example, the Netherlands has implemented a policy of long-term storage. Radioactive waste is stored above ground, allowing retrieval at all times for a period of at least one hundred years. This example is being followed by Spain and Italy. Long-term waste management decisions are being left to future generations.

In 2007, the Radioactive Waste Management Committee (RWMC) of the OECD/NEA initiated a multi-year project in geological disposal, with the aim of screening approaches to reversibility and retrievability (NEA/ OECD 2012).[8] One of the major findings was that: "The development of any geological repository for radioactive waste will take place over many decades and should be open to progress in science and technology, to evolving societal demands, and to fixing potential implementation errors. In this regard, selecting technologies that are as reversible as possible is a prudent approach" (NEA/ OECD 2012: 3).

A growing number of countries are considering the option of depositing their radioactive wastes in such a way that they are readily retrievable from repositories and some countries are looking into reversibility. The reasons for this are manifold, but among them are the expectation of finding a safer technological solution in the future and that for future generations waste could even be a resource.

Countries pursuing retrievable DGD are Canada, France, Sweden, Switzerland, and the United States. In France, the so called "Planning Act" of 2006 requires that HLW disposal should not only be retrievable but also reversible.[9] There are different interpretations as to whether reversibility applies to a nuclear waste management strategy or also to policy decisions. The latter refers to the ability to modify or reverse a decision and take a different course of action, so long as a repository has not been sealed. Which explanations can be found for these new debates and different approaches, as well as for the erosion of the DGD paradigm?

---

organisations, European Commission, municipalities, waste management organisations, utilities, and other members of the nuclear industry as well as NGOs.

8 Retrievability means that nuclear waste is deposited retrievably and can be monitored and observed. Reversibility of nuclear waste disposal will allow future generations to reverse previous decisions and steps taken as new technological options become available. For details, see NEA (2012)

9 For details see the chapter on France in this volume.

Science, society, and politics are not well prepared for dealing with these complex circumstances, with the regulatory challenges of nuclear waste disposal, its link with nuclear power, the risks and uncertainties, conflicting values and preferences of stakeholders, and still unresolved socio-technical and political issues. New analytical approaches are needed in order to understand the strongly "politicized" problems and the conflicts posed by nuclear waste disposal. The editors of this volume postulate that nuclear waste storage and disposal should be understood as a wicked problem that must be examined from a multi-level governance perspective (Brunnengräber et al. 2012; 2014).

2 Nuclear waste as a wicked problem

The search for suitable sites and the management of final repositories have become major political dilemmas for governments and companies worldwide at the EU level, as well as locally. The final disposal of radioactive waste must address social, political, and economic concerns and conflicts as well as techno-scientific requirements. "Waste management decisions involve the allocation of uncertain risks and benefits to different regions of the country, to different generations, and to different social groups. Many of these decisions are linked to the national debate over the role of nuclear energy and the future of nuclear weapons" (National Research Council 1984: 1).

A clear distinction between the environmental, societal, political, economic and technological dimensions of nuclear waste is difficult, as these domains are entangled in complex ways. The way that competing information and knowledge are processed and put to use by different actors and in different political and cultural contexts also plays an important role in siting decisions. Certainly, without knowledge about geological formations and their corresponding (hydro-logical) morphologies, no long-term solution to the disposal of HLW will be possible. But the process that leads to a selection of clay, salt or granite as host-rock for DGD is hardly only technical. Numerous socio-political factors also come into play in selecting a site. Regulatory structures, legal frameworks and at least in democratic states, societal acceptance, are necessary in siting processes. In many countries, where non-transparent top down-approaches have been used in site selections, latent conflicts have turned up and open conflicts have grown often over the decades. These conflicts in turn help explain why only "clumsy solutions" have so far been possible (Verweij and Thomson 2006).

There are also many ethical and justice issues to be taken into consideration. Waste siting decisions have both inter- and intra-generational ethical implica-tions. Future generations have no say in the process; they simply inherit problems that were not of their making. Often radioactive waste siting is focused on less-well off communities that are then compensated for taking on nuclear waste. Nearby communities may also be impacted (for example due to waste transport) but do not benefit from such compensation.

Neither state control and steering of the kind that shaped the ideas of political-administrative systems in many countries in the 1980s and continue to do so in some countries today, nor ambitious multi-national approaches based on public-private-partnerships (PPP), such as the ones governing the international fusion for energy (ITER) project, provide suitable models for the challenges linked to the various operational phases of radioactive waste management. This is because they tend not to pay enough attention to societal concerns. An international approach, such as is found in environmental regimes like the Kyoto Protocol, could result in basic rules regarding safety principles or monitoring strategies. Such an approach, however, does not take sufficiently into considerations the many specific national, regional and local issues and conflicts that influence the search for a final disposal. Building upon and supplementing Bergmans et al. (2008), we identify various characteristics of nuclear waste governance that make it such a "wicked" challenge:

- The strengths and weaknesses of a process leading to a final repository can only be understood *ex post*, i.e. when the search process is completed and the final repository has been built. Specific problems are unpredictable and may only emerge during the search process for a site, during a facility's construction or at the time of, or after disposal. Even then, given the long-term nature of disposal and the complexities involved, the process will never be fully understood.
- Problems will never be solved completely and satisfactorily, since residual risks remain in all phases of nuclear waste governance.
- There can be no ideal solution. Categories such as "right" or "wrong" do not apply to this kind of issue. A global standard solution is unlikely given national and regional differences.
- Solutions and concepts cannot be fully tested experimentally or in a laboratory; there are limits to learning by trial and error especially given that many things will not be known until disposal begins and even then there will be many unanswered questions.
- Efforts to solve one aspect of the problem (whether societal, technical or political) may end up creating new problems.
- The search for final solutions is shaped by highly cultural, social, political and discursive contextual dependences and by various economic, ideological, environmental, cultural, and other interests.

- There can be no objective findings given the many uncertainties related to nuclear waste.
- Solutions can never be justified, for example, simply by virtue of technical-material merits. Problems are "socially produced" and their (possible) solution depends on how the problem is framed. The definition of problems also depends on the potential solution being considered or taken.

Finding a long-term nuclear waste disposal option requires iterative learning, addressing societal conflicts and the possibility of readjusting strategies. In the end, a decision will have to be reached, as no decision leaves future generations to deal with even greater problems which they played no role in creating. Although there may be resistance to more participatory and open planning structures, new and more democratic approaches to nuclear waste management are likely to be necessary to move forward, even if the process is often slow and cumbersome.

3 The international framework of nuclear waste governance

Nuclear waste governance is characterised by a complex, intertwined relationship and interaction between the different political-territorial levels from the global to the local. Current formal state legislations are also based on international conventions and directives. Because of the transnational safety aspects and nuclear non-proliferation issues, the management of radioactive waste and spent fuel is an area for regulation at the international and supranational levels. These regulations are laid down in international agreements and conventions.

At the global level, the most important international organisations are the International Atomic Energy Agency (IAEA), the Nuclear Energy Agency (NEA) of the Organisation for Economic Cooperation and Development (OECD) as well as the International Commission on Radiological Protection (ICRP). Since 2001, all countries with nuclear technologies are subject to the "Joint Convention on the Safety of Spent Nuclear Fuel Management and on the Safety of Radioactive Waste Management".[10] The Joint Convention applies to the spent fuel and radioactive wastes generated by both civilian and military programmes. The aim of the convention is to ensure the safety of spent fuel and radioactive waste. The principles and requirements of the Joint Convention are recognised internationally, but are not mandatory and there exist no sanctions in the case of

10 For more information see http://www.iaea.org/Publications/Documents/Conventions/jointconv.html, last accessed 16 September 2014.

non-compliance. Additionally, the safety requirements of the IAEA are not legally binding. However, most countries have incorporated these standards in their legislation on a voluntary basis.

At the EU level, the competences and responsibilities regarding spent fuel and radioactive waste are governed by the treaty which established the European Atomic Energy Community (Euratom Treaty). Under the Council Directive 2011/70/Euratom, Member States are required to establish, implement and keep updated "national programmes" for the management of spent nuclear fuel and HLW waste by 2015.[11]

The requirement of Directive 2011/70/Euratom of 19 July 2011 for mandatory implementation at the national level has been a major accelerator for the development of national nuclear waste disposal plans. Article 4 on General Principles calls for the disposal of radioactive waste in the Member State where it was produced unless an agreement is reached. Despite the August 23, 2015 deadline, delays in making and implementing decisions, selecting appropriate sites, and constructing repositories for HLW make it unlikely that all Member States will be in compliance with the directive.

This volume examines NWG in 10 Euratom countries and offers the possibility to compare approaches. Additionally, the nuclear waste management in Switzerland and the USA is analysed. In most of the other Euratom countries the site selection procedures have either not started yet, are on-going, or are controversial and therefore there is a deadlock situation. The case studies addressed here represent a rather heterogeneous sample as far as their energy policy, size of nuclear programmes, share of nuclear power in domestic electricity generation, political decisions to phase out nuclear plants, and state of advancement of decisions and procedures for the siting process for nuclear waste repositories are concerned. Variance also exists with regard to their relevant national institutional, legal, and industrial frameworks. Moreover, there are marked geographical, demographic, socio-economic, cultural, and political differences.

The nuclear waste produced from years of military nuclear activities (nuclear weapons material production, submarine reactors, military research, reprocessing plants, etc.) is also an extremely important topic but is not part of the discussion here. Nuclear waste management for the military is dealt with in a separate (and not very transparent) system. These wastes are referred to as "legacy wastes" and represent a liability not covered by conventional nuclear waste legislation and funding arrangements.

11 For details on the international organisations and the Euratom Directive, see the separate chapter in this volume.

Within the countries analysed, institutional settings and governance modes vary from top-down to deliberative participative governance. Progress in relation to the search for DGD sites for HLW has been limited and several states have opted instead for storage facilities for the next decades or even centuries, putting off decisions about DGD sites.

The implementation of the Euratom directive varies considerably. In most Member States legal and institutional frameworks are now in place. Licensing requirements and procedures for site selection and safety criteria have been established, and the responsibilities of diverse stakeholders defined. The major actors involved are: nuclear plant operators, nuclear waste producers, nuclear waste management organisations, regulatory bodies, civil society and the affected public, and policy makers at the national, regional and local levels.

In the cases analysed there are some similarities in the subdivision of responsibilities between waste producer and waste management organisation. Functional separation between "operators" and "regulators" is also the norm in most cases. Regulatory authorities are in charge of overseeing safety requirements and standards. Differences are most prominent in relation to the ownership structure of relevant institutions, some of which are state agencies and others are in the hands of waste producers.

Financing structures vary. Usually the "polluter pays principle" is applied. The costs for disposal are to be paid by the ones who generate the waste. Most countries can count on ad-hoc disposal funds which are financed by the "nuclear" utilities, but are de facto paid by the consumers through levies on their electricity bills. These funds cover the financing of a wide spectrum of tasks, ranging from feasibility studies to decommissioning to operating costs. Often these revenues are invested in growing capital funds. There are, however, many vagaries, such as how financing of the monitoring, reporting and maintenance costs tied to disposal that will be required long into the future (centuries and conceivably millennia) will be secured.

Societal acceptance of nuclear energy varies. Public opinion surveys made in Europe before the Fukushima nuclear accident revealed that public acceptance of nuclear power would increase significantly if the waste disposal problem could be solved (Eurobarometer 2010). Regardless of their views of nuclear power, people realise that spent fuel and radioactive waste generated by existing nuclear facilities must be dealt with. In most countries the debate is no longer confined to scientific and techno-political actors, but also includes many other relevant stakeholders, including actors from civil society and social movements. There are some marked differences among Euratom countries in the processes they are using to come to decisions on how to deal with highly radioactive nuclear waste. Voluntary procedures have been used in Sweden, Finland, and

Spain and partly in the United Kingdom and Belgium, whilst consultative processes with the participation of the affected communities (*débat public*) have been implemented in France. Top-down decision making is prevalent in the Czech Republic despite the fact that there have been 28 local referenda.

4 Structure of this volume

Various organisational logics can be used in looking for patterns across cases. Here the editors decided to group the case studies according to three main categories:

a) countries that have opted for geological disposal after first reprocessing their nuclear fuel;
b) countries that are pursuing geological disposal of radioactive waste and spent nuclear fuel without first reprocessing; and
c) countries that have opted for long-term surface storage as a temporary solution for dealing with high level radioactive waste.[12]

a) Geological disposal after reprocessing nuclear fuel
Whilst the majority of countries in the OECD (and elsewhere) have opted for direct disposal, there are still a number of states that reprocess most of their spent fuel. China, France, India, Japan, the Russian Federation, and the United Kingdom reprocess spent nuclear fuel on a commercial basis and plan to dispose of their radioactive waste after reprocessing. None of these countries has a completed nuclear waste disposal facility for HLW, although with varying degrees of urgency they are exploring sites for final disposal facilities. It is interesting to note that several of these countries (though not Japan) are nuclear weapons possessing states.

France and the United Kingdom have stopped their fast breeder programmes; China, India, Japan, and the Russian Federation are still on the

12 The World Nuclear Association (WNA) uses category a) and b) in its classification. They include also countries that once did reprocessing, but subsequently introduced a moratorium on reprocessing under the category reprocessing countries (e.g. Belgium and Switzerland). For details, see WNA (2013) http://www.world-nuclear.org/info/Nuclear-Fuel-Cycle/Nuclear-Wastes/Radioactive-Waste-Management/, last accessed 26 May 2014.

advanced reactor track.[13] In the past, other states were also engaged in reprocessing, but for various reasons stopped their programmes.[14]

Plutonium recovered from used fuel is either stored or used for nuclear weapons production. A small part is recycled into MOX fuel (WNA 2013). The global MOX production capacity is currently around 200 tonnes per year, most of which is in France (WNA 2013).[15] Apart from the military necessity to separate plutonium for nuclear weapons, over the last 50 years the principal reasons for reprocessing used fuel have been to "close the fuel cycle". This option was given up by operators because reprocessing is the most expensive way to treat spent fuel.

Of the countries planning geological disposal after reprocessing of their nuclear fuel, we focus on the Euratom countries with the largest operating reprocessing facilities: France and the United Kingdom. They also treat spent fuel from most of the other countries analysed in this volume. The Chinese, Japanese and Russian cases will be dealt with in a subsequent volume.

b) Direct disposal of radioactive waste and spent nuclear fuel

Direct disposal of spent nuclear fuel without nuclear reprocessing is being given the most attention as a long-term solution. The financial advantages for direct disposal over reprocessing are that this solution avoids the necessity of establishing a full nuclear fuel cycle or sending spent fuel abroad for reprocessing. Under the national cases with direct disposal of spent nuclear fuel are two groups of countries: those that originally reprocessed but do not do any longer and those that never pursued a complete fuel cycle.

In Belgium there has been a moratorium on reprocessing since 1993. In Germany, following the decision to phase out nuclear energy (the amended Atomic Energy Law which went into force in 2002), the operators had to stop spent fuel reprocessing in July 2005. In Switzerland reprocessing is considered as part of spent fuel management, but in 2006 the parliament introduced a 10 year moratorium on the shipment of spent fuel for reprocessing to the United Kingdom and France. In the United States civil reprocessing plants have not been in operation since the middle of the 1970s. In 1976, the United States abandoned reprocessing, especially because the Gerald Ford administration

13 Also Germany and the USA stopped their fast breeder reactor programmes.

14 In the Euratom countries, Italy abandoned reprocessing in the 1980s, Belgium placed a moratorium on reprocessing in 1993 and Germany stopped spent fuel reprocessing in July 2005. Also Switzerland introduced in 2006 a ten-year moratorium on the export of spent fuel for reprocessing. Additionally, Argentina, Korea, and Taiwan gave up their reprocessing plans.

15 Further details can be found in http://www.world-nuclear.org/info/Nuclear-Fuel-Cycle/ Fuel-Recycling/Processing-of-Used-Nuclear-Fuel/, last accessed 17 July 2014.

feared that this could lead to the proliferation of nuclear weapons. Later, the William Carter administration banned reprocessing. Although this ban was removed by the Ronald Reagan administration, commercial reprocessing has never been resumed. Countries that have not attempted to establish a complete fuel cycle include Finland, Sweden and the Czech Republic. The two main categories of long-term disposal designs under construction or in various planning stages are direct disposal with no option for retrievability (e.g. Finland) and designs that include a retrievability option (e.g. Sweden, Switzerland and the United Kingdom). Finland, in contrast, is pursuing deep geological disposal and is forgoing retrievability.

c) Surface long-term storage

Several countries are putting off decisions about long-term disposal of HLW and are instead opting for long-term storage facilities. The Netherlands, Italy, and Spain have postponed the search for DGD sites and are instead pursuing storage in specially designed surface repositories. These countries vary significantly in terms of the volume of HLW they produce and the role of nuclear power in their national energy policies. The decisions of both Italy and Spain to follow this path appear to have been influenced by the Netherland's pioneering role.

5 On the contributions in this volume

Achim Brunnengräber and *Miranda Schreurs* explain the many challenges facing the nuclear energy industry, including competition from other energy sources, costs, the legacies of nuclear accidents, and the need to find a solution to nuclear waste. They present a brief overview of national policies on nuclear energy to show that despite both government and industry plans, there has been little new construction and a decline in global nuclear power capacity. They show that the gap between nuclear rhetoric and the nuclear reality is widening. The chapter then turns to a discussion of nuclear waste governance and of *wicked problems*. They explain why nuclear waste management represents a *wicked problem* and touch upon some of the difficulties countries are facing in siting and developing DGD facilities. Various approaches being taken by countries to nuclear waste governance and public participation processes in siting decisions are introduced. The chapter considers some of the lessons learnt in relation to failed and successful siting instances and argues that in most countries there is still a long way to go before socially acceptable, technically safe, and secure geological repository and storage sites will be identified and facilities constructed.

Maria Rosaria Di Nucci and *Ana María Isidoro Losada* focus on the international and supranational dimensions of spent fuel and radioactive nuclear waste disposal. Their chapter concentrates in particular on the conditions set for European nuclear waste policy by Directive 2011/70/Euratom and provides critical comments on several salient articles. Additionally the authors present the possibilities for shared repositories and discuss what kind of impact the Directive 2011/70 may have on the attempts to develop regional disposal activities, such as the shared solutions being investigated by the European Repository Development Organisation (ERDO).

Gordon MacKerron turns attention to the *United Kingdom* with its particularly large and complex stock of nuclear wastes. The UK Government has only recently begun to tackle some of the most pressing issues involved in NWG. According to MacKerron, there are three major substantive challenges with UK NWG. The first is the long-term question of developing a clear management strategy for higher level radioactive wastes and the continuing pursuit of a DGD site. A second and urgent challenge is the need to clean up old and hazardous waste facilities at the Sellafield site. The third is to find a satisfactory way to dispose of the large UK stockpile of separated plutonium. In recent years there have been activities in all three areas and institutional structures have progressively changed in order to introduce a more open and deliberative governance style. However, as the chapter shows, progress has been limited, especially in the search for a geological disposal site. The UK government is also planning new construction of its nuclear capacity. It anticipates that the resulting waste will be placed in the same repository that is being planned for legacy wastes.

Markku Lehtonen examines the case of Europe's largest nuclear power state: *France*. After a decades-long search for a suitable site for its long-lived intermediate and high-level radioactive wastes, France is advancing towards the construction of a DGD in clay formations underneath the village of Bure, situated in a remote, sparsely populated, and economically declining region in the eastern part of the country. Provided that the government and the nuclear safety authorities give their approval, the national radioactive waste management agency will start constructing the repository in 2017. The plan is to deposit the first waste packages in 2025. French legislation requires both retrievability of the waste packages and the reversibility of decisions concerning the project. Detailed conditions of reversibility will be stipulated in a specific Act in 2016. Lehtonen argues that in view of numerous uncertainties and complexities, as well as persistent local opposition to DGD siting, there could well be delays in what could become one of the largest industrial projects ever undertaken in Europe.

Belgium has a relatively large nuclear programme and quite advanced low- and intermediate-level radioactive waste management programmes. *Jantine Schröder, Anne Bergmans* and *Erik Laes* give an overview of the Belgian radioactive waste management situation and discuss some of the challenges facing the country related to radioactive HLW management. They argue that Belgium's nuclear waste management governance is challenged by the limitations of national decision-making authority and capacity. The radioactive HLW management programme is ambiguous; on the one hand it has specialised, advanced and dedicated national research, but on the other it has a rather impassive, fragmented and lagging policy structure. There is advanced technical knowledge and know-how which finds recognition within the international research community, but this knowledge remains isolated vis-à-vis society and politics. The challenge is to develop a process that will outline how a complex, long-term project like geological disposal can be implemented, taking into account a mixture of demands such as transparency, safety, flexibility, control, justice, and accountability.

Peter Hocke and *Sophie Kuppler* introduce the Swiss case. *Switzerland* started an ambitious nuclear programme in the 1960s. Following the nuclear accidents at Three Mile Island and then Chernobyl, phase-out scenarios were discussed. Civil society was successful in mobilising against the civilian use of nuclear power and in blocking the industry's plans to install repositories for radioactive waste without substantial participation of the public from the affected regions. Actors in different policy areas reacted by adopting New Governance approaches developed by think tanks and governmental organisations. Questions of participation and deliberative democracy became increasingly important. Hocke and Kuppler claim that collective actors were successful in "modernising" the radioactive waste management concept before such attempts could be blocked by central actors in politics, industry, and civil society. They show in this chapter that the Swiss government and competent authorities introduced an alternative planning strategy and a number of procedural changes, which gave a new range of flexibility to nuclear waste management. These changes may have paved the way for a potentially successful siting process.

Germany foresees geological disposal for LLW, ILW, and HLW each in different kinds of repositories. Siting activities for a DGD for heat generating waste have been on-going for decades and have faced harsh opposition from societal and political actors. Site selection processes were brought to a standstill. In 2013, a political compromise for a "restart" of the siting procedure was found and stipulated in the "Repository Site Selection Act". This act defines a stepwise, criteria based, comparative selection process. In this chapter, *Beate Kallenbach* and *Peter Hocke* portray the law as a milestone. Their analysis focuses on the intertwined political expectations and societal criticisms and how they are influencing governance concepts for the disposal of heat generating waste over the long term. The authors conclude that the act may be regarded as an important step in overcoming the deadlock in DGD. The process, however, is starting amidst substantial skepticism on the part of societal groups, making it unclear how willing they will be to compromise. The siting process is on-going and it remains to be seen whether the integration of contrasting positions in an open and democratic procedure and the formulation of a consensus on how to proceed will be possible in the near future.

Sweden is the focus of *Tomas Kåberger* and *Johan Swahn's* chapter. Like Finland, Sweden has gained international attention for having found a site for the disposal of radioactive waste. Its approach to governance and management, as well as its legislation governing radioactive waste disposal, is often seen as a model by other countries. The authors suggest that there is considerable transparency in the system and public participation processes are adequately developed. In addition, significant interest is focused on the financing system for the final disposal of radioactive waste which is controlled by the Swedish Government. Kåberger and Swahn express serious doubts that the financing model can be sustained and point out that the financing system is under strain. They see the risk that the radioactive waste fund will be insufficient and ask whether it is too late for a new model based on an industry-financed safe NWG. The authors recommend increasing the leeway for the regulator and the government to revise the financing system in order to reduce the risk of future generations having to pay the costs of clean-up, maintenance, and decommissioning. They also identify problems with meeting modern safety standards and legal obligations. The most challenging issue in their eyes remains the on-going licence review for a repository to be located near the Forsmark NPP and the work necessary before for the formal review can be started.

Finland has attracted worldwide attention, as it is already in an advanced stage of DGD construction. The construction of the Onkalo nuclear waste repository at Olkiluoto started in 2004; the HLW disposal facility is scheduled to begin operations by 2020. The facility is still in a development and testing phase. *Burkhard Auffermann, Pertti Suomela, Jari Kaivo-oja, Jarmo Vehmas* and *Jyrki Luukkanen* present the Finnish policy on the disposal of HLW waste and analyse specific domestic conditions, problems, and solutions as well as their institutional and political context. The authors offer a comprehensive overview of Finland's socio-technical and political system. They point out that while the institutional and legal frameworks are in place and well-defined, there are still some open questions relating to how to determine payment levels into the Nuclear Waste Fund, what the implications are of the practically limited liability of the nuclear energy companies, and what coverage is provided by the Nuclear Liability Act in relation to the final disposal phase.

Since the mid-1980s, the *Czech Republic* has been trying to find a site for spent nuclear fuel from its nuclear power plants. Two intermediate storage facilities for nuclear waste and two repositories for waste produced by R&D and industry are currently in operation. *Martin Bursik* considers inconsistency and the democratic deficit as the greatest weaknesses of the Czech repository planning process. The Czech government was forced by the pressure of public opinion to suspend its exploration work for five years, beginning in 2004. Thanks to an amendment to the Atomic Act in 2011, a new policy of compensation instead of a dialogue with the municipalities was put in place. Bursik claims that the process of searching for the ultimate repository has not been transparent and has not followed any clear criteria, which is reflected in the fact that there has been a continuous change in the list of potential sites. The response of the inhabitants of affected municipalities is unambiguous: in all of the 28 local referenda realised between 2003 and 2013, the inhabitants refused the exploration work connected with the repository siting. The Czech government has therefore changed its strategy: instead of increasing involvement of municipalities in decision-making or the findings of safety and environmental impact assessments, "enforceability" has become the main criterion used for selecting possible sites for a repository.

Richard Forrest describes the management regime for HLW in the *United States*, a unique and complex case due to the long history and large-scale nature of the country's military and civilian nuclear power programmes. Currently, HLW is stored mainly at reactor sites in spent fuel pools, dry casks, and underground storage tanks. The system of governance in the United States offers multiple opportunities to overturn or block the implementation of policy decisions. A clear determination was made in 1987 to transfer and place the country's large volume of HLW within a permanent DGD at Yucca Mountain, Nevada. Approximately $15 billion was spent on related preparations, but in 2009 the Obama administration halted funding for the repository. No detailed rationale for this decision was provided, but there were clearly political factors at play. Several years later, Yukka Mountain is once again being looked at due to a 2012 ruling by the Ninth Circuit Court reprimanding the National Regulatory Commission for not doing more to identify a suitable site for a DGD site. The final nuclear waste repository remains a problem in need of a solution.

Nuclear waste governance has been on the Dutch government's political agenda ever since the start of their nuclear energy programme. Nuclear energy was and still is a highly contested energy option in *the Netherlands*. Despite the nuclear ambitions of some political groups in the early 1980s, the Netherlands has only one operational nuclear power plant. In addition, there are some nuclear research facilities. The current radioactive waste policy has been accompanied by some societal resistance. *Maarten Arentsen* claims that NWG in the Netherlands is characterised by a "mixture of politics and technocracy, with periodical clashes between politics, industry and society." The governance model is central coordination under government regulation, with a single publicly owned company responsible for transport and storage of all categories of radioactive waste. Arentsen argues that the feasibility of nuclear power in a small, densely populated country like the Netherlands is questionable and this is probably one of the reasons for the transparency of the domestic NWG. The dominant approach is retrievable permanent storage to allow reprocessing of stored nuclear waste, should the future state-of-the-art technology provide low hazard treatment and storage options. Additionally, a research programme is concerned with exploring geological repositories in Dutch salt and clay layers.

Maria Rosaria Di Nucci looks at nuclear waste management in *Italy* and describes it as a highly politicized issue. Three decades of fierce opposition and inconsistent policies on nuclear research and nuclear energy have provoked a stalemate and rendered the mandatory search for a national site more problematic than is the case in most other European countries. Following a popular revolt against a designated disposal site in southern Italy in 2003, the DGD alternative has been abandoned. Instead, a solution similar to the Dutch model is being considered: the construction of a central surface repository for the storage of both low and medium activity waste, also housing temporarily HLW. The siting selection has not started yet and one of the major challenges is to adopt a procedure on the basis of socio-technical criteria which incorporates inclusive and decentralised forms of decision making and interaction among players and stakeholders at all levels of (risk) governance. Di Nucci argues that the main governance bodies can be partly seen as the very source of the problem and criticises the absence of a strong, independent regulator. In order to make progress it is necessary to reinforce competent and independent authorities that have the trust of the majority of stakeholders and are deserving of public confidence. But, most of all, it is the implementers that must be perceived by the public as trustworthy. A prerequisite for this, Di Nucci argues, is that decisions are made in an understandable and transparent way. They must be based on an open process in which stakeholders and the affected local population know the "rules of the game."

Spain has opted for the installation of a centralised interim storage facility for its radioactive HLW and spent nuclear fuel. As a consequence of the high priority given to the interim storage solution, the development of strategies for a long-term disposal have been deferred and related research efforts on DGD disposal have been postponed for at least 15 years. A final DGD facility is not expected before 2050. An interim storage facility will be constructed beginning in 2016 with a design similar to the Dutch facility. It is being designed to hold HLW for at least 100 years. The criticism expressed by the anti-nuclear organisations and local groups points out that even though the government and "the establishment" have made a decision, the lack of societal consensus will represent a major challenge for the implementation of the project. *Ana María Isidoro Losada* claims in this chapter that in view of the low level of transparency in the decision-making process – and considering the failing broad societal support related to the siting procedures for the centralised interim storage – the scheduled timetable is not likely to be realized. The author argues that NWG in Spain is characterized by a notable lack of transparency. Decisions tend to be taken by politicians in consultation with a limited group of involved stakeholders behind closed doors. For future decision-making processes it would be necessary to assure the involvement of all key political-territorial tiers and authorities and to establish transparent and open communication and information systems among stakeholders.

References

Bergmans, A., and Schröder, J. (2012). Review of initiatives addressing socio-technical challenges of RWM & geological disposal in international programmes. http://www.insotec.eu/publications/topicalreports, last accessed 17 June 2014.

Bergmans, A., Elam, M.; Kos, D.; Polič, Marko, Simmons; P., Sundqvist, G., and Walls, J. (2008). Wanting the Unwanted: Effects of Public and Stakeholder Involvement in the Long-Term Management of Radioactive Waste and the Siting of Repository Facilities (Final Report CARL-Project). http://uahost.uantwerpen.be/carlresearch/docs/20080222112500ZGYI.pdf, last accessed 16 June 2014.

Brunnengräber, A., Mez, L., Di Nucci, M. R., and Schreurs, M. (2012). Nukleare Entsorgung – ein „wicked" und höchst konfliktbehaftetes Gesellschaftsproblem. *TATuP – Technikfolgenabschätzung – Theorie und Praxis*, 59-64. http://www.tatup-journal.de/tatup123_brua12a.php, last accessed 16 June 2014.

Brunnengräber, A., Di Nucci, M. R., Häfner, D., and Isidoro Losada, A. M. (2014). Nuclear Waste Governance – ein wicked problem der Energiewende. In Brunnengräber, A., Di Nucci, M. R. (2014), *Im Hürdenlauf zur Energiewende. Von Transformationen, Reformen und Innovationen*, Wiesbaden: Springer VS, 389-399.

European Commission (2010). Europeans and Nuclear Safety Report. Eurobarometer 324, Brussels.

European Nuclear Energy Forum (ENEF). Working Group "Risks" (2009). Roadmap to successful implementation of geological disposal in the EU, 29 October 2009.

http://ec.europa.eu/energy/nuclear/ forum/risks/doc/2010_02_05_roadmap.pdf, last accessed 16 September 2014.

Feiveson, H., Mian, Z., Ramana, M.V., and von Hippel, F. (Eds.) (2011). Managing Spent Fuel from Nuclear Power Reactors. Experience and Lessons from Around the World. International Panel on Fissile Materials – IPMF, Princeton. https://www.princeton.edu/sgs/publications/ipfm/Managing-Spent-Fuel-Sept2011.pdf, last accessed 30 September 2014.

International Atomic Energy Agency (IAEA) (2009). Geological Disposal of Radioactive Waste: Technological Implications for Retrievability, IAEA Nuclear Energy Series No. NW-T-1.9, Vienna. http://www-pub.iaea.org/MTCD/publications/PDF/Pub1378_web.pdf last accessed 16.09.2014.

Landström, C., and Bergmans, A. (2012). Socio-Technical Challenges to Implementing Geological Disposal: a Synthesis of Findings from 14 Countries. http://www.insotec.eu/publications/topicalreports, last accessed 17 September 2014.

National Research Council (Panel on Social and Economic Aspects of Radioactive Waste Management, Board on RWM) (1984). Social and Economic Aspects of Radioactive Waste Disposal: Considerations for Institutional Management, National Academy of Science, National Academy Press. Washington D.C. http://www.nap.edu/catalog.php?record_id=316, last accessed 17 September 2014.

Nuclear Energy Agency (NEA) (2008). Moving forward with Geological Disposal of Radioactive Waste, OECD, Paris, France. http://www.oecd-nea.org/rwm/reports/2008/nea6433-statement.pdf, last accessed 17 September 2014.

NEA (2012): Reversibility and Retrievability in Planning for Geological Disposal of Radioactive Waste. Proceedings of the "R&R" International Conference and Dialogue. 14-17 December 2010, Reims. OECD, Paris.

Rittel, H., and Webber, M. M. (1973). Dilemmas in a General Theory of Planning, *Policy Sciences*, 4, 155-169.

Verweij, M. and Thompson, M. (2006). *Clumsy Solutions for a Complex World: Governance, Politics, and Plural Perceptions*, Palgrave MacMillan.

World Nuclear Association (WNA) (2013). Radioactive Waste Management (updated November 2013). http://www.world-nuclear.org/info/Nuclear-Fuel-Cycle/Nuclear-Wastes/Radioactive-Waste-Management/, last accessed 17 September 2014.

II. The International Dimension

Nuclear Energy and Nuclear Waste Governance
Perspectives after the Fukushima Nuclear Disaster

Achim Brunnengräber and Miranda Schreurs

1 Introduction[1]

In the 1950s and 1960s, when commercial nuclear energy was first being developed, it was portrayed as an almost miraculous and limitless form of energy that would in the future be able to meet the world's growing energy demands. While nuclear energy did indeed grow to become an important element of electricity systems in some countries, it has been plagued by many problems and challenges. Today, the nuclear energy industry is facing challenging times that are linked to past failures in nuclear reactors, in energy utilities' planning, competition from alternative sources of energy, concerns about safety, and the Achilles' heel of nuclear waste.

The global use of nuclear power has been declining since 2006. There has been a drop in both the number of operational nuclear power plants (NPPs) worldwide and in total global nuclear power capacity. The industry's woes have been exacerbated by governmental announcements in several countries of plans to abandon nuclear energy all together. According to the World Nuclear Industry Status Report 2012 the industry is "suffering from the cumulative impacts of the world economic crisis, the Fukushima disaster, ferocious competitors and its own planning and management difficulties" (Schneider and Froggatt 2012: 5; 2013; 2014). The factors behind the industry's problems emerged well before the catastrophic incident at the Fukushima Dai-ichi nuclear power facility in March 2011. There are many processes which are changing the energy world order dramatically and threatening further nuclear energy development.

1 This chapter is a contribution of the Environmental Policy Research Centre (Forschungszentrum für Umweltpolitik, FFU) of the Freie Universität Berlin to the project, "Multi Level Governance-Perspective on Management of Nuclear Waste Disposal. A Comparative Analysis: Actors, Instruments and Institutions" that is part of the inter-disciplinary and inter-institutional ENTRIA project funded by the German Federal Ministry of Education and Research (BMBF/ 02S9082B). ENTRIA is the acronym for the German language equivalent of: "Disposal options for radioactive residues: Interdisciplinary analyses and development of evaluation principles" (www.entria.de).

To explain the ongoing decline of nuclear energy in many, if not all parts of the world this chapter reflects on challenges facing the nuclear energy industry, including changes in the architecture of the world energy system, new economic realities, societal scepticism, and court orders and regulatory decisions requiring that solutions are found for high level radioactive waste (HLW) disposal. A brief overview of national nuclear power policies with a focus on the five biggest nuclear electricity producers is followed by brief overviews of nuclear energy developments in other countries, ranging from plans for new nuclear power stations to plans for complete nuclear phase out are also presented.

A second major focus of the chapter is the emerging responses to the global crisis in nuclear waste management. For decades, decisions makers have proved unable or unwilling to take on the nuclear waste challenge. They have allowed nuclear waste storage to continue for decades in temporary facilities that were not designed as long term waste repositories. With global stockpiles of HLW accumulating, there is growing pressure to address this unresolved problem.

For those countries newly considering adding nuclear energy to their power supply, the failure of early nuclear energy states to tackle this problem from the start should be a powerful warning lesson. For those countries continuing with nuclear energy, developing safe and secure HLW storage must be given priority. For countries that are planning to phase out nuclear energy, the nuclear energy "problem" cannot be considered solved until the nuclear waste challenge has been addressed in a satisfactory manner.

Nuclear energy can be considered a *wicked problem* (Bergmans et al. 2008; Brunnengräber et al. 2012; 2014) that can not be untangled from highly complex safety and security risks, complicated financial questions, and the major societal challenges of trying to identify nuclear waste sites and building adequate facilities for low, intermediate and especially high level radioactive waste (respectively, LLW, ILW, HLW). Siting processes tend to be accompanied by conflicts, disagreements, delays and financial calculations that underestimate actual costs. Most significantly, there is no perfect solution. One of the biggest dilemmas with nuclear waste is its longevity. Highly radioactive waste remains dangerous for tens of thousands of years. Thus, it is necessary to store it safely for periods of time that are essentially beyond human comprehension. What will be necessary under such difficult condititons is to develop the best possible solutions – from technical and safety perspectives as well as in terms of societal acceptance given current understandings and state of knowledge.

There is still considerable public mistrust regarding nuclear waste management in most countries. Governments are partly to blame. Governments pursued nuclear power development but pushed off the waste question into the future. This, however, cannot be done indefinitely. The unwelcome truth is that the international community is still a long way off from having adequate solutions to the backend problem of nuclear energy use. Worldwide there has been little progress in site selection due to various wicked challenges. As the chapter will show however, efforts to find solutions are intensifying.

2 Talk of a nuclear renaissance

The massive explosions at the Chernobyl NPP in the former Soviet Union (now, the Ukraine) in April 1986 led to a global slowdown in the building of NPP. The explosions at the Chernobyl nuclear facility sent radioactive plumes across parts of the Soviet Union and Europe. In Europe, this led to decisions to strengthen nuclear safety standards and institutions, such as the International Atomic Energy Agency (IAEA). Internationally, the World Association of Nuclear Operators (WANO) was established to improve global operating standards and "to achieve the highest possible standards of nuclear safety".[2]

The nuclear accident sent a chill through the industry, but at the same time, was portrayed as a problem of Soviet-era technology. Many assurances were given that a similar accident could not happen in the West. These assurances began to have some effect. Nuclear energy began to win renewed support from governments, interest groups, and publics. The nuclear industry began to talk of a nuclear renaissance and from 1995 on received support at annual United Nations conferences on climate change and from some national governments looking for ways to implement climate mitigation plans. Nuclear energy was portrayed as a "clean" energy and promoted as a good way to combat greenhouse gas (GHG) emissions. This clever campaign made inroads, and even some non-governmental organizations (NGOs) – especially in the United States – began to support nuclear energy out of their concerns about climate change. The Intergovernmental Panel on Climate Change (IPCC) has lent support to this ongoing debate, calling nuclear energy a zero- and low-carbon energy technology, which "does not contribute to direct GHG emissions" (IPCC 2014, chapter 7: 23, see below). According to the "Energy Technology Perspectives" published by the International Energy Agency in June 2008 for the G8 summit, if in addition to other measures, 32 new NPPs were built per year, they could reduce harmful GHG emissions by 50% by 2050 (International Energy Agency (IEA) 2008).

Plans began to be set in several countries to strengthen, revitalize or newly launch nuclear power programs. As we will show below, however, for a variety

2 See http://www.wano.info/en-gb/aboutus/, last accessed 20 September 2014.

of reasons, including but not limited to the Fukushima nuclear accident in 2011, this nuclear renaissance has failed to materialize (Mez 2011). According to the World Nuclear Industry Status Report 2014, the 388 operational reactors in the world as of 1 July 2014 are 50 fewer than their peak number, which was in 2002.[3] Total installed capacity reached a high of 367 GWe in 2010, but then a process of decline began. Annual nuclear electricity generation which was at 2,660 TWh in 2006 was only at 2,359 TWh in 2013. There was a stabilization of the situation in 2013 (+0.6 percent) after two consecutive years of significant decline (-4% in 2011, -7% in 2012) (Schneider and Froggatt 2014: 7). While about three-quarters of the drop between 2006 and 2013 is due to the nuclear shut downs in Japan, 16 other countries have decreased their nuclear generation (Schneider and Froggatt 2013: 6). The reasons for the decline's start in 2006 are many-fold, as will be discussed later.

National rates of dependence on nuclear energy vary considerably. The "big five" nuclear generating countries in installed capacity are the United States of America (99 GWe), France (63 GWe), Japan (44 GWe), the Russian Federation (24 GWe), and the Republic of Korea (21 GWe). Together they represented 68% of total global nuclear capacity as of September 2013 (IAEA 2013; Schneider and Froggatt 2014: 8). In the United States there are 100 NPPs in operation (with five under construction) (World Nuclear Association (WNA) 2014b), 58 in France (WNA 2014c), 33 in Russia, 23 in South Korea and 21 in India; most countries with nuclear power have fewer than 10 NPPs. Nuclear energy generates about 75% of France's electricity, 19% of the United States' and 15% of Germany's (Power Reactor Information System (PRIS) 2014).[4] These countries are also the major suppliers of nuclear energy technologies, dominate nuclear energy research and development, and are the largest exporters of nuclear energy equipment. Below developments related to nuclear energy and waste in these countries are briefly considered.

3 The "big five" nuclear generating countries

United States of America

The United States is the world's largest producer of nuclear power, accounting for about one-third of global nuclear energy output (PRIS 2014, WNA 2014b).

---

3 The IAEA's list of 437 operational nuclear reactors, includes the reactors in Japan, even though in 2014 they were not in use see: http://www.iaea.org/PRIS/WorldStatistics/OperationalReactorsByCountry.aspx, last accessed 10 October 2014.

4 See http://www.iaea.org/PRIS/WorldStatistics/OperationalReactorsByCountry.aspx, last accessed 13 October 2014.

Its reactors are aging as there has been little new NPP construction since the 1979 Three Mile Island and 1986 Chernobyl nuclear reactor accidents. In the 1990s, acceptance of nuclear energy did begin to resurface due to energy security and climate change concerns. The first signs of change began with the granting of contract extensions to nuclear reactors, beginning with the application by Baltimore Gas & Co. for its two reactors at Calvert Cliffs in Maryland in 1998. This approach to maintaining nuclear capacity has since spread across the country. As of 2014 three-quarters of the USA's reactors had been granted extensions to their operating licenses (United States Nuclear Regulatory Commission (U.S.NRC) 2014). The extensions mean that at least some reactors of this existing capacity could be operating until the middle of the century.

With growing concerns about long-term energy security, the Energy Policy Act of 2005 offered financial incentives (a 2.1 cents/kWh tax credit for the first 6,000 MW of capacity in the first eight years) for the construction of new NPPs and federal loan guarantees. The United States Department of Energy put the program in place in 2008 and in the ensuing months received applications for 21 nuclear reactors. In January 2012, the U.S.NRC approved permits for two reactors at a new nuclear power facility to be built in eastern Georgia, the first new approvals since 1978. Subsequently, several other applications were approved.

Yet, with the energy situation rapidly changing in the United States, several of the construction plans have been put on hold or withdrawn for economic reasons. Some plans for extensions of operating licenses were also withdrawn and a few plants are beginning to be shut down. The Vermont Yankee NPP which began operations in 1972 and received a 20-year operating time extension in 2011, will shut down on cost competitiveness grounds (The Guardian 2013). Wisconsin's Kewaunee reactor has been shut down for economic reasons but its owners are holding off on decommissioning plans.

Democratic Congressman Edward J. Markey has raised a point of concern: "Once these old nuclear reactors shut down – as we're seeing now – it will take 60 years and hundreds of millions of dollars to decontaminate them. Taxpayers should have assurances that these nuclear relics don't outlive their corporate owners and their ability to fund nuclear cleanup costs, leaving ordinary Americans to foot the bill" (Wald 2013). Two other plants (San Onofre in California and Crystal River in Florida) are being shut down as repair costs are considered too high. Several more plants could be shut down in the coming period for similar reasons. The combination of the Fukushima meltdowns and low natural gas prices tied to the extensive extraction of shale gas has dampened expectations in the industry. Thus, despite very strong industry and political support at the highest levels, including from President Barack Obama, the outlook for the U.S. nuclear industry is shaky.

France

France is one of the leading producers and consumers of nuclear energy, with nearly three-quarters of its electricity generated in nuclear reactors. In 2013 403.7 TWh of electricity was produced by nuclear power, 73.3% of total electricity generation (PRIS 2014). The French nuclear group, Areva, has strong backers in the French government, both within the conservative and the socialist parties. Nevertheless, the government of President François Hollande is pursuing a policy of diversification for the electricity system. During very cold or hot periods demand exceeds supply due to the lack of more flexible generating plants, and France needs to import electricity from its neighbors, often at very high prices. The Fukushima nuclear accident was a powerful warning to France. Diversity is therefore being pursued. Goals have been set to raise the share of renewable electricity to 23% in 2020 and 40% by 2030 in order to enhance energy security. And after François Hollande's victory in the 2012 presidential election, a partial nuclear phase-out and the closure of the oldest 24 reactors by 2025 is on the agenda. Plans for the shut down of specific plants remain vague; currently only Fessenheim 1 and 2 at the German border, have been identified for shut down at the end of 2016. Yet, one-third of France's nuclear reactors will have been in operation for 40 years or more in 2022, thus even if operating-time extensions are granted, France, like the United States, will have an increasingly old nuclear fleet (Broomby 2014a). Currently, France is building a new generation reactor in Flamanville; grid connection is scheduled for December 31, 2016. Like Finland's reactor Olkiluoto 3 of similar design, Flamanville 3 is suffering from cost overruns and repeated delays. France will face costly decisions in the coming years as it is forced to decide whether to build new nuclear plants or to instead shift to other power sources. Decisions about extending the operating life-time of the existing NPP will only be made at the end of this decade.

Japan

Japan invested heavily in nuclear energy in previous decades and became the world's third largest producer of nuclear power after the United States and France. As a major component of its efforts to combat climate change, the Japanese government envisioned sharply increasing the share of nuclear energy in the electricity mix. Its 2010 Basic Energy Plan set out plans to expand nuclear energy to meet 50% of electricity production by 2030. Those plans were badly shaken by the Fukushima nuclear accident. Four reactors at the Fukushima

Dai-ichi NPP were severely damaged by the earthquake, tsunami, subsequent hydrogen explosions and nuclear meltdowns and permanently shut down in May 2011. It is estimated that their dismantling and decommissioning will take several decades and cost the country trillions of yen. Due to the dire situation in Fukushima, the other two reactors at the Dai-ichi Nuclear Facility in Fukushima were also permanently shut down in December 2013 (PRIS 2014) and designated for decommissioning.

Of the remaining operational 48 NPPs, all have been shut down for safety checks and are awaiting decisions about their restart under new safety guidelines.[5] At the time of the Fukushima nuclear accident in March 2011, nuclear power supplied 27% of Japan's electricity. After the Fukushima catastrophe, the nuclear share dropped to 1.7% of total electricity production in 2013 due to shut downs (PRIS 2014). Between March 2011 and the end of 2014, Japan has functioned on a greatly reduced share of nuclear energy and long periods when there was nearly and even no nuclear energy connected to the grid. The loss of electricity production from nuclear reactors has been compensated by energy efficiency improvements, energy conservation and greater reliance on imported coal and gas. Japan's carbon dioxide emissions have risen sharply, but its energy demand has also dropped substantially (Schreurs and Yoshida 2013).

In December 2013, a newly elected conservative government led by Shinzo Abe presented a new energy plan. The plan states that nuclear energy remains an important energy source, but that Japan will strive to decrease dependency on nuclear energy while expanding reliance on renewable energy. The plan does not give a percentage for nuclear energy in the future as it is unclear how many of Japan's remaining NPP will pass new safety standards established after the Fukushima nuclear accident. The Japanese nuclear industry, the powerful Ministry of Economy Trade and Industry, and other nuclear supporters have argued that Japan will face soaring energy costs, rising GHG emissions, and economic decline without nuclear energy. Nuclear opponents argue that Japan can survive without nuclear energy by promoting renewable energy and energy conservation and point to the country's ability to function after the Fukushima nuclear accident with no nuclear energy connected to the grid.

While the government of Shinzo Abe is eager to restart the country's NPPs, opposition in the country remains strong. Plans to restart two reactors at the Oi NPP in Fukui were set back by a local court decision that said the plant's owner had not adequately shown the plants were safe given their location in a possible earthquake zone (Muroya 2014). In the meantime, renewable energy is back on the agenda in Japan. In the summer of 2012, a feed-in-tariff was

5 See http://www.iaea.org/pris/CountryStatistics/CountryDetails.aspx?current=JP, last accessed 10 October 2014.

introduced to incentivize renewable energy, and especially photovoltaics (PV). A target of 33 GW of PV and 9 GW of wind was set for 2020. Japan has become an attractive photovoltaic market, and by 2013 had over 10 GW of installed capacity.

The Japanese government and the nuclear industry envision a closed nuclear fuel cycle despite significant international opposition. The Rokkasho Reprocessing plant, with the capacity to reprocess 800 tons of uranium or 8 tons of plutonium, is scheduled to open in late 2014. The start up of the plant is highly controversial due to concerns about nuclear proliferation (Tirone and Adelman 2014).

Russian Federation (Russia)

Despite the meltdowns at the Chernobyl reactor and the Fukushima Dai-ichi NPP, Russia has indicated its intentions to expand nuclear energy capacity by 2020, develop new reactor technologies, and export NPPs. Russia presently has 33 operational reactors totalling 24,164 MWe, 10 reactors under construction and 5 in permanent shutdown (PRIS 2014); in addition, most reactors are being given operating extensions (WNA 2014d). Russia's nuclear power production peaked in 2010 but the Russian industry has expanded its export policy with financial backing from the state. Russia is cooperating with countries that are building reactors for the first time (Belarus, Turkey and Bangladesh) and with countries like Finland and Hungary that already have NPPs. A deal for a Russian loan to finance a nuclear plant expansion project in Hungary was approved by the Hungarian parliament in June 2014 (although the plan awaits consideration by the European Union). Under the agreement, Russia's state-owned Rosatom will build 2,000 MW of additional capacity at Hungary's state-owned MVM Paksi Atomeromu (Gulyas 2014). Russia has also signed a deal to build a NPP in Bangladesh (Russia Beyond the Headlines 2014). Despite these potential markets, Russia is currently only constructing reactors at home and in China, and Belarus (Schneider and Froggatt 2014: 48).

Republic of Korea (South Korea)

In South Korea, 23 nuclear reactors with an installed capacity of 20.7 GWe provided almost one-third of South Korea's electricity in 2013. Five reactors are under construction (PRIS 2014). Korea's aim is to provide about half of the electricity from over 30 units by 2022 (WNA 2014e). A year before the Fukushima accident, South Korea sold four nuclear power reactors to the United Arab Emirates and in January 2010, the South Korean Ministry of Knowledge Economy launched NuTech 2030, a plan to develop indigenous reactor technology (Innovative, Passive, Optimised, Worldwide Economical Reactor

(I-Power)). The goal is to export 80 NPPs by 2030 and to make South Korea the third largest supplier of nuclear technology in the world. Six new reactors are to be completed by 2016. The government's plans, however, are under scrutiny. The natural disaster and ensuing nuclear crisis in Japan have opened a debate about Korea's energy future and led to calls to invest more in energy efficiency improvements and renewable energy development.

Comparing across the Big Five Producers

Despite the Fukushima nuclear accident, there is political support for continuing with nuclear energy in the big five nuclear energy producers. Yet, the impacts of the accident can be felt in these countries as well. The industry is struggling and the high growth expectations of the past have been dampened. In the United States, both the Republican and Democratic parties continue to back nuclear energy, but the economics of nuclear energy could make it difficult to do more than continue with existing plant capacity for the foreseeable future. New construction may not be able to offset the decline in operational nuclear NPPs. The Japanese government wishes to continue the use of nuclear energy, but there is strong public apprehension. Japan is the first country in the world to have a large nuclear energy fleet that was forced to shut down entirely for an extended period of time due to safety concerns. France is planning a major cut back in the share of nuclear energy in its electricity mix. South Korea and Russia have the most ambitious plans for nuclear energy expansion of the big five, although in South Korea there is now more opposition to nuclear energy than there was in the past. All five countries are in competition with each other in terms of exporting nuclear energy technologies, but there are relatively few buyers.

4 Challenges facing the nuclear energy industry

There are many challenges facing the nuclear energy industry. The most obvious is the lack of public trust in the safety of nuclear energy after the nuclear accidents at Three Mile Island, Chernobyl, and Fukushima. They occurred despite government and industry assurances of safety. After each major accident, the industry has suffered major setbacks with loss of public confidence in the industry's assurances of the technology's safety (Schreurs 2013b). After the Fukushima accident, the European Commission ordered stress tests for all NPPs operating in the European Union. Peer reviews took place in 2011 and 2012. Many other countries and territories also conducted comprehensive nuclear risk and safety assessments, based on the EU stress-test model. These included Switzerland and the Ukraine (both of which fully participated in the EU stress tests), Armenia, Turkey, Russia, Taiwan, Japan, South Korea, South Africa and Brazil. The peer review performed by the European Commission and the European Nuclear Safety Regulators' Group (ENSREG) identified weak spots and called for action to make the NPP more robust.[6]

An aging nuclear fleet

Apart from a few countries that have continued to build NPPs (like South Korea) or that have only in the last decade started to heavily invest in nuclear power (like China), many of the world's nuclear facilities are getting older. With age come greater concerns about plant safety and added costs for upgrades. Few countries in the west have built new NPPs since the Chernobyl nuclear accident; a majority of the nuclear facilities around the world are within one or two decades of the end of their original licensed operating times. NPPs in most countries are only allowed to operate for 30 or 40 years, even though some are getting operating extensions for an additional decade or two. Worldwide there are 39 reactors that have run for over 40 years. As there has been little new construction, the global NPPs fleet is aging. At the end of 2014, the average age of operating NPPs in Europe is 30 years, in the United States, 34 years (United States Energy Information Agency (EIA) 2013) and globally, 28.5 years (Schneider and Froggatt 2014: 6). Clearly, the global nuclear power fleet is aging and this leads to additional costs and risks (Haverkamp 2014).

A rapidly changing global energy structure

Nuclear energy is but one of many forms of energy in a rapidly changing global energy structure. As the International Energy Agency (IEA) has noted: "the rise of unconventional oil, gas and renewables is transforming our understanding of the distribution of the world's energy resources" even if the situation differs between countries and regions (IEA 2013a: 3). Demand for energy is expected to grow by as much as one-third in the coming decades (through about 2035), triggered mainly by the enormous thirst for energy in China, India and the Middle East. These regions account for about 60% of the growth in energy demand globally. While one might expect these tendencies to speak in favor of nuclear energy, the situation is quite complex. On the one hand, energy security concerns have led countries like China and India to pursue all forms of energy development, including nuclear power. Growing energy demand and the desire for greater energy independence are also attracting other countries to consider developing nuclear energy, such as Bangladesh, Belarus, Poland, Turkey and Vietnam, or to maintain nuclear energy, such as the Czech Republic.

6 For more information see http://ec.europa.eu/energy/nuclear/safety/stress_tests_en.htm, last accessed 20 March 2014.

On the other hand, other energy developments – and especially new possibilities related to unconventional sources of oil and gas as well as cheaper coal and renewable energies – challenge the outlook for the nuclear industry. In many Organization for Economic Cooperation and Development (OECD) countries, there has been a pronounced shift towards gas and renewables (IEA 2012a) and a declining investor interest in nuclear energy. This is the case in Europe and North America. Although energy demand in OECD countries is expected to be quite flat, the IEA predicts that EU reliance on imported oil will increase from around 80% today to more than 90% by 2035. This increases the EU's vulnerability to supply and energy price shocks (European Commission 2014: 11). Yet, except for in a few countries (most notably, the United Kingdom and Poland), the EU is not responding to this situation with a push for new nuclear energy development, but rather with energy efficiency plans and renewable energy development. New supply strategies are being developed, import diversification is being promoted, and resource diversification is being given new priority. The changes come in response to technological advancements in the renewable energy sector as well as non-conventional fossil fuels, societal demands, concerns about climate change, and new oil and gas discoveries.

Paradoxically, the United States, which did not ratify the Kyoto Protocol, has been taking some steps towards decoupling greenhouse gas emissions from energy consumption. Carbon dioxide emissions dropped by 10% between 2005 and 2012 and by 3.4% from 2011 to 2012, due to a variety of factors. These include the so-called shale gas boom and the resulting shift by many industries from coal to gas, energy efficiency measures introduced by President Obama, and the economic crisis which dampened economic activity. Although carbon dioxide emissions rose by 2% in 2013 as a result of the pick up of the economy, projections are that the United States will be able to meet its Copenhagen pledge to reduce US carbon dioxide emissions by 17% of 2005 levels by 2020.[7]

The IEA has called this the "Golden Age of Gas" (IEA 2012a). It is estimated that fossil fuel reserves will last for decades and possibly centuries to come. The extraction of unconventional oil and gas means that contrary to outlooks of just a few years ago that saw the United States facing major domestic energy supply shortages, new estimates are that the United States will become energy self-sufficient by around 2035 (IEA 2013a; 2012b). This means that the demand for new domestic sources of energy (and especially nuclear power, which carries with it various risks and high up front investment costs) will be lower than might otherwise have been the case (Smith 2012). Of course, "fracking" for shale gas and oil brings with it its own environmental and health

7 For more details see http://www.epa.gov/climatechange/ghgemissions/usinventoryreport.html, last accessed 25 September 2014.

concerns, which is one reason why its introduction in densely populated Germany was advised against by the German Environment Advisory Council (Sachverständigenrat für Umweltfragen (SRU) 2013: 43). Yet, despite some opposition to fracking in the United States, shale gas and oil extraction is likely to continue, dimming the prospects for nuclear energy.

Renewable energy development

Worldwide renewable energy development continues its rapid expansion in both capacity and generation. In 2013, three of the world's four largest economies – China, Germany, Japan and India – generated more power from renewables than from nuclear power (Japan due to the shut down of its NPP after the Fukushima nuclear accident). In China, solar electricity generation grew by 400% in one year, 2012 (Schneider and Frogatt 2013: 11). In the European Union, renewable energy targets of 20% of final energy consumption have been set for 2020 and there is a European Commission proposal for a 27% target for 2030.[8] In many states in the United States, renewable energy targets have also been established (e.g. 33% of electricity supply by 2020 in California and 30% of electricity consumption by 2015 for New York).

Renewable energy use is growing at astounding rates across much of the world (REN21 2014). Liberalization of electricity markets and state subsidies for renewable energy have significantly changed the prospects and the attractiveness of nuclear power in the world. In several countries, power companies which did not see or react to this new trend are in trouble. They face difficulties keeping their fossil-based industries profitable. Investing in nuclear energy in this context is becoming an increasingly high risk option from an economic perspective.

Global climate change and nuclear energy

Of course, there is the argument that nuclear energy is important in a climate constrained world. The Intergovernmental Panel on Climate Change (IPCC) suggested in their Fifth Assessment Report on Mitigation of Climate Change (Working Group III AR5, 2014) that nuclear energy is one way to reduce GHG emissions and to protect the climate: "Nuclear energy is a mature low GHG emission source of base-load power, but its share of global electricity generation has been declining (since 1993). Nuclear energy could make an increasing contribution to low carbon energy supply, but a variety of barriers and risks exist (*robust evidence, high agreement*). Those include: operational risks, and the associated concerns, uranium mining risks, financial and regulatory risks, unresolved waste management issues, nuclear weapon proliferation concerns,

8 See http://ec.europa.eu/clima/policies/2030/index_en.htm, last accessed 26 September 2014.

and adverse public opinion (*robust evidence, high agreement*). New fuel cycles and reactor technologies addressing some of these issues are being investigated and progress in research and development has been made concerning safety and waste disposal" (IPCC Summary for Policymakers 2014: 21f).

With memories of the Chernobyl nuclear accident fading in the period before the Fukushima nuclear accident, there were some voices, especially within the industry but also in various non-governmental organizations (NGOs) concerned with climate change, who began to speak of the need for and the inevitability of a renaissance of the nuclear industry. Notable statements of political support for nuclear energy came from the "big five" (Japan, South Korea, the United States, Russia and France) as well as from the United Kingdom; all linked a revival of the industry to their climate change policy plans. Even Germany, which in 2002 had passed a law to phase out nuclear energy once a designated amount of operating time was guaranteed for its fleet of nuclear reactors, made plans in 2010 to extend the operating time of its nuclear energy facilities for an additional 8-14 years.

The argument that nuclear energy has a low carbon footprint, however, can be challenged. Greenhouse gas emissions cannot be calculated solely on the basis of actual operations, but must take the entire production cycle and nuclear waste treatment into account. Nuclear power plants have to be planned and constructed; uranium has to be mined, processed and transported; and waste has to be treated, stored for decades, and transported to a permanent geological disposal side. Nuclear power plants have to be decommissioned. All these steps in the life cycle of nuclear energy produce carbon emissions. When this is taken into account, it is clear that NPPs are not greenhouse gas-free production systems, as presented in the advertisements of the operating companies. It is estimated that nuclear technologies emit twice as much carbon as solar photovoltaic technologies, and six times as much carbon as onshore wind farms (Sovacool 2008; Mez 2012b).

While climate protection is an argument often brought forth by the nuclear power industry and some environmentalists to justify nuclear energy development, there has been little global movement on climate mitigation. With the climate negotiations deadlocked, climate change has become a less powerful and visible playing field for nuclear energy supporters. The United Nations Conference on Climate Change in 2010 in Cancún, Mexico, agreed to commit to a maximum increase in the planet's average temperature of two degrees above the pre-industrial level. Rhetorically, the United Nations still supports this target, but little additional progress has been made at the subsequent climate negotiations held in Durban (2011), Doha (2012), and Warsaw (2013) with little expected progress in Lima (2014). The Paris 2015 climate negotiations are targeted for the development of a new global climate change agreement, but little more than a lowest common denominator agreement is expected. It is obvious that the United Nations Framework Convention on Climate Change (UNFCCC), signed in 1994, and the Kyoto Protocol, which came into effect in 2005 with an initial commitment period through 2012, have not had the power to bring many governments to reduce their emissions substantially (Altvater and Brunnengräber 2011a; 2011b). Between 2000 and 2012, global greenhouse gas emissions increased by an annual average of 2.9%. According to the International Energy Agency, "despite many countries taking new actions, the world is drifting further and further from the track it needs to follow" (IEA 2013: ii). While nuclear energy might be considered as a response to this problem, even putting nuclear risk factors aside, NPP approval and construction is a very lengthy and costly process. It is difficult to imagine that nuclear energy could make much of a dent in global GHG emissions. Moreover, energy efficiency improvements, energy saving, and renewable energy offer safer and cheaper means of combating climate change.

5 The impacts of the Fukushima nuclear accident

While there are some countries, especially in East Asia (and in particular China and South Korea), the Middle East, and in central and eastern Europe (especially countries that have been heavily reliant on Russia for gas) that are planning new NPPs, in much of the world, the industry is struggling to maintain its historic share of electricity production or slow its decline. First, the global financial crisis that began in 2008 hit financial institutions hard and has made it difficult for the nuclear industry to convince investors that the time is ripe for a nuclear renaissance. Second, the Fukushima nuclear disaster largely shattered any remaining hope the industry had for a renaissance in the foreseeable future. Immediately after Fukushima, several countries made decisions to turn away from nuclear energy.

Germany is in the process of closing down its remaining 9 NPPs by 2022 (Mez 2012a, Schreurs 2013a). For many Germans, Fukushima brought back memories of Chernobyl. Chancellor Angela Merkel responded to these developments in March 2011 by announcing a three-month moratorium on the nuclear plant running-time-extension law. She further ordered that the seven oldest nuclear reactors be taken off line (adding to one that has been off line for several years due to fire that broke out in the facility in 2007 and 2009). In Italy, in a referendum held in June 2011, over 95% of respondents voted against restarting a nuclear energy program. The Berlusconi government gave up on its nuclear ambitions after the referendum (Kennedy 2011).

In May 2011, in the face of growing unease about nuclear energy, the Swiss cabinet recommended the country's five nuclear power stations to be shut down, starting in 2019 and finishing by 2034. In addition, it was decided that no new NPPs are to be built. Switzerland's parliament approved the plan in the summer of 2011 and is now developing plans for an energy transition that will require both a sharp reduction of greenhouse gas emissions and a gradual nuclear energy phase-out. In Belgium a long debate initiated by the Belgian Greens resulted in a nuclear phase-out law, which was adopted in 2003, shelved in 2010, and revived in 2012. Belgian's seven nuclear reactors are to be shut down between 2015 and 2025. In 2014, the Doel-3 and Tihange-2 nuclear reactors were temporarily shut down (until further notice) due to technical problems discovered during safety tests (Federaal agentschap voor nucleaire controle (FANC) 2014).

Despite the phase out decisisions in these countries, the production of nuclear waste will continue for years or decades. At the same time, the amount of HLW will increase globally, even though there is still no country with an operating deep geological disposal (DGD) facility for HLW from the operation of commercial NPPs. In fact, there are few middle-term HLW storage facilities (100 to 300 years) and no long-term facilities anywhere in the world. In the absence of long-term waste strategies in many countries, highly radioactive spent fuel rods are being stored in cooling pools and dry casks for periods far longer than was initially planned for. Some countries are now turning to dry cask surface storage of HLW and interim storage facilities at reactor sites until decisions on long-term waste disposal are resolved and geological storage sites are built.

6 (No) long-term radioactive waste management

Radioactive waste is produced during various stages of the fuel cycle, including in the open pit- and underground-mining of uranium. Uranium ore is ground to produce uniform-sized particles and then uranium is extracted with a chemical leaching process. The milling process yields a dry powder-like material consisting of natural uranium that has sarcastically been called "yellowcake". There are various health and environmental concerns linked to uranium mining. Uranium ore emits radon gas, a serious health concern. Contaminated residual materials, soil, sludge and dust are also produced and must be disposed of properly. There are concerns that economically disadvantaged and marginalized groups are most at risk, as they are often the ones involved in uranium mining.

Good waste management should begin before waste is generated. New concepts include *cradle to* cradle or *zero waste* initiatives. The starting point should be to avoid or reduce waste to a minimum at its source. In the case of nuclear waste, neither of these ideas can be realized. Although it was obvious from the beginning of the nuclear industry that safe and secure disposal is needed, the industry and government moved forward with nuclear energy development but with no concrete plans for what to do with the nuclear waste. This is a little like building an airplane without a runway or, as it has been said elsewhere, building a condominium without a toilet (Japan Times 2014a). From the beginning, questions were raised about how nuclear waste was to be handled. Experts and citizens voiced concerns that civilian nuclear energy development could contribute to nuclear proliferation.

This concern stood as a challenge to the concept of the Atoms for Peace Initiative, the idea of pursuing atomic energy for peaceful economic and conventional energy production. Take Japan as an example: it is estimated that there is sufficient plutonium waste in Japan to produce thousands of nuclear bombs (Takubo and von Hippel 2013: 6). The United States has pressured Japan to return weapon's grade plutonium to the United States to reduce the threat of the waste falling into terrorists' hands. China has raised concerns that Japan is pursuing reprocessing in order to have the material necessary to produce nuclear weapons (Japan Times 2014b). There are also concerns that HLW in temporary storage facilities could become the target of a terrorist attack. Security forces in Germany were alarmed by a call put out by a German Salafist from Syria in August 2014 for those who are like-minded to attack a U.S. nuclear weapon's storage facility in Rheinland-Pfalz (Focus Online 2014).

In the past, the Soviet Union dumped radioactive waste and decommissioned nuclear reactors into Arctic seas (Digges 2012) and Russia is currently surveying whether some of the waste (17,000 containers, 19 vessels with radioactive waste, and 14 nuclear reactors) can be safely removed from the Kara sea, which has become of interest for possible offshore fossil fuel exploration (Peter 2013). In 2011, the Russian Duma passed the Radioactive Waste Management Law, which introduced a unified state radioactive waste management system and brought Russia into compliance with the United Nations Joint Convention on the Safe Management of Spent Nuclear Fuel (Nuclear Threat Initiative (NTI) 2014).

Governments and the nuclear industry did place hope in the development of a closed nuclear fuel cycle. This has led some to reprocess nuclear waste, the case in the United Kingdom and France, and others to pursue this know how (e.g. Japan). Reprocessing means that fissionable plutonium is separated and recovered from irradiated (spent) nuclear fuel. This can then be used in MOX fuel in thermal nuclear reactors, but as noted above, brings with it various security risks (particularly concerns about proliferation). These security risks are why the United States does not reprocess. Moreover, while reprocessing reduces the volume of waste, by itself it does not reduce radioactivity or heat generation. There is HLW produced, meaning that DGD and a long-term management strategy are still necessary. What progress can be observed?

The following sections are based on the findings of the chapters in this book, the authors' own research and reports on nuclear waste management in different countries. The material has been put into a comparative structure in order to assess patterns across states in their efforts to manage high level radioactive waste.[9]

6.1 Countries with deep geological disposal (DGD) after reprocessing nuclear fuel (France, Japan, United Kingdom)

France and Japan have chosen to reprocess spent fuel and to use the recovered uranium in NPPs. These countries see DGD as the preferred option for their radioactive waste, and are currently exploring possible site locations. No final site selections have been made. In France, the search for a site has been going on since the late 1970s. Site investigations conducted in the late 1980s generated intense local opposition, prompting the government in 1990 to declare a one-year moratorium on the search for a site. After extensive consultation, the parliament adopted the country's first law on nuclear power in 1991, which reopened the search for a waste solution. The law also marked the beginning of a more participatory approach to waste management policy. Since 1998, reversibility has been the cornerstone of French radioactive waste management policy. The concept was confirmed after a public consultation process organized by the National Commission of Public Debate in 2005-2006. In 1998, the government selected Bure, a small village on the border between the departments of Meuse and Haute-Marne in the northeast of the country, as the site for an underground research laboratory (URL) for DGD. Since then the Bure site has been the primary candidate for hosting the final disposal facility. If all goes according to plan, the construction of the repository should begin in 2017 and be finished in 2025, at which time the installation should be ready to receive the first waste

9 See also the country reports from the research project International Socio Technical Challenges for implementing geological disposal (InSOTEC), www.insotec.eu/publications/file-cabinet, and the Nuclear Energy Agency (NEA)/ OECD, https://www.oecd-nea.org/rwm/, last accessed 9 May 2014.

containers. The project is not without detractors, however, as was made apparent during the public consultations that took place in 2013.

For Japan, dealing with the nuclear waste from the Fukushima nuclear accident has become a wicked problem of new and huge proportions. Across contaminated regions, lightly radioactive soil and other materials are being temporarily stored under plastic sheets in the countryside while plans are made for longer-term storage. Discussions are underway regarding what to do with the highly contaminated radioactive waste both from the Fukushima nuclear accident and from spent fuel produced in Japan's many NPPs. The government solicited requests from communities to host a facility in exchange for large-scale compensation, but no communities have made themselves available. Now the government is following a new strategy and is looking for suitable sites as part of a national search.

Since 2000, disposal of HLW originating from nuclear reactor operations in Japan has been governed by the Designated Radioactive Waste Final Disposal Act. The Nuclear Waste Management Organization of Japan, which is under the Ministry of Economy, Trade, and Industry and was set up in the same year, is responsible for implementing DGD of HLW and LLW that contains long-lived nuclides. It aims to have a licensed and operating site by 2040.

In 2001, an Advisory Board on HLW Repository Safety issued a report entitled "Safety Communication on Geological Disposal." This report recognized the importance of establishing a safety communication system to enable stakeholders or their representatives to participate in waste decision processes in order to win their long-term confidence in the safety of long-term geological disposal. It was seen as essential that both the public and the technical community are confident in the engineering feasibility as well as in the long-term safety of DGD. This is particularly important given that the Japanese government and nuclear industry are challenged by a lack of public trust as a result of the Fukushima nuclear accident. Due to the volcanic nature of the Japanese archipelago, finding a suitable disposal site is particularly complicated.

In 2002, NUMO started to solicit applications for communities interested in conducting feasibility studies to host the country's radioactive waste site. The only town to come forward was Toyo in Kochi Prefecture, but the application was withdrawn after the mayor who had supported it lost an election to a candidate that ran on a platform against the idea. A report by the Science Council of Japan issued in September 2012 notes the difficulties of achieving public consensus, criticizes using financial incentives as an approach, and argues for the need to address the nuclear waste issue within the broader context of nuclear energy policy (Edahiro 2012). In November 2013, the Japanese government issued a plan that outlines a new approach, whereby the national government will list candidate disposal sites and consider measures to support potential host communities (The Asahi Shimbun 2013).

In the United Kingdom, many nuclear facilities are nearing the end of their operating lifetimes, and there is the need for decommissioning, site clean-up and restoration. The UK government announced that the thermal oxide reprocessing plant at Sellafield will be closed in 2018. The Magnox nuclear fuel reprocessing plant in Sellafield celebrated its fiftieth birthday in 2014, and is now twice as old as it was originally designed to be. Decommissioning of nuclear facilities falls under the responsibility of the Nuclear Decommissioning Authority (NDA), which became operational in 2005. The Managing Radioactive Waste Safety White Paper, published in June 2008 by Defra, BERR and the devolved administrations of Wales and Northern Ireland, sets out how the government policy for geological disposal of HLW will be implemented.[10] The paper was accompanied by an invitation to communities to express interest in entering into open discussions with the UK government about the possibility of hosting a geological disposal facility for higher activity radioactive waste.

The government aimed for a voluntarist and participatory approach, in which potential host communities would have a right to withdraw from the process up to a pre-determined point. Dialogues were held with the planning authority, potential host communities, other interested parties and the general public. Local authorities in West Cumbria initially expressed interest in researching the suitability of the region for a deep disposal site, but the Cumbria County Council voted against further research to the disappointment of the UK Department for Energy and Climate Change (Harrabin 2014). The search for a suitable site for a geological disposal facility for HLW in the UK remains a big challenge. After the Cumbria decision, the UK government has moved to take the veto right of county governments away and to move toward a policy that would simply require a positive "test of community support" for the project. Councils are to be paid to join in the site investigation process (Broomby 2014b).

6.2 Countries with direct deep geological disposal of spent nuclear fuel (Belgium, Switzerland, Sweden, Finland, Germany, United States, Czech Republic)

Belgium has developed an underground research laboratory for experimental research on geological disposal for HLW named HADES, after the Greek god of the underworld. HADES is located in a Boom clay layer, which serves as a host

10 See https://www.gov.uk/government/uploads/system/uploads/attachment_data/file/228903/7386.pdf, last accessed 13 October 2014.

rock, and is one of the first such laboratories to be built worldwide. In 2006 the government selected Dessel for surface disposal of LLW and ILW. For HLW, Belgium is pursuing advanced research on geological disposal, but no formal political decisions have been taken so far. Currently no specific procedures – other than impact assessment regulations – have been developed. The aim of various consultation processes has been to draw up a 'map' of public concerns; the instruction given to facilitators has been to avoid polarised debates and to look for consent wherever possible. Public consultations have taken place, but have also elicited critique.

In Switzerland, the concept of "controlled geological disposal" for high-level radioactive waste was established by the "Expertengruppe Entsorgungskonzept für radioaktive Abfälle" (EKRA) and adopted in the Nuclear Energy Act. A pilot facility is to be built next to a main facility to allow for monitoring for a certain period. A siting process for a DGD for LLW and ILW began in Wellenberg in the Swiss canton of Nidwalden in 1986 but failed two times when put before the public, in 1995 and again in 2002. The Canton opposed the license for the construction of an exploratory drift. In response to these failures new procedures were introduced. Since 2002, the "Sectoral Plan for geological repositories" has been organized in a stepwise siting process that involves stakeholder and public participation. In the first step of this process, the National Cooperative for the Disposal of Radioactive Waste (Nagra), chose six potentially suitable sites for geological disposal of LLW and ILW and three for HLW. These locations were approved for research by the Swiss Federal Council. On 13 February 2011, a cantonal referendum on a proposed geological repository in Nidwalden, however, turned into a fiasco for its proponents. Eighty percent of the voters refused the nuclear waste disposal project. Potential sites for surface facilities are now to be found based on recommendations of Regional Conferences established in the potential host regions.

Direct deep underground geological disposal facilities for spent fuel are presently being built in both Finland and Sweden. They are still in the development and test phases, and are not yet approved for operation. The two companies that own the Finnish *Fortum Oyj and Teollisuuden Voima Oyj* (TVO) NPPs, in 1996 formed a joint company, Posiva Oy, to manage nuclear waste. The site and the project for the final disposal solution is called Onkalo, and is located at the Olkiluoto NPP, in the southwest of Finland. The facility's construction started in 2004; it is to begin operations around 2020. Onkalo is proceeding with very little public debate; the influence of non-governmental organizations has been very limited. Initially six areas were analyzed for their suitability and a list of four candidate regions selected. The decision for Olkiluoto was made by a group of experts and ministerial department heads. The

Olkiluoto site was chosen both because of its geological and technical suitability and the public's willingness to accept the facility. It is important to note that Olkiluoto already hosts a radioactive waste facility started in 1992 for LLW and ILW.

In Sweden, the Swedish Nuclear Fuel and Waste Management Company (SKB) is responsible for managing nuclear and radioactive waste. Nationwide test-drillings in search of suitable sites for a geological depository resulted in widespread local protests and opposition (Åhäll et al. 1988). This led to a decision to turn to a voluntary siting process in which all municipalities in Sweden were invited to host initial feasibility studies. After local referenda blocked potential sites in northern Sweden, the focus shifted to communities hosting nuclear waste facilities. Two towns, Oskarshamn and Östhammar, expressed interest and competed with each other to become host to a high-level radioactive waste site in exchange for considerable financial compensation (even the loser, Oskarshamn came out ahead) (Knight 2009). The SKB decision for Forsmark in the municipality of Östhammar was made in 2009. The facility, when it is completed sometime in the 2020s, will take HLW from all Swedish nuclear facilities. As stipulated by environmental law since 2004, funds have been made available from the Nuclear Waste Fund to enable environmental groups and other NGOs to participate in the evaluation and public examination of Swedish nuclear waste management policy.

In Germany, siting for a geological repository for heat generating waste has been going on for almost 40 years. The process has been highly controversial and contributed to the rise of a powerful anti-nuclear-movement. In parallel with the decision after the Fukushima nuclear accident to phase-out nuclear power by 2022, Chancellor Angela Merkel stressed the importance of addressing long-term HLW disposal. In contrast with an earlier nuclear phase out decision made in 2001, which was resisted by Germany's conservative political parties, the 2011 phase-out decision won support across the political party spectrum. This improved the chances for mapping-out and implementing a site-selection process with wider political and societal support. The "Repository Site Selection Act," passed in 2013 by the two chambers of the German Parliament, stipulates a stepwise, criteria based, comparative selection process. The site selection is to be done with the guidance of a Commission for the Disposal of High-active Waste, which started its tenure in May 2014. The Commission represents different groups in society. Its mandate is to discuss basic questions of nuclear waste management, to evaluate the Site Selection Act, and to propose criteria for siting.

Although Germany has plans to abandon nuclear energy, the process of site selection remains a significant challenge. Strong local opposition that emerged in response to site explorations in Gorleben in 1977 and the years of sometimes violent protest associated with Castor transports mean that some in the public are distrustful of government and industrial actors when it comes to nuclear waste questions. The decades of controversy have contributed to the wicked nature of nuclear waste siting in Germany.

In the United States, politicians have failed for over four decades to find a solution to the country's growing nuclear waste problem. As of 2012, about 67,500 metric tons of high-level radioactive waste was being temporarily stored in about 75 closed and still-operating nuclear plant sites (Jeffrey and Johnsson 2012). The Nuclear Waste Policy Act Amendments of 1987 (NWPAA) directed the Department of Energy to study Yucca Mountain, Nevada to determine if it was suitable to serve as a nuclear waste repository. In 2002, Congress designated Yucca Mountain as a long-term nuclear waste repository, but the decision was highly contested. Strong local opposition and a campaign against the designation eventually led President Barack Obama to stop funding for research at the site in 2010.

The search for a repository site continues. In 2012, four states, a Native American community, and several environmental groups filed suit in the U.S. Court of Appeals in Washington, D.C., petitioning for a review of the Nuclear Regulatory Commission's "waste confidence decision", which determined that a permanent repository will be available "when necessary" (and not by a specific date) and that spent nuclear fuel can be safely stored at plants for at least 60 years beyond the licensed life of the plant. The U.S. Court of Appeals ruled in favor of the plaintiffs, determining that the Nuclear Regulatory Commission had failed to fully evaluate the environmental risks associated with storing spent nuclear fuel on site and to properly examine future dangers and consequences.
The finding of the court is significant: "Due to the government's failure to establish a final place for spent fuel, spent nuclear fuel (SNF) is currently stored on site at nuclear plants. This type of storage, optimistically labeled 'temporary storage,' has been used for decades longer than originally anticipated. The delay has required plants to expand storage pools and to pack SNF more densely within them. The lack of progress on a permanent repository has caused considerable uncertainty regarding the environmental effects of temporary SNF storage and the reasonableness of continuing to license and relicense nuclear reactors." The NRC was ordered to draft new rules (Environmental Law Report (ELR) 2012).

In 2013, the powerful U.S. Court of Appeals of the District of Columbia Circuit rebuked the Nuclear Regulatory Commission for its failure to find a long-term disposal site for HLW. This ruling put renewed pressure on the Nuclear Regulatory Commission to study the Yucca Mountain site. In August 2013, the U.S. Court of Appeals ordered the NRC to review a license application for the

Yucca Mountain site. The state of Nevada has appealed the decision with the Federal Appeals Court. In 2013, in a Hearing before the U.S. Congress on the 2013 Nuclear Waste Administration Act, Secretary Ernest J. Moniz of the Department of Energy noted: "Promising experiences in other countries indicate that a consent-based process, developed through engagement with states, tribes, local governments, key stakeholders, and the public, offers a greater probability of success than a top down approach to siting."[11]

The United States not only has a large stockpile of nuclear waste from its commercial NPP, but also a stockpile related to its defense activities. The U.S. Department of Energy's Waste Isolation Pilot Plant (WIPP), located in Carlsbad, New Mexico, was established for DGD of transuranic (TRU) nuclear waste—items, such as rags and clothing, that have been contaminated with plutonium or other radioactive materials during Cold War era atomic research. TRU waste has been disposed of at WIPP since 1999. WIPP opened after decades of controversy and is the world's only currently operating facility for DGD of nuclear waste. The WIPP experienced a waste drum explosion in February 2014 that led to a radiation leak and concerns about the condition of other waste drums in the facility (Butler 2014).

The Czech Republic has set its eyes on irreversible, irretrievable DGD of HLW in granite. The government focused on a half-dozen locations for testing as potential sites. In response to the government's actions, NGOs and local communities organised 28 valid local referenda between 2003 and 2013. In all of these referenda, the inhabitants of the concerned municipalities refused the exploration for a nuclear repository in their region. Although there is to be monetary compensation for communities where there is exploration for siting for a nuclear waste repository, the municipalities remain opposed. A "Working group for dialogue on the deep repository involving representatives of municipalities" was set up in 2010 with the aim of improving communication with the communities concerned, but later was disbanded by the Ministry of Industry and Trade This belated attempt to involve the municipalities in the decision-making process failed. Here, as in Germany communication, mistakes of the past continue to make site selection difficult.

11 See Hearing Before the Committee on Energy and Natural Resources, United States Senate, 113th Congress, first session, to receive testimony on S. 1240, The Nuclear Waste Administration Act of July 30, 2013, http://www.gpo.gov/fdsys/pkg/CHRG-113shrg 85875/pdf/CHRG-113shrg85875.pdf, last accessed 25 September 2014.

6.3 *Countries with surface long-term storage for HLW (The Netherlands, Italy and Spain)*

The Netherlands, Italy and Spain are pursuing surface long-term storage for HLW in centralized facilities. The Spanish government has postponed making a decision on final DGD of nuclear waste with the idea that technological innovations may provide alternative possibilities for managing radioactive waste. A decision on a DGD site is not expected before 2050. Since 1992, LLW and ILW has been stored at El Cabril in a near-surface repository. In 2006, a political consensus was reached in parliament regarding the need to develop an interim storage facility. After a delay, at the end of 2011, the government designated the municipality Villar de Cañas (Castilla la Mancha) in the province of Cuenca, a locality with 455 inhabitants, to host the site for the ATC. Villar de Cañas was one of the 14 towns which volunteered to host the facility. There were Not In My BackYard (NIMBY) movements and complaints from autonomous communities about non-transparent and non-inclusive decision-making procedures, but despite this opposition, the facility is to be constructed with a start date foreseen for 2016. The design will be similar to the *Hoogradioactief Afval Behandelingsen Opslag Gebouw* (HABOG facility) at the NPP in Borssele in the Netherlands.

In the Netherlands, both nuclear energy production and the treatment of nuclear waste have been controversial. Compared with other countries examined in this book, the Netherlands' volume of nuclear waste is relatively small, as there is only one operational NPP. There has been strong local resistance to research and experimentation with DGD. Test drilling in the north was stopped due to local opposition, although local and regional authorities have little authority under the Dutch nuclear energy law. The Dutch government favors retrievability. In the meantime, it is using the centralized, long-term interim storage facility at Borssele. The buildings are to be robust enough to last at least 100 years. During this storage time, funds will be saved and research in international collaborative programs on waste technologies will be carried out. The hope is that future technological innovations will offer low hazard treatments that will simplify finding a final national or international disposal method.

Due to harsh opposition, Italy has also abandoned its original plans for a deep geological repository and opted instead for a centralised surface storage facility for LLW and ILW, which should also be able to temporarily house HLW. A technology park with a centre for R&D and innovation in the field of decommissioning and radioactive waste management is also planned at the repository. There is still no information on potential sites, only a set of criteria and guidelines for site selection issued by the organisation acting as regulator.

The non-transparent top down decision for a centralized surface storage unit is based on the governmental assumption that transferring the waste there will ensure maximum safety for the population and the environment and minimize costs. There is little information available on the storage time frames being considered and no independent cost assessments. There is a strong implementer in charge of the decommissioning of NPPs but no dedicated regulator. These functions are undertaken by the Institute for Environmental Protection and Research ISPRA.

7 Conclusions: dealing with wicked challenges

Nuclear energy has had a turbulent history. As is often the case with technologies, the excitement, power and promise of the new technology persuaded publics and politicians alike to move forward with building nuclear energy facilities. The hope and belief of many was that nuclear energy would offer large supplies of cheap ("too cheap to meter"), safe, near endless and climate-friendly energy. While there are certainly still many countries that continue to support nuclear energy, there are growing numbers of countries that are exhibiting more caution and restraint, and a substantial number that have chosen to abandon or to not begin with nuclear energy. There are many reasons for this shift in nuclear energy's fortunes – they range from concerns about safety to problems with dealing with nuclear energy's back end. Here we focus in particular on the wicked problem of nuclear energy and nuclear waste governance, the Achille's heel of nuclear energy production. Regardless of whether a country chooses to continue with nuclear energy production, reprocess spent nuclear fuel, or shut down its nuclear energy production capacities, nuclear waste must be dealt with – and for periods of time that extend beyond all human experience. High level radioactive waste must be managed safely for at least tens of thousands of years. Some questions regarding this time dimension of nuclear waste disposal simply cannot be answered with certainty.

Wicked problems are complex, imperfectly understood and have no easy solution. Problem solving is exacerbated by the reality that those who are asked to bear the costs of nuclear waste management are often not the ones who have benefitted from its use. In the case of nuclear waste, this is true both intra- and inter-generationally. As has been stated by the World Social Science Report WSSR: "The question of how societies manage (or fail to manage) this imbalance of private goods and public 'bads' forms the central problem for environmental and sustainable governance" (ISSC and UNESCO 2013: 413). Different countries have different local and regional wicked problems to manage and solve related to nuclear energy production and waste. In terms of nuclear energy production, there are societal acceptance, cost, and risk factors that posechallenges to the nuclear industry. Nuclear energy is produced for today's use, but the costs of its production impacts future generations in both foreseeable and unforeseeable ways. There is also the concern for nuclear proliferation. It is difficult to fully estimate the cost of the treatment of nuclear waste. These costs have not been fully included in nuclear energy production cost calculations. All of these factors make nuclear waste management an immensely complex process. Moreover, there are many unanswered and unanswerable questions, e.g. regarding new disposal technologies, the safety of transporting nuclear waste and long-term containment.

Decision makers also must deal with the uncertainty of political processes and public acceptance. On the societal and political levels, nuclear waste is often highly politicized. Protest-movements are common and in many countries governments and the nuclear industry have made their own challenges greater by underestimating the importance of public involvement. Top-down decision making regarding facility siting, which was common in the past, has recently elicited wide-spread frustration and protests and contributed to an atmosphere of distrust. Many factors influence the politics of nuclear waste siting, including the country's political system, its geological context, the levels of available technical expertise, the priorities of local municipalities, the strength and outlook of social movements, legal requirements and informal rules and procedures, as well as the country's specific nuclear history. Highly radioactive nuclear waste management is a field where we still have little positive experience. It will require new democratic standards, rules, and procedures as well as long-term monitoring and enforcement to ensure the highest possible levels of safety. Only in a small number of cases, most notably Finland and Sweden, has it been possible to select sites on a voluntaristic basis with relatively broad local support. This was done with public debate and discussion in areas already hosting nuclear facilities, but also with assurances of economic benefits to the region.

In the past, governments tended to employ top-down approaches to identifying sites for nuclear energy facilities. For too long, nuclear energy policy making has been dominated by clientelism, patronage, and a *fait acompli politics*. As demands for democracy have intensified in many parts of the world, both government and industry have been forced to change their strategies and to accept more public participation. Decision-making processes that were viewed as non-democratic, non-inclusionary, and non-transparent played a role in the rise of a variety of NGOs, anti-nuclear-movements, local government associations and political parties that have become more deeply engaged in nuclear waste siting questions. Many of these have developed substantial scientific and societal expertise and wield considerable power and influence.

Learning from the past, governments are moving towards more inclusive, open, and participatory processes. There has been a shift from a more traditional *decide, announce and defend* model to a broader model of *communicate, participate and influence*. Without intense dialogue, participation, information availability and the *right of veto,* progress will be difficult. The promise of jobs, financial compensation and local investments could help ameliorate local opposition, but it could also lead to local resistence. Public debate and participation may still result in negative siting decisions and enormous delays in the building of repositories, but this is part of a democratic political process. Building consensus and acceptance takes time. Socio-ecological research for robust political and social action is required for humanity to stay within planetary boundaries and ensure socially just and sustainable development (Brand and Brunnengräber 2013). There is considerable work to be done in this regard. In many countries there are no reliable nuclear waste inventories, the legal and institutional framework is not yet set up and the implementation of the Euratom Directive is far from being accomplished.

Many questions regarding appropriate forms of public participation in the siting and monitoring processes, as well as technological, financial and economic matters, must be addressed. New regulatory and institutional frameworks that address operational safety and the appropriate management of nuclear waste still need to be developed. The international exchange of lessons learned will be extremely important.

References

Åhäll, K.-I., Linström. M., Holmstrand, O., Helander, B., and Goldstick, M. (1988). Nuclear Waste in Sweden – The Problem is Not Solved! The Peoples' Movement Against Nuclear Power and Weapons. http://www.folkkampanjen.se/nwfront.html#Contents, last accessed 11 August 2014.

Altvater, E., and Brunnengräber, A. (Eds.) (2011a). *After Cancún: Climate Governance or Climate Conflicts*. Wiesbaden: VS Verlag für Sozialwissenschaften, VS Research Energiepolitik und Klimaschutz.

Altvater, E., and Brunnengräber, A. (2011b). With the Market Against Climate Catastrophe – Can That Succeed? In Altvater, E. and Brunnengräber, A. (Eds.), *After Cancún: Climate Governance or Climate Conflicts*. Wiesbaden: VS Verlag für Sozialwissenschaften, VS Research Energiepolitik und Klimaschutz, 11-21.

Asahi Shimbun, The (2013, November 20). Government changes tack on finding disposal site for radioactive waste. http://ajw.asahi.com/article/behind_news/ politics/AJ201311200060, last accessed 8 August 2014.

Bergmans, A., Elam, M., Kos, D., Polič, M., Simmons, P.; Sundqvist, G., and Walls, J. (2008). Wanting the Unwanted: Effects of Public and Stakeholder Involvement in the Long-Term Management of Radioactive Waste and the Siting of Repository Facilities (Final Report CARL-Project). http://uahost.uantwerpen.be/ carlresearch/docs/20080222112500ZGYI.pdf, last accessed 16 September 2014.

Brand, U., and Brunnengräber, A. (2013). Debating transformation in multiple crises. In ISSC and UNESCO (Eds.), *World Social Science Report 2013. Changing global environments*. OECD Publishing and UNESCO Publishing. 479-484.

Broomby, R. (2014a, January 11). France struggles to cut down on nuclear power. *BBC*. http://www.bbc.com/news/magazine-25674581, last accessed 8 August 2014.

Broomby, R. (2014b, July 24). Local politicians to lose nuclear waste site veto right. *BBC*. http://www.bbc.com/news/science-environment-28463042, last accessed 11 August 2014.

Brunnengräber, A., Mez, L., Di Nucci, M. R., and Schreurs, M. (2012). Nukleare Entsorgung – ein „wicked" und höchst konfliktbehaftetes Gesellschaftsproblem. *TATuP – Technikfolgenabschätzung – Theorie und Praxis*, 59–64. http://www.tatup-journal.de/tatup123_brua12a.php, last accessed 16 September 2014.

Brunnengräber, A., Di Nucci, M. R., Häfner, D., and Isidoro Losada, A. M. (2014). Nuclear Waste Governance – ein *wicked problem* der Energiewende. In Brunnengräber, A., and Di Nucci, M. R. (Eds.), *Im Hürdenlauf zur Energiewende. Von Transformationen, Reformen und Innovationen*, Wiesbaden: Springer VS, 389-399.

Brunnengräber, A. (2014). Bridging the gap with agrofuels. Energy hunger, energy scarcity and climate change in the European Union. In Dietz, K., Engels, B., Pye, O., and Brunnengräber, A. (Eds.), *The Political Ecology of Agrofuels*, Routledge.

Butler, D. (2014, May 25). Waste facility on high alert over risk of new explosions. *Nature.com*. http://www.nature.com/news/nuclear-waste-facility-on-high-alert-over-risk-of-new-explosions-1.15290.

Dauvergne, P., and Neville, K. J. (2009). The Changing North-South and South-South Political Economy of Biofuels, *Third World Quarterly* 30 (6) 1087-1102.

Digges, Ch. (2012, August 28). Russia announces enormous finds of radioactive waste and nuclear reactors in Arctic seas. *Bellona*. http://bellona.org/news/ uncategorized/ 2012-08-russia-announces-enormous-finds-of-radioactive-waste-and-nuclear-reactors-in-arctic-seas, last accessed 11 August 2014.

Edahiro, J. (2012). Science Council of Japan Releases Policy Recommendation on High-Level Radioactive Waste Disposal. *JFS Newsletter*, No. 124, December 2012. http://www.japanfs.org/en/news/ archives/news_id032508.html, last accessed 11 August 2014.

European Commission (2014): A policy framework for climate and energy in the period from 2020 to 2030, COM(2014)15 final. Brussels.

Environmental Law Report (ELR) (2012). New York v. Nuclear Regulatory Commission, 42 (17), 20124, No. 11-1045 (D.C. Cir., 8 June 2012). https://elr.info/litigation/42/ 20124/new-york-v-nuclear-regulatory-commission, last access 16 September 2014.

Federaal agentschap voor nucleaire controle (FANC) (2014): Dossier Pressure Vessel Doel 3 & Tihange 2. http://www.fanc.fgov.be/nl/page/dossier-pressure-vessel-doel-3-tihange-2/1488.aspx?LG=2, last accessed 9 August 2014.

Focus Online (2014, August 7). Salafist droht mit Anschlägen in Deutschland. http://www.focus.de/politik/deutschland/terrorgefahr-in-deutschland-salafist-ruft-zu-anschlaegen-auf-auch-gegen-atombombenlager_id_4043909.html, last accessed 9 August 2014.

Guardian, The (2013, August 27). Vermont Yankee nuclear power plant to close by end 2014. http://www.theguardian.com/world/2013/aug/27/vermont-yankee-nuclear-plant-close-2014, last accessed 8 August 2014.

Gulyas, V. (2014, June 23). Hungary Parliament Approves Nuclear Plant Financing Deal with Russia. *The Wall Street Journal*. http://online.wsj.com/articles/hungary-parliament-approves-nuclear-plant-financing-deal-with-russia-1403547549, last accessed 9 August 2014.

Harrabin, R. (2014, January 30). Cumbria nuclear project rejected by councilors, *BBC*. http://www.bbc.com/news/uk-england-cumbria-21253673, last accessed 11 August 2014.

Haverkamp, H. (Ed.) (2014). Lifetime extension of ageing nuclear power plants: entering a new era of risk, Greenpeace Switzerland, March 2014. http://www.greenpeace.org/sweden/Global/sweden/karnkraft/dokument/2014/Outofage.pdf, last accessed 8 August 2014.

International Atomic Energy Agency (IAEA) (2013a). The Power Reactor Information System (PRIS) and Its Extension to Non-Electrical Applications, Decommissioning and Delayed Projects Information. Vienna, Austria. http://www.iaea.org/PRIS/WorldStatistics/OperationalReactorsBy Country.aspx, last accessed 05 May 2014.

International Energy Agency (IEA) (2012a). *World Energy Outlook 2012*. Paris: International Energy Agency.

IEA (2012b). Golden Rules for a Golden Age of Gas, World Energy Outlook. Special Report on Unconventional Gas. Paris: International Energy Agency. http://www.worldenergyoutlook.org/media/weowebsite/2012/goldenrules/WEO2012_GoldenRulesReport.pdf, last accessed 6 December 2013.

IEA (2013). World Energy Outlook Special Report 2013: Redrawing the Energy Climate Map, Paris: International Energy Agency. http://www.worldenergyoutlook.org/ media/weowebsite/2013/energyclimatemap/RedrawingEnergyClimateMap.pdf, last accessed 9 October 2014.

Intergovernmental Panel on Climate Change (IPCC) (2014). *Climate Change 2014: Mitigation of Climate Change*. Prepared by Working Group III contribution to the IPCC Fifth Assessment Report, final draft, Genf. http://mitigation2014.org/report, last accessed 7 May 2014.

IPCC (2011). IPCC Special Report on Renewable Energy Sources and Climate Change Mitigation. Prepared by Working Group III of the Intergovernmental Panel on Climate Change, Cambridge University Press, Cambridge, United Kingdom and New York, NY, USA. http://srren.ipcc-wg3.de/report, last accessed 9 May 2014.

International Social Science Council/ United Nations Educational, Scientific and Cultural Organization (ISCC/ UNESCO) (Eds.) (2013). *World Social Science Report 2013, Changing Global Environments*. Paris: OECD Publishing/ UNESCO Publishing.

Japan Times (2014a, January 21). Japan's Nuclear Waste Problem. http://www.japan times.co.jp/opinion/2014/01/21/editorials/japans-nuclear-waste-problem/#.U-Z2O8a KiFI, last accessed 9 August 2014.

Japan Times (2014b, March 24). Japan to Return Weapon Grade Plutonium to the US. http://www.japantimes.co.jp/news/2014/03/24/national/japan-to-return-weapons-grade-plutonium-to-u-s/#.U-Z518aKiFI, last accessed 9 August 2014.

Jeffrey, D., and Johnsson, J. (2012, June 9). Nuclear Waste Storage Rules Thrown Out by U.S. Court. *Bloomberg*. http://www.bloomberg.com/news/2012-06-08/nuclearwaste-storage-rules-thrown-out-by-u-s-court.html, last accessed 16 September 2014.

Kennedy, D. (2011, June 14). Italy nuclear: Berlusconi accepts referendum blow. *BBC*. http://www.bbc.co.uk/news/world-europe-13741105, last accessed 12 May 2014.

Kim, S. J. (2014). *A Comparative Analysis of Nuclear Energy Policy between Germany and South Korea: Crisis Tendencies of Nuclear Energy*, Ph.D. Dissertation, Freie Universität Berlin, 20 June 2014.

Knight, S. (2009, September 18). How two Swedish towns vied for nuclear waste. *Financial Times Magazine*. http://www.ft.com/intl/cms/s/0/e8ea6602 -a322-11de-ba74-00144feabdc0.html#ax zz3A7cIf3yk, last accessed 11 August 2014.

Mez, L. (2011). The future of nuclear energy in the wake of Fukushima. In Netzer, N., and Steinhilber, J. (Eds.), *The end of nuclear energy? International perspectives after Fukushima*, Friedrich-Ebert-Stiftung, Berlin, 13-22.

Mez, L. (2012a). Germany's merger of energy and climate change policy, *Bulletin of the Atomic Scientists*, 68(6) 22–29.

Mez, L. (2012b). Nuclear energy – Any solution for sustainability and climate protection? *Energy Policy*, 48 (2012) 56–63.

Muroya, H. (2014, July 13). Ruling banning nuclear reactor restarts translated into English, Korean, Chinese. *The Asahi Shimbun*. http://ajw.asahi.com/ article/behind_ news/social_affairs/AJ201407130025, last accessed 16 September 2014.

Nuclear Threat Initiative (NTI) (2014): *National Operator for Radioactive Waste Management* (NO RAO). http://www.nti.org/facilities/920/.

OECD (2010). *Partnering for Long-Term Management of Radioactive Waste: Evolution and Current Practice in Thirteen Countries*, OECD Publishing. http://www.oecd-ilibrary.org/nuclear-energy/partnering-for-long-term-management-of-radioactive-waste_9789264083707-en, last accessed 7 May 2014.

Peter, L. (2013, January 25). Russia explores old nuclear waste dumps in Arctic. *BBC*. http://www.bbc.com/news/world-europe-21119774, last accessed 18 October 2014.

Power Reactor Information System (PRIS) (2014). Operational & Long-Term Shutdown Reactors by countries. http://www.iaea.org/PRIS/WorldStatistics/Operational ReactorsByCountry.aspx, last accessed 13 October 2014.

Russia Beyond the Headlines (2014, April 18). Russia, Bangladesh initial construction contract for Rupper NPP. http://rbth.com/news/2014/04/18/russia_bangladesh_ ini tial_construction_contract_for_ruppur_npp_36010.html, last accessed 8 August 2014.

Schneider, M., and Froggatt, A., with Hazemann, J. (2012). The World Nuclear Industry Status Report 2012. http://www.worldnuclearreport.org/IMG/pdf/2012MSC-WorldNuclearReport-EN-V2.pdf, last accessed 25 September 2014.

Schneider, M., and Froggatt, A., with Hosokava, K., Thomas, S., Yamaguchi, Y., and Hazemann, J. (2013). *The World Nuclear Industry Status Report 2013*. Retrieved from http://www.worldnuclearreport.org/IMG/pdf/20130716msc-worldnuclearreport2013-lr-v4.pdf, last accessed 25 September 2014.

Schneider, M., and Froggatt, A., with Ayukawa, Y., Burnie, Sh., Piria, R., Thomas, S., and Hazemann, J. (2014). *The World Nuclear Industry Status Report 2014*. http://www.worldnuclearreport.org/IMG/pdf/201408msc-worldnuclearreport2014-lr-v3.pdf, last accessed 25 September 2014.

Schreurs, M. A. (2013a). Orchestrating a Low-Carbon Energy Revolution Without Nuclear: Germany's Response to the Fukushima Nuclear Crisis. *Theoretical Inquires in Law* 14 (1), 83-108.

Schreurs, M. A. (2013b). The International Reaction to the Fukushima Nuclear Accident and Implications for Japan. In Schreurs, M. A.m and Yoshida, F. (Eds.), *Fukushima: A Political Economic Analysis of a Nuclear Disaster*, Sapporo: Hokkaido University Press, 1-20.

Schreurs, M. A. and Yoshida, F. (Eds.) (2013). *Fukushima: A Political Economic Analysis of a Nuclear Disaster*. Sapporo: Hokkaido University Press.

Schreurs, M. A. (2014a). The Ethical Dimension of Energy Policy Decisions: The Role of Advisory Bodies. *Zeitschrift für Politikwissenschaft*, Sonderband 2013, 119-138.

Schreurs, M. A. (2014b). The Ethics of Nuclear Energy: Germany's Energy Politics after Fukushima. *The Journal of Social Science*, 77, 9-29.

Smith, R. (2012, March 15). Cheap Natural Gas Unplugs U.S. Nuclear-Power Revival. *The Wall Street Journal*. http://online.wsj.com/news/articles/SB10001424052702304459804577281490129153610?mg=reno64-wsj&url=http%3A%2F%2Fonline.wsj.com%2Farticle%2FSB10001424052702304459804577281490129153610.html, last accessed 07.05.2014.

Sovacool, B. (2008). Valuing the greenhouse gas emissions from nuclear power: A critical survey. *Energy Policy* 36, 2950–2963.

Sachverständigenrat für Umweltfragen (SRU) (2013): Fracking for Shale Gas Production. A contribution to its appraisal in the context of energy and environment policy, The German Advisory Council on the Environment, Statement No. 18. Berlin. http://www.umweltrat.de/EN/TheGermanAdvisoryCouncilOnTheEnvironment/thegermanadvisorycouncilontheenvironment_node.html;jsessionid=32E6C2FF54D065BFDAB2E63A4EFBE439.1_cid325, last accessed 5 December 2013.

Takubo, M., and von Hippel, F. (2013). Ending Reprocessing in Japan. An Alternative Approach to Managing Japan`s spent nuclear fuel and separated plutonium, International Panel on Fissile Materials (IPMF). http://fissilematerials.org/ library/rr12.pdf, last accessed 13 October 2014.

Tirone, J., and Adelman, J. (2014, March 24). Japan's Plutonium Plans Stoke China Tensions on A Bomb Risk. *Bloomberg*. http://www.bloomberg.com/ news/2014-03- 23/japan-s-plutonium-potential-stokes-china-tensions-on-a-bomb-risk.html, last accessed 9 August 2014.

United States Energy Information Agency (EIA) (2013). How old are U.S. nuclear power plants and when was the last one built? http://www.eia.gov/tools/faqs/faq.cfm?id=228&t=21, last accessed 08 August 2014.

United States Nuclear Regulatory Commission (U.S.NRC) (2014). Fact Sheet on Reactor License Renewal. http://www.nrc.gov/reading-rm/doc-collections/fact-sheets/fs-reactor-license-renewal.html, last accessed 8 August 2014.

Wald, M. L. (2013, May 7). As Price of Nuclear Energy Drops, a Wisconsin Plant is Shut. *The New York Times*. http://www.nytimes.com/2013/05/08/ business/energy-environment/kewaunee-nuclear-power-plant-shutsdown.html?_r=0, last accessed 8 August 2014.

World Nuclear Association (WNA) (2014a). Nuclear Power in the World Today. http://www.world-nuclear.org/info/Current-and-Future-Generation /Nuclear -Power-in-the-World-Today/, last accessed 24 March 2014.

WNA (2014b). Nuclear Power in the USA. http://www.world-nuclear.org/ info/Country-Profiles/Countries-T-Z/USA--Nuclear-Power/, last accessed 8 August 2014.

WNA (2014c). Nuclear Power in France. http://www.world-nuclear. org/info/Country-Profiles/Countries-A-F/France/, last accessed 8 August 2014.

WNA (2014d). Nuclear Power in Russia. http://www.world-nuclear.org/ info/Country-Profiles/Countries-O-S/Russia--Nuclear-Power/, last accessed 25 September 2014.

WNA (2014e). Nuclear Power in South Korea. http://www.world-nuclear.org/info/ Country-Profiles/Countries-O-S/South-Korea/, last accessed 25 September 2014.

Welsh, I. (1993). The NIMBY Syndrome: It's Significance in the History of the Nuclear Debate in Britain. *The British Journal for the History of Science*, 26(1): 15-32.

An Open Door for Spent Fuel and Radioactive Waste Export?
The International and EU Framework

Maria Rosaria Di Nucci and Ana María Isidoro Losada

1 Introduction: The international and supra-national frameworks[1]

Worldwide, the majority of countries with nuclear activities have – or plan to have – national legislative, regulatory, and organisational frameworks. There is an implicit ethical principle or even a common understanding that each country should be responsible for the safe disposal of its own spent fuel elements and radioactive waste. Due to safety and proliferation concerns, however the disposal of radioactive waste should not be considered a national matter. National frameworks and policies are designed in accordance with standards and guidelines set at, and agreed to at the international level. These are laid down in international agreements and conventions.

At the international level, the most relevant organisations are the International Atomic Energy Agency (IAEA), the Nuclear Energy Agency (NEA) within the Organisation for Economic Cooperation and Development (OECD), as well as the International Commission on Radiological Protection (ICRP). The IAEA was established in 1957 as a United Nations organisation to carry out the "Atoms for Peace" programme. The Agency develops nuclear safety standards and promotes the achievement and maintenance of safety levels, as well as the protection of human health and the environment against ionizing radiation. Another major field of activity is the verification of compliance with the Non-Proliferation Treaty (NPT) and other agreements on the peaceful use of nuclear energy. The OECD Nuclear Energy Agency (NEA) was created in 1958 and currently counts 31 OECD countries as members. In 1975, the NEA Radioactive Waste Management Committee (RWMC) was created. It represents one of the

---

1 This chapter is a contribution of the Environmental Policy Research Centre (Forschungszentrum für Umweltpolitik – FFU) of the Freie Universität Berlin to the project "Multi Level Governance-Perspective on Management of Nuclear Waste Disposal. A Comparative Analysis: Actors, Instruments and Institutions" that is part of the inter-disciplinary ENTRIA research platform funded by the German Federal Ministry of Education and Research (BMBF/ 02S9082B).

early multinational initiatives to promote international co-operation and exchange of information in radioactive waste management (RWM). NEA is organised as an informal, international network so as to discuss issues of common concern, in particular the regulatory framework. Both the IAEA and the European Commission (EC) participate in the work of the RWMC committee. Within this committee, there are different stakeholder groups, made up of representatives from regulatory authorities, radioactive waste management and decommissioning organisations, policy making bodies, and R&D institutions from the NEA countries. The work of the RWMC is conducted in close cooperation with national high-level advisory bodies and transnational bodies such as the International Commission on Radiological Protection (ICRP) and Governance directorate of the OECD. In 2014 the RWMC established an expert group on the management of radioactive waste with a focus on aspects related to (long term) storage activities and transportation of spent fuel (SF) and radioactive waste (RW) (NEA 2014a). Additionally, in 2014 a joint IAEA and EC working group "Status and Trends" (STJWG) was formed under the auspices of the IAEA, with the aim of periodically informing on issues related to spent fuel and radioactive waste management (NEA 2014b).

The Joint Convention on the Safety of Spent Fuel Management and on the Safety of Radioactive Waste Management is the most significant international agreement in the nuclear waste management policy field. This convention was agreed upon in 1997 by a diplomatic conference convened by the IAEA. It became effective in June 2001 after various ratifications. The principles and requirements laid down in the Joint Convention are recognised internationally, but are non-binding and, in the case of non-compliance, there are no sanctions. The safety requirements of the IAEA are also not legally binding. However, most countries have incorporated these standards into their regulatory frameworks on a voluntary basis.

Due to the transnational safety issues, the management of radioactive waste and spent fuel is an area for regulation at the EU level. The competences and responsibilities regarding SF and RW are governed by the Treaty establishing the European Atomic Energy Community (Euratom Treaty), which is one of the so-called founding treaties of the European Community. It followed the Coal and Steel Community (CSC) Treaty, which was signed in 1951, and was established in 1957 by the same original six Member States – Belgium, France, Germany, Italy, Luxembourg, and the Netherlands. Unlike other treaties, such as the CSC or the European Economic Community (EEC) Treaty, which was also established in 1957, the terms of the Euratom Treaty have remained basically unchanged.

It aims to:

- promote the formation and development of Europe's nuclear industry;
- guarantee high standards of safety for the health of workers and the general public with respect to the dangers of ionizing radiation; and,
- ensure that nuclear materials are not diverted from civil to military use (European Nuclear Society 2007).

Euratom is party to international conventions. Although in the course of time many institutions have either been suppressed, as the CSC in 2002, or replaced, as the EEC (following the Lisbon Treaty in 2009), the Euratom has remained as a separate Community with a separate treaty and legal identity. This means that the principle of subsidiarity does not apply to the Euratom.

In 2009, the Council Directive 2009/71/Euratom on the safety of nuclear installations entered into effect.[2] The legislative procedure which gave life to the nuclear waste directive was lengthy and difficult. The Commission had initially prepared a proposal in 2003, and revised it in 2004, however it was not possible to reach a consensus within the Council (Council Directive COM (2004) 526). Finally, in 2010, a proposal was adopted (European Commission 2010) and the Directive 2011/70/Euratom became effective in August 2011.

In the following section, the conditions which were set for European nuclear waste policy by the Directive 2011/70/Euratom will be analysed. This analysis will be complemented by a discussion of the possibilities for international, multinational, and regional shared repositories and of potential advantages and disadvantages of these shared solutions. Finally, our focus turns to the impact of Directive 2011/70 on the attempts to develop regional disposal activities, such as the cooperative approach culminating in the European Repository Development Organisation (ERDO).

2 The main provisions of the Directive 2011/70/Euratom

The Council Directive 2011/70/Euratom of 19 July 2011 establishes a community framework for the responsible and safe management of spent fuel

2 In December 2013, a revision of the Basic Safety Standards, dating from 1996, was adopted in the New Council Directive. The new Directive 2013/59/Euratom (BSS Directive) reflects recent international developments, responds to technological advances, benefits from new recommendations of the ICRP, and broadens the application to the whole range of radiation sources and categories of exposure. For details, see http://eur-lex.europa.eu/LexUriServ/LexUriServ.do?uri=OJ:L:2014:013: 0001:0073:FR:PDF.

and radioactive waste within the European Union. The primary foundations for the Directive are the IAEA Safety Standards and the Joint Convention. Member States (MS) are mandated to put into effect the regulations and administrative provisions necessary to comply with the Directive by August 2015. The Directive imposes obligations on all Euratom countries and asserts in its Preamble (Recital 24) that it is an 'ethical obligation' that EU Member States "avoid any undue burden on future generations in respect of spent fuel and radioactive waste including any radioactive waste expected from decommissioning of existing nuclear installations". Furthermore, the Preamble (Recital 25) of the Directive states that the: "principle of national responsibility, as well as the principle of prime responsibility of the licence holder for the safety of spent fuel and radioactive waste management under the supervision of its competent regulatory authority, should be enhanced and the role and independence of the competent regulatory authority should be reinforced by this Directive".

Chapter 1 of the Directive (Art. 1-4) concerns scope, definitions, and general principles. Subject matter, scope, and definitions consistent with the existing EU legislation and IAEA safety glossary are covered by Art. 1, Art. 2, and Art. 3 respectively. The Directive ensures that MS provide for national arrangements for a high level of safety, supplementing the basic standards referred to in the Euratom Treaty (Art. 4.1). The Directive covers all stages of the management of spent fuel and radioactive waste from civilian activities. It endorses the ultimate responsibility of MS for the management of SF and RW generated in their territories. Member States are obliged to establish and maintain national policies on SF and RW based on principles specified in Art 4. These principles include that: the generation of RW is to be kept to the minimum; interdependencies between all steps in SF and RW generation and management need to be taken into account; SF and RW are safely managed in the long term with passive safety features; SF and RW producers cover the costs for their management; and, the decision making process is to be evidence-based and documented.

Chapter 2 of the Directive concerns the obligations of the Euratom countries. Art 5 requires that MS establish and maintain a national legislative, regulatory, and organisational framework for spent fuel and radioactive waste management that allocates responsibility and provides for coordination between relevant competent bodies. The frameworks should include:

- a national programme for the implementation of SF and RW management (Art. 5a);
- national arrangements for the safety of SF and RW management (Art. 5b);

- a system of licensing of SF and RW management activities and facilities (Art. 5c);
- a system of appropriate control, a management system, regulatory inspections, documentation, and reporting obligations for SF and RW management activities and facilities, including appropriate measures for the post-closure periods of disposal facilities (Art. 5d);
- enforcement actions, including the suspension of activities and the modification, expiration, or revocation of a licence together with requirements, if appropriate, for alternative solutions that lead to improved safety Art. 5e;
- an allocation of responsibility to the bodies involved in the different steps of SF and RW management and an attribution of responsibility for SF and RW to their generators or, under specific circumstances, to a licence holder to whom this responsibility has been entrusted by competent bodies (Art. 5f);
- national requirements for public information and participation (Art. 5g); and,
- a financing scheme(s) for SF and RW management (Art. 5h) in accordance with Article 9.

Art. 6 requires that MS establish and maintain a competent regulatory authority in the field of safety of spent fuel and radioactive waste management (Art 6.1).

Art. 7 rules that responsibility for the safety lies with the licence holder (Art 7.1) and that – under the regulatory control of the competent regulatory authority – the licence holder shall assess, verify, and improve the safety of the SF and RW management facility or activity in a demonstrable way.

Art. 9 requires that adequate financial resources are available for the implementation of the national programmes for SF and RW management, taking due account of the responsibility of SF and RW generators.

Art. 10 regulates transparency and obliges the regulatory authority to inform the public of its competence (Art. 10.1). Moreover, it is required that the public participates "effectively" in the decision making process for SF and RW management "in accordance with national legislation and international obligations" (Art. 10.2).

Art. 11, 12 and 13 concern the national programmes to be implemented and regularly reviewed and updated (Art. 11), as well as procedures regarding the notification of these programmes and subsequent changes (Art. 13). Art. 12 lays down the major contents of the national programmes, i.e. overall objectives of the national policy; milestones and timeframes; inventory, including estimates for future volumes of waste; concepts or plans and technical solutions for the post-closure period of a disposal facility; RD&D; and key performance indicators to monitor progress, cost assessment practices, and the financial scheme(s) in effect. Finally, Art 12 1.k envisages agreements with a MS or a third country on SF and RW management, including the use of a disposal facility abroad.

Art. 14 lays down the procedures concerning periodical reporting. Member States are obliged to organise international peer reviews of their national framework at least every 10 years.

Lastly, Chapter 3 encompasses final provisions. Whereas Art. 15 regards the transposition of the Directive and requires MS to submit the first reports to the Commission on the implementation of the Directive by 23 August 2015, Art. 16 states the date the Directive goes into effect and Art.17 specifies the addressees.

The 2011/70/Euratom Directive is not immune to criticism. Already the revised proposal for a Council Directive on the management of spent fuel and radioactive waste – and the respective consultation process – has been criticised, especially since it resulted in a "watering-down" of the initial proposals. In a number of technical and juridical commentaries related to the Council Directive 2011/70/Euratom, the requirements are regarded as insubstantial. For example, they are said to be solely a common framework for a minimal consensus. The directive has been criticised for being too "soft'", since "it was not possible to reach a majority in the Council in a more harmonised approach" (Södersten 2012: 1). The requirements largely replicate the international framework and are considered to have a limited value added. Nevertheless, it should be conceded that this Directive, together with the Directive on Nuclear Safety of 2009, sets for the first time a binding regulation in the fields of nuclear waste. It also obliges MS to present timelines for the siting and construction of disposal facilities. Both Directives are so-called framework Directives, which similarly to "laws" are binding, but leave discretionary autonomy for the MS in setting and implementing their own national frameworks.

3 A loophole for the "export" of spent fuel and radioactive nuclear waste?

During the first reading of the Directive, there were differences in the positions of Council members regarding the export of radioactive waste. Both Commission and the European Parliament were in favour of a ban on the export of SF and RW to non-EU countries.[3] In the end, the Council agreed that under certain

3 "In its initial directive proposal, the Commission had advocated a complete export ban in the new directive. On 23 June 2011, the European Parliament in its plenary session voted in favour of a complete export ban as proposed by the Commission. As the legal basis for this directive is the Euratom Treaty, the European Parliament was only consulted and the opinion is therefore

restrictions nuclear waste could continue to be shipped to other countries. The possibilities for exporting, shared facilities, and multi-national repositories remain much debated and controversial issues.

Article 4.1 of the 2011/70/Euratom Directive specifies that "each MS *shall* have ultimate responsibility for management of the spent fuel and radioactive waste generated in it" (italics added for emphasis). Moreover, Article 4.2 stipulates that: "[w]here radioactive waste or spent fuel is shipped for processing or reprocessing to a Member State or a third country, the ultimate responsibility for the safe and responsible disposal of those materials, including any waste as by-product, *shall* remain with the Member States or third country from which the radioactive waste was shipped" (italics added for emphasis).

Article 4.4 includes the requirement that "[p]rior to a shipment to a third country, the exporting Member State shall inform the Commission of the content of any such agreement". This Article specifies the restrictions on the export of radioactive waste, but allows an exemption: a MS may export radioactive waste for disposal abroad when "at the time of shipment, an agreement, taking into account the criteria established by the Commission in accordance with Article 16(2) of Directive 2006/117/Euratom, has entered into force between Member State concerned and another Member State or a third country to use a disposal facility in one of them."[4] Hence, Article 4.4 can be interpreted as an opening for SF and RW disposal in third countries and as a possible alternative to a national disposal site. This interpretation tacitly implies the possibility of extending the export of radioactive waste to other countries.

In this context, Article 16 of the Council Directive 2006/117/Euratom of 20 November 2006 on the supervision and control of radioactive waste and spent fuel (the so-called Shipments Directive) prohibits MS from exporting to destinations "south of latitude 60° south", to "a State which is party to the Partnership Agreement between the members of the African, Caribbean and Pacific Group of States of the one part (Cotonou ACP-EC Agreement) which is not a Member State".[5] Furthermore it is prohibited to export to those third

not binding. The final decision was made only by the Council," European Commission press release, see http://europa.eu/rapid/press-release_IP-11-906_en.htm, last accessed 18 September 2014.

4 Further exemptions are included in Art. 2 of the Directive 2011/70, which refers to the authorisation to repatriate "disused sealed sources to a supplier or manufacturer" (Art. 2.3).

5 The Commission Recommendation 2008/956/Euratom of December 2008 points out that the decision to authorise shipments of radioactive waste and spent fuel to third countries falls into the responsibility of the exporting MS. Contrary to EU policy statements, current European legislation allows – under specified conditions – export out of the EU.
Since 2000 this has represented the framework for the European Community and its MS, with 79 countries from Africa, the Caribbean and the Pacific.

countries which, in view of the competent authorities of the MS of origin, do not meet or fulfil the requirements concerning the "administrative and technical capacity and regulatory structure to manage the radioactive waste or spent fuel safely" (European Council 2006: 28).

The Directive does not apply for shipments of:

- sources which are returned to a supplier, manufacturer, or authorised installation;
- radioactive substances covered through reprocessing and destined for different use;
- natural radioactive substances which do not result from treatment (European Council 2006: 23; Art. 1.3-1.5).

The MS opposing export and the anti-nuclear organisations, in particular, have criticised the Directive for not establishing a general ban on exports of RW and SF in a third country. Although export control requirements are specified, the possibility for export remains. For this reason, Sweden, Austria, and Luxembourg abstained from voting for the adoption of the Directive and expressed their disapproval in a joint declaration (European Council 2011a: 3). The exceptions envisaged in Article 4.4 "open up the possibility for exports of radioactive waste without sufficient mechanisms in place to ensure compliance by the recipient country with standards equivalent to those required by the Directive within the EU" (European Council 2011a: 2). The absence of an explicit ban on export is partly due to the fact that, within the framework of the Global Threat Reduction Initiative,[6] some MS requested exemptions for the return of spent fuel and advocated for the possibility of exporting.[7] The exception for export is also justified with the argument that some MS consider the development and use of shared repositories for the storage or disposal of radioactive waste and spent fuel as a potentially beneficial, economical and reliable option. In this sense, the Directive enables the possibility of action in this direction.

---

6 The Global Threat Reduction Initiative was launched in 2004 by the National Nuclear Security Administration, "a semi-autonomous agency within the U.S. Department of Energy [...which] is responsible for enhancing national security through the military application of nuclear science" (see www.nnsa.energy.gov/aboutus) in order "to reduce and protect vulnerable nuclear and radiological material located at civilian sites worldwide." For further details, see http://nnsa.energy.gov/aboutus/ourprograms/dnn/gtri, last accessed 16 September 2014.

7 This applies especially to signatories of the Global Threat Reduction Initiative that have running agreements with the Russian Federation and the United States.

4 Shared repositories and multinational/regional cooperation as a possible option?

The Preamble of the Joint Convention on the safety of SF and RW management and on the safety of radioactive waste management does not exclude the possibility that countries could develop a multinational or regional disposal concept. The Joint Convention states in its Preamble "that any State has the right to ban import into its territory of foreign spent fuel and radioactive waste" but concedes that "in certain circumstances, safe and efficient management of spent fuel and radioactive waste might be fostered through agreements among Contracting Parties to use facilities in one of them for the benefit of the other Parties" (IAEA 1997). This provision was meant for countries with limited volumes of spent fuel or radioactive waste or for those without adequate geological sites. In these cases, multinational disposal options would represent a more adequate and cost-effective solution.

In 2004 the Director General of the IAEA, El Baradei, set up a Multinational Nuclear Approaches Expert Group (MNA) to develop, among other things, proposals for possible multilateral arrangements for radioactive waste disposal. Their conclusion concerning spent fuel disposal was: "At present there is no international market for spent fuel disposal services, as all undertakings are strictly national. The final disposal of spent fuel is thus a candidate for multilateral approaches. It offers major economic benefits and substantial non-proliferation benefits, although it presents legal, political and public acceptance challenges in many countries. […] The final disposal of spent fuel (and radioactive waste as well) in shared repositories must be looked at as only one element of a broader strategy of parallel options. National solutions will remain a first priority in many countries. This is the only approach for States with many nuclear power plants in operation or in past operation. Small countries should keep options open (national, regional or international), be it only to maintain a minimum national technical competence necessary to act in an international context" (MNA 2004 cited in IAEA 2011: 19).

As already mentioned, the Directive 2011/70/Euratom leaves room for cooperative solutions (Art. 4) and "regional" storage and disposal facilities represent a legal possibility also within the European Union. The option for disposal outside the European Union is only contemplated when the country of destination has radioactive waste management and disposal programmes with a high level of safety comparable to that established by the Directive (Art. 4.4b). Additionally, it is possible when "the disposal facility in the country of destination is authorised for the radioactive waste to be shipped, is operating prior to the shipment, and is managed in accordance with the requirements set down in the radioactive waste management and disposal programme of that country of destination" (Art. 4.4c). In Recital 32 and 33 of the Preamble it is stated that "[c]ooperation between Member States and at an international level could facilitate and accelerate decision-making through access to expertise and technology" and that: "[s]ome Member States consider that the sharing of facilities for spent fuel and radioactive waste management, including disposal facilities, is a potentially beneficial, safe and cost-effective option when based on an agreement between the Member States concerned."

4.1 The challenge of sharing repositories

Organisations such as the IAEA and the EU have a positive attitude towards the concept of shared repositories, whereas the NEA has not formulated an official position on this issue. In the last two decades the IAEA has encouraged research on regional, international, and multinational disposal options. As a result, a number of initiatives which give serious consideration to concepts focusing on multinational approaches to the back-end of the nuclear fuel cycle have been promoted. These initiatives claim to represent a viable alternative which would ensure non-proliferation and mitigate security and energy concerns. Nevertheless, the ambition to promote binding international norms that regulate such international and multinational undertakings has failed so far. National sovereignty considerations play a relevant role in this. Unresolved and critical issues include how and under whose guidance the required legal and contractual frameworks should be designed, and how the relations between the national entities and European Commission of the EU and the IAEA should be organised (Štefula 2006: 4). Amongst the many relevant open questions, International Panel on Fissile Materials (IPFM) asks about: "how the host facility would be funded and managed; the rights of the host country to terminate foreign use of the facility and to return the spent fuel; the rights of foreign users to withdraw their spent fuel before it is finally disposed; and the coverage of IAEA safeguards if the facility is shared by weapon states and non-weapon states" (2011: 119).

Multinational repositories, i.e. repositories where more than one country disposes its waste in a shared facility, must be understood as a complement to national disposal programmes. Some studies have argued that multilateral storage facilities and repositories can benefit from economies of scale and be more cost-effective than a system of separate national facilities. These studies have outlined important factors that would have to be addressed in the development of multinational facilities (IAEA 2005; IPMF 2011).[8] The advocates of this solution, however, are aware of the need to resolve several legal and institutional issues connected to multilateral facilities, as well as the importance of addressing political, social, and public acceptance issues. Countries such as Belgium, Bulgaria, Hungary, Italy, the Netherlands and Norway manifested interest in multinational repository solutions because of geologic factors or the size of their national nuclear programmes (IAEA 2004: 1). It is often adduced that such a solution can be justified on grounds of safety and security. A reduction in the number of disposal facilities (under multinational control) would decrease the risk that in some countries there could be inadequate safety provisions. As already mentioned, this solution could also be justified on non-proliferation grounds. In spite of the potential advantages however, the major challenge to shared facilities remains the international transportation of spent fuel. According to IPFM, "there has not been a thorough analysis of this issue. The IAEA study on multilateral approaches does emphasize the obligation of countries to ensure that any transport will be done safely. There is, of course, a considerable history of international transport of both spent fuel and high-level reprocessing waste due to France, Russia and the UK providing reprocessing services to other countries" (2011: 118).

4.2 IAEA scenarios

The International Atomic Energy Agency considers three different scenarios for developing and implementing multinational repositories. These depend on the specific circumstances in potential host countries.

Add-on scenario

In this scenario, a host country decides to expand its existing national repository and accepts the import of spent fuel and radioactive waste from other countries. The disposal facility remains a repository under *national* responsibility. The "add-on" scenario requires that the geological conditions in the hosting country are favourable for the development of a disposal facility and that this solution is financially and technically feasible. Moreover, sufficient political will is an essential prerequisite. In this scenario the facility would first be implemented as a national repository, and then later on, disposal services would be offered to other interested countries. Adequate candidates for hosting repositories would be large countries with a significant radioactive waste inventory. On the one hand

8 For further details related to the viability of multinational initiatives see IAEA (1998; 2004; 2011).

countries “with small nuclear waste inventories and/or complex geological environments”, for example Slovenia, Taiwan and Switzerland, and countries with only spent fuel from research reactors, for example Australia, Denmark and Norway represent the most obvious partners (IAEA 2004: 19). The motivation for getting involved in such activities can be attributed to various reasons, including: “straightforward business initiative; desire to share repository development costs; willingness to help neighbours [...]; interest in reducing global security risks; commitment to reduce the number of disposal sites worldwide; opportunity to trade its offer to take radioactive waste from its partners for some other national goal to which all partner countries can contribute” (Margeanu 2012: 50).

Cooperation

In this scenario, a group of countries makes a mutual agreement to develop a shared disposal concept and identifies a potential host, either within the territory of the participating states or in a third country.
Three variants can be outlined:

a. Several countries with small nuclear programmes agree to share a disposal facility in a host state which meets the relevant essential geological and technical requirements. This scenario represents a potential approach to spent fuel within the European Union. The socio-political aspects are not explicitly mentioned.

b. Countries with small volumes of RW and SF and which face difficulties in implementing a domestic disposal concept cooperate to ensure that in one country of the group the necessary technology is available and the necessary infrastructures are implemented. Amongst the potential countries are those operating research reactors or those that have only a limited nuclear power plant park. In 2003 and 2004, competent authorities and stakeholders from Austria, Bulgaria, Croatia, the Czech Republic, Hungary, Slovakia, and Slovenia met in order to discuss the possibility of multinational, regional radioactive waste disposal solutions (Newman 2013).

c. Countries implementing disposal facilities for specific types of wastes with the possibility to combine them with commercial arrangements for international exchange of different types of waste, e.g. the collection of spent sealed sources for disposal or the replacement of LILW with lower quantities of HLW. Examples of this option are the exchange-agreement between Sweden and Germany involving German spent mixed-oxide (MOX) fuel and the transport of 120 drums of low-level waste from Spain to the United States in August 2000 (Newman 2013).

International/ supranational
In relation to the collaborative disposal options, the IAEA distinguishes between the concept of a "regional repository" and of an "international repository", whereby the *regional* concept refers to multinational facilities in which the host country and the collaborating countries belong to the same geographic region, while an *international* concept implies that the radioactive waste disposal facility or a network of facilities are organised under the supervision of a supra-national body e.g. the United Nations (IAEA 2004: 5). In this context the IAEA states: "Willingness to cede land for a full fuel cycle nuclear park might be more easily found since this could bring more high-tech activities into a country than is the case with disposal alone" (IAEA 2011: 33). The implementation of this model implies a partial renouncement of national sovereignty and delegation to an international body. Currently this is considered to be the most unlikely scenario.

4.3 Attempts for multinational repositories and ongoing initiatives

The idea of shared or multinational repositories dates back to the 1970s, but hardly any endeavours have since proved feasible. In the mid-1990s, multinational efforts to develop an international repository on the Marshall and Wake/Palmyra Islands in the Pacific and the Pangea-Project in Western Australia had to be abandoned due to opposition from the United Stated government and strong public and political opposition (IAEA 2004: 13). To date, there has been very little progress in presenting viable concepts. Amongst the initiatives developed with the participation of Euratom countries, Arius and ERDO are particularly worth mentioning. Table 1 below summarises a number of such proposals and attempts.

The Association for Regional and International Underground Storage (Arius), is a non-commercial organisation founded in 2002 and based in Switzerland. Founding members include Belgium (ONDRAF Waste Agency), Bulgaria (Kozloduy Power Plant), Hungary (PURAM Waste Agency), Japan (Obayashi Corporation) and Switzerland (Colenco Power Engineering, backed by two of the Swiss nuclear power utilities). The Association aims to "promote concepts for socially acceptable, international and regional solutions for environmentally safe, secure, and economic storage and disposal of long lived radioactive wastes".

Figure 1: Overview of the development of proposals for shared multinational waste management facilities

Date	Proposals	Proponents	Description
1970s and 1980s	International Spent Fuel Management	IAEA/ Expert group	Discussion of key elements of international agreements to be drawn up for an international SF venture. It ran from 1979 to 1982.
	International Plutonium (Pu) Storage	IAEA/ Expert group	Proposition that all separated Pu exceeding current requirements would be stored under international control.
	Austrian fuel to China or to the former U.S.S.R.	Verbundgesellschaft (responsible for the Zwentendorf reactor)	Contracts were signed with China and the U.S.S.R. for receiving all RW. Following the Austrian decision not to open the plant, these agreements were revoked.
1990s	Spent fuel and plutonium storage (IMRSS)	Germany and USA	German and US institutions launched concept for an "International Monitored Retrievable Storage". Discussed at international conferences; no negotiations followed.
	Disposal of SF and RW; reve-nues to be used for N test site remediation	Marshall Islands to host	Strong opposition from Pacific countries and USA; feasibility study started. Initiative terminated in 1999.
	Non- proliferation; SF and Pu storage; fuel leasing	USA Fuel and Security (private) as host, including a Russian partner	Proposed storage of SF and excess Pu on Palmyra and later, Wake Island. Opposed by the US Government; supported by Minatom (Russian Federation); abandoned in favour of Non-Proliferation Trust.
	International storage and SF disposal	IAEA and then an International Working Group	IAEA started an initiative. Work continued by experts from Australia, China, Germany, South Africa, and Switzerland. Feasibility report published.
	Storage of SF and RW in Asia	Taiwan, South Korea and Japan	Various concepts discussed, but no formal initiative.
		China a potential host to Taiwan SF	Letter of intent but no formal agreement.
	Russian Federation to host SF and RW	A. Suzuki, Global Peace Initiative	Storage and disposal in the Russian Federation.
		Bunn Norwark	Storage in the Russian Federation.
		Kurchatov Institute	Emphasis on storage in Krasnoyarsk.

Date	Proposals	Proponents	Description
	Multinational disposal of SF and RW	Pangea: commercial initiative	Investigation of suitable geological regions worldwide. Western Australia identified as preferred region. Initiative ceased in 2000.
2000s	Storage and return of SF, or storage and reprocessing without return of Pu and HLW. Fuel leasing.	Russian Federation Minatom as host	Official Russian government policy; enabling legislation. Russian offer to customers. SF reprocessing plans.
			Lease fuel and take back SF to the Russian Federation.
	LLW storage and disposal	Kazakhstan Government	Proposal for a repository in a disused uranium mine; Kazakhstan Government in favour; debate in parliament.
	Develop framework for multinational disposal	IAEA/Working Group	WG met with representatives from Czech Republic, Hungary, Germany, Slovenia, South Africa, Switzerland and USA. This was followed by a Technical Committee meeting to review the draft of TECDOC 1413.
	Promotion of multinational storage disposal	Arius (Belgium, Bulgaria, Hungary, Japan, Italy and Switzerland)	Promoting concepts for international or regional solutions to the disposal of HLW (ongoing initiative).
	Promotion of cooperative regional disposal options in the EU	ERDO-WG (Austria, Bulgaria, Ireland, Italy, Lithuania, Netherlands, Poland, Romania, Slovakia, and Slovenia); first meeting in 2009.	Regional solutions to the disposal of long lived radioactive waste (ongoing initiative).

Source: adapted from IAEA (2004: 8) and
European Repository Development Organisation (ERDO) (n.d.).

Within the 6th Framework Programme of the European Arius and DECOM Slovakia, carried out a pilot study called Support Action: Pilot Initiative on European Regional Repository (SAPIERR) in 2003, which focused on a regional approach to waste disposal. This initiative was also supported by a large number of EU MS. As a follow-up to this pilot study, the EC funded the SAPIERR project (Strategic Action Plan for Implementation of European Regional Repositories) from 2006 to 2009; the project investigated the feasibility of a European solution for shared, regional geological repositories. A strong focus was placed on legal, liability, and economic issues as well as safety and security aspects.

As a primary outcome, 14 countries decided to set up the Working Group European Repository Development Organisation, ERDO WG, which provides a forum for countries interested in a cooperative solution. The ERDO-WG secretariat is provided by Arius and the administration is provided by COVRA, the Dutch radioactive waste agency. At the end of 2011, the ERDO WG presented a possible structure and financing options for a formal, multinational, European waste management agency, the ERDO. The respective documents were submitted to interested EU governments and information was also provided to countries that pursue national radioactive waste disposal solutions, such as Sweden, Finland, and France. In a model under development, a second organisation, ERO is considered. The European Repository Organisation should be tasked with the implementation and operation of the repository (ERDO n.d.).

In recent years, Arius has also started to investigate whether the ERDO model can be adapted for use in other regions with existing or potentially new nuclear power countries like the Arabian Gulf, South East Asia, North Africa, and Central and South America.

5 Conclusions

The implementation of the 2011/70 Euratom Directive obliges Member States to present national plans for the final disposal of spent fuel and radioactive waste, identify sites and set up timelines for the construction of the facilities. The Euratom Directive has been criticized for a lack of specific targets and for the fact that it does not explicitly ban exporting nuclear waste to other countries. Until now, no country has accepted the storage or disposal of (highly) radioactive materials. In France and the United Kingdom, bilateral contracts stipulate that the high level wastes (recovered uranium and plutonium) from reprocessing have to be sent back to the country of origin.

In Russia, the national legislation is vague and does not include an explicit ban on the import of foreign SF and RW. The domestic spent fuel management policy has to be seen in the context of the broader national strategy to achieve a closed fuel cycle. As a legacy of the Soviet-era and as a tacit policy, Russia still takes back the SF which originated from Soviet-era designed nuclear power plants for reprocessing (IPFM 2011: 18). Currently only Bulgaria transfers its spent fuel to the Russian Federation.[9] Considering the critical situation between

9 It is unclear if the annexation of Crimea by Russia in April 2014 may lead to changes in this context. For more details see IPFM (2011) in particular the chapter on Russia, pp. 70-77.

the Russian Federation and the Ukraine, it is questionable and uncertain that the Ukrainian spent fuel will still be transferred to the Russian Federation.

The fact that a number of countries are opting for interim solutions, such as centralised (near) surface storage facilities for both low and medium radioactive waste and HLW, can be interpreted as an indicator of a possible interest in shared facilities and the export option. However, countries pursuing a national strategy oppose these shared solutions. Thus, Finland´s Nuclear Energy Act, amended in 1994, states that radioactive waste generated in Finland shall be handled, stored, and permanently disposed of in the country and excludes the handling, storage, or permanent disposal of SNF produced anywhere else. This means that an import of nuclear waste from other countries is forbidden – and this was agreed upon in the membership negotiations before Finland joined the EU in 1995.

As already discussed, there are many hurdles to implementing regional, international, and multinational solutions. At least within the EU, it can be assumed that the construction of a regional repository could encounter exactly the same problems with siting and public acceptance as the process has at a national level. Apart from acceptance issues, which make the search for an adequate siting a difficult enough problem at a national level, there are normative, ethical, and environmental justice criteria that, individually and together, speak against such attempts. But there are not only acceptance problems and difficult technical arrangements associated with the RW and SF transport logistics; there is also the fact that an especially complex legal, financial, and political framework would be necessary for such multinational facilities. Much work is still necessary in the next decade. In spite of the possible loopholes, the Euratom Directive clearly states that MS have an ethical responsibility to dispose of their own SF and RW.

In order to establish a jointly owned European Waste Agency and comply with the EU Council Directive 2011/70/Euratom, a concise development plan and detailed work schedule by the respective participating countries is necessary. Consequently, the MS that would be potentially involved in such an initiative would need to pursue a "dual track" approach. This consists of setting up, on the one hand, the framework for a national plan, and on the other, a shared programme. The task for the next decade is to investigate suitable sites, and then in the period 2020-2025, under the future organisational structure of a "European Repository Organisation", to identify a preferred site and formulate binding agreements (Margeanu 2012: 56).

The ERDO initiative is something to be followed closely; however in the short to medium run it appears that no country is prepared to house a "regional" or multinational spent fuel and radioactive waste facility. As far as third-party, non-EU countries are concerned, the matter could be even more complex. Compensation is likely to play an important role in finding a multinational solution and that less privileged countries would be willing to host repositories. From an ethical and environmental justice perspective, however, this option can hardly be taken into consideration by industrialised countries.

References

European Commission (2010). Commission Staff Working Document, Accompanying document to the revised proposal for a Council Directive (Euratom) on the Management of Spent Fuel and Radioactive Waste, SEC/2010/1289 final – NLE 2010/0306.

European Council (2006). Council Directive 2006/117/EURATOM of 20 November 2006 on the supervision and control of shipments of radioactive waste and spent fuel. *Official Journal of the European Union* – OJ L 337, 20 November 2006.

European Council (2009). Council Directive 2009/71/Euratom of 25 June 2009 establishing a Community framework on the nuclear safety of nuclear installations. *Official Journal of the European Union* – OJ L 172, 2 July 2009.

European Council (2011a). Proposal for a Council Directive (Euratom) on the management of spent fuel and radioactive waste – Adoption. 8 July 2011, Brussels, 12248/11., last accessed 31 May 2014.

European Council (2011b). Council Directive 2011/70/Euratom of 19 July 2011 establishing a Community framework for the responsible and safe management of spent fuel and radioactive waste. OJ L 199/48, 2 August 2011.

European Nuclear Energy Forum (ENEF) Working Group "Risks" (2009). Roadmap to successful implementation of geological disposal in the EU. Sub Working Group "Waste Management", 29 October 2009. EUR 24301EN. Luxembourg: Office for Official Publications of the European Communities.

European Nuclear Society (2007). 50 years of the Euratom Treaty: reflecting on the past, safeguarding the future, No. 16 Spring, http://www.euronuclear.org/e-news/e-news-16/euratom-treaty.htm, last accessed 18.09.2014.

European Repository Development Organisation (ERDO) (2011). Impact of EC Waste Directive 2011 on Regional Disposal Initiatives. http://www.erdo-wg.eu/Document_files/Impact of 2011 EC Waste Directive on Regional Disposal Intiatives 2.pdf.

ERDO (n.d.). European Repository Development Organisation Model Structure and Plan, last accessed 18 September 2014.

ERDO (n.d.). Shared Solutions for Spent Fuel and Radioactive Wastes responding to EC Directive 2011/70/EURATOM. http://www.erdo-wg.eu/Documents_files/ERDO%20Strategy%20Document.pdf, last accessed 18 September 2014.

Feiveson H., Mian, Z., Ramana, M.V., and von Hippel, F. (IPFM) (2011). Managing Spent Fuel from Nuclear Power Reactors Experience and Lessons from Around the World. IPFM Princeton.

International Atomic Energy Agency (IAEA) (1997). Joint Convention on the Safety of Spent Fuel Management and on the Safety of Radioactive Waste Management. INFCIRC/546, IAEA, Vienna.

IAEA (1998). Technical, institutional and economic factors important for developing a multinational radioactive repository. IAEA-TECDOC-1021, Vienna.

IAEA (2004). Developing multinational radioactive waste repositories: infrastructural framework and scenarios of cooperation. IAEA-TECDOC-1413, Vienna.

IAEA (2005). Technical, economic and institutional aspects of regional spent fuel storage facilities. IAEA-TECDOC-1482, Vienna.

IAEA (2011). Viability of sharing facilities for the disposal of spent fuel and nuclear waste. IAEA-TECDOC-1658, Vienna.

IAEA (n.d.). Storage and Disposal of Spent Fuel and High Level Radioactive Waste, http://www.iaea.org/About/Policy/GC/ GC50/GC50InfDocuments/English/gc50inf-3-att5_en.pdf.

Margeanu, C. A. (2012). Overview on the Multinational Collaborative Waste Storage and Disposal Solutions. Presentation at the "XI Radiation Physics & Protection Conference", Nasr City – Cairo/Egypt, 25-28 November 2012.

Nuclear Energy Agency/ Radioactive Waste Management Committee (NEA/ RWM) (2014a). Mandate of the Expert Group on Predisposal of Radioactive Waste (PMRW). NEA/RWM (2014)9, OECD Paris. http://www.oecd.org/official documents/publicdisplaydocumentpdf/?cote=NEA/RWM%282014%299&docLanguage=En, last accessed 18 September 2014.

NEA/ RWM (2014). Mandate of the Expert Group on Waste Inventorying and Reporting Methodology (EGIRM). NEA/RWM (2014)7, OECD Paris. http://www.oecd.org/offcialdocments/publicdisplaydocumentpdf/?cote=NEA/RWM%282014%297&doc Language=En, last accessed 18 September 2014.

Newmann, A. (2013). Multinational Spent Fuel Disposal: Nonproliferation Challenges and Opportunities. Presentation at the INMM 54th Annual Meeting, Palm Desert – California, 14-18 July 2013.

Södersten, A. (2012). The EU and Nuclear Safety: Challenges Old and New. *European Policy Analysis* 2012/10, September, Stockholm.

Štefula, V., and McCombie, Ch. (2004). SAPIERR paves the way towards European Regional Repository. Presentation at the 5th International Conference on Nuclear Option in Countries with Small and Medium Electricity Grids Dubrovnik, Croatia, 16-20 May 2004.

Štefula, V. (2006). SAPIERR – Support Action: Pilot Initiative for European Regional Repositories. Final report EUR 22400, European Commission – Directorate-General for Research Euratom. ftp://ftp.cordis.europa.eu./pub/fp6-euratom/docs/sapierr-projrep_en.pdf, last accessed 16 September 2014.

World Nuclear Association (WNA) (2014): Nuclear Power in the European Union. http://www.world-nuclear.org/info/Country-Profiles/Others/European-Union/, last accessed 16 September 2014.

III. Countries with Geological Disposal after Reprocessing Nuclear Fuel

Multiple Challenges
Nuclear Waste Governance in the United Kingdom

Gordon MacKerron

1 Introduction: the challenges

The United Kingdom has a long history of both military and civilian activity in nuclear technology. In the early days of nuclear development, especially the 1950s, there was neglect of waste management, and this neglect has serious consequences in the present. It leads directly to the first major challenge for UK nuclear waste governance: extracting, characterising and safely packaging large quantities of legacy wastes, especially those held at Sellafield. Much more publicity attaches to the second challenge, which is the search for an acceptable long-term management route for wastes, especially the search for a suitable and acceptable site for deep geological disposal of higher activity wastes. The third challenge is to find a way of managing the large stockpile of separated plutonium that has resulted from the long-standing UK practice of reprocessing spent fuel. The fourth challenge is institutional and involves the need to rationalize the funding and management arrangements for nuclear wastes, given that there are currently three separate systems for managing wastes: one for publicly-owned legacy wastes; another for privately-owned wastes from existing reactors; and a new, third system to apply to wastes derived from any new-build reactors.

2 Current policy for waste management

2.1 The inventory

The question of what constitutes the inventory of radioactive waste in the United Kingdom is not straightforward. Spent fuel, separated plutonium (Pu), plus depleted and reprocessed uranium (U) were always officially regarded as potential resources rather than wastes, at least until 2003.[1] However when the

1 In this paper the terms radioactive waste and nuclear waste are used interchangeably. Note that the small quantities of wastes that derive from medical and other small-scale uses are not covered: the focus is on wastes that have their origin in research, development and

independent Committee on Radioactive Waste Management (CoRWM) was set up in that year, the UK Government asked it to consider what would be the implications if these materials were re-classified as wastes. Consequently this paper includes all these materials as potential wastes and the UK Government is currently considering the best way to disposition separated plutonium. The UK Government has also announced the closure of the two UK reprocessing facilities by 2018, and some spent fuel from the current stock of operating reactors will not be reprocessed. Spent fuel has therefore explicitly started to become, at least *de facto,* a waste form.

The UK Government in seeking a disposition route for separated plutonium now explicitly recognises that it constitutes a waste form. It is also now explicit that if/when there is new nuclear build in the United Kingdom, the reprocessing of spent fuel will not take place. For the rest of this paper, therefore, spent fuel, plutonium and uranium will be treated as part of the waste inventory. UK radioactive waste classifications recognize three categories:

- low level waste (LLW, including very low level waste (VLLW)) virtually all of which is of limited radioactivity and decays relatively quickly. It is disposed in a shallow, engineered repository near Drigg (close to Sellafield), with capacity to at least 2050. There is little controversy here, so while LLW is dominant in terms of waste volumes, it is not considered further in this paper;
- intermediate level waste (ILW) defined as wastes of higher activity than the limit for LLW but which do not need heat to be taken into account in the design of storage or disposal facilities; and,
- high level waste (HLW) where radioactive decay heat has to be taken into account in the design of management facilities. Where there is reprocessing, HLW consists of highly active liquors which are then vitrified. When reprocessing does not take place, HLW consists of spent fuel.

The inventory of wastes as expressed in CoRWM (2006) is shown in Table 1. As this inventory is intended to account for all wastes liable to arise by 2120 (assuming no new build) it is unlikely to vary significantly under current plans. The total volume of waste is expected to be circa 478,000 cubic metres.

commercialization related to nuclear power, plus limited attention to military-origin wastes, where data is scarcer but waste quantities relatively limited.

Table 1: Inventory of United Kingdom wastes (including potential wastes but excluding LLW)

Material	Volume (%)	Activity (TBq)	Activity (%)
HLW	less than 0.3	39 million	50
ILW	73.9	2.4 million	3
Separated Pu	0.7	4 million	5
Uranium	15.7	3 million	below 0.01
Spent fuel	1.7	33 million	42

Source: CoRWM (2006: 20).

2.2 Current policy

Policy for waste management can be divided into time periods. In the short-term, and corresponding to the first challenge outlined above, there is high priority given to reducing the hazards caused by past poor practice in the management of wastes. The public agency, the Nuclear Decommissioning Authority (NDA), has an annual expenditure on decommissioning and waste management of around £2.4bn, of which in 2012/13, £1.6bn was spent at Sellafield (NDA 2013a: 4 and 20), which is home to a very high proportion of the UK's higher activity wastes, measured by radioactive content.

Expenditure outside Sellafield is primarily focused on the early stages of reactor decommissioning and cleaning up redundant research sites. At Sellafield itself there are a number of old facilities, known as Legacy Ponds and Silos, dating back 50 years and more, and which were neglected for decades until the mid–2000s. These represent high hazard levels, and the budget for managing them is currently constrained only by the rate at which clean-up activities can be technically achieved (NDA 2013a). Other major activities at Sellafield include the attempt to continue the process of vitrification of highly active liquors resulting from spent fuel reprocessing, including the current installation of a new evaporator. In undiscounted terms the expected total budget to clean up the Sellafield site is now £67.5bn (House of Commons Public Accounts Committee 2013) and expected to rise yet further as many waste streams are yet to be characterized. The total expected cost of managing all public sector liabilities is now £104bn (NDA 2013a).

A second short-term policy (the third challenge outlined above) concerns the management of separated plutonium. Reprocessing spent fuel has been a continuous activity in the United Kingdom since the early 1960s. Two large reprocessing plants are still working at Sellafield, though both are scheduled to close by 2018. In addition to reprocessing UK-origin spent fuel a significant quantity of overseas spent fuel has also been reprocessed at the THORP reprocessing plant. As none of the resulting separated plutonium has ever been re-used, the United Kingdom now has a stockpile at Sellafield of 120 tonnes of plutonium, of which 96 tonnes (as at 31 December 2012) was UK-owned. The UK Government is currently considering three options: converting it into mixed oxide fuel (MOX) (for use in light water reactors), which is the official current preference; consuming it in new reactors (e.g, the GE PRISM fast reactor); or immobilising it in ceramic form (DECC 2013a). No final decision is likely until at least 2015.

In relation to the long-term, and corresponding to the second challenge above, there is the issue of policy for long-term waste management. The UK Government and nuclear industry, in common with most other countries with waste to manage, have long advocated deep geological disposal (DGD) as the end-point for higher activity wastes. However policymaking essentially came to a halt in 1997, following rejection by a planning inquiry into a Rock Characterisation Facility near Sellafield.

In 2003 an independent Committee on Radioactive Waste Management was formed to recommend to Government the best long-term management option for wastes. It reported in 2006, affirming that DGD was the best long-term option for legacy waste (CoRWM 2006). However it also recommended a more robust long-term interim storage policy in the interim and – most important of all – strongly argued for a voluntarist and participatory approach, granting potential host communities a right to withdraw from the process up to a pre-determined point.

The UK Government accepted these recommendations in a White Paper of 2008 (Defra 2008). It began a search process to find communities who would agree to engage in preliminary 'Expressions of Interest' to host a repository, but without long-term commitment. Local authorities around Sellafield alone expressed initial interest. After a long partnership-based negotiation under this policy, Cumbria County Council exercised its right of withdrawal. This means that Government has agreed to disengage from any further activity in that area (DECC 2013b). Policy remains to seek a volunteer, but at present no community or its local authority has made any further Expression of Interest and the Government is considering how to move the voluntarist process forward (DECC 2013c).

Scotland, where radioactive waste management policy is devolved from the UK Government to the Scottish Government, has a different policy for waste management, involving the principles that waste should be stored near the surface, and close to its existing locations. However this is not a universal policy, as spent fuel from the two operating reactor sites in Scotland, at Hunterston and Torness, is currently sent to Sellafield under long-term contractual arrangements.

2.3 Waste locations and financing

Wastes are held at 36 sites across England, Scotland and Wales, most of them reactor or research locations (NDA/ DECC 2010). The largest concentration by far in terms of radioactivity is at Sellafield, which holds all high level waste (reprocessing products) and the great bulk of spent fuel (NDA 2013a). Only the reactor site at Sizewell, aside from Sellafield, currently holds substantial quantities of spent fuel. This is because the private owners of the Sellafield B PWR (now EDF) decided against reprocessing their spent fuel and must consequently store it for an indefinite period in an on-site store.

There are currently two financial schemes for managing wastes: In the first, all public sector wastes are now owned and managed by the NDA. This is a non-departmental public body set up in 2005 by the Energy Act 2004. It took ownership of all public sector nuclear sites from their previous owners, British Nuclear Fuels (BNFL) and the Atomic Energy Authority (UKAEA), and BNFL was subsequently wound up. The NDA has no financial reserves but receives money from the UK Government out of current tax revenues on an annual basis within a series of four-year Spending Reviews and Plans (MacKerron 2012). The NDA also earns diminishing amounts of money from commercial activities – operating one Magnox reactor site and the two Sellafield reprocessing plants, all of which will close within the next four years. The NDA spends around £2.4bn annually on waste management and decommissioning, around 66% of it at Sellafield (NDA 2013a). This expenditure represents more than half of the annual budget of the UK's Department of Energy and Climate Change.

The total cost of managing public sector nuclear waste liabilities (including decommissioning) is officially estimated at £104bn (undiscounted) or £59bn. (discounted) (NDA 2013a: 21). While the expected costs of cleaning up Sellafield have risen consistently from 2006/7 to 2012/13 for reasons explained above, the expected costs of managing wastes at other sites have, by contrast, been falling slightly in the last two years. Overall however the estimated future costs of cleaning up all UK public sector sites have risen sharply between 2004/5

(£57bn.) and 2012/3 (£104bn) (NDA 2013a: 20, all values expressed at 2013 prices).

The second financing scheme applies to all private sector wastes. These are currently owned by EDF and arise from its seven advanced gas-cooled reactors (AGRs) and one pressurised water reactor station. This involves an entirely different financial system. The Nuclear Liabilities Fund (NLF), a trust set up under Scottish law, is responsible for managing moneys to pay for all EDF decommissioning and waste management expenses. It is an external segregated fund, currently worth a little over £8bn (NLF 2013). This endowment is largely made up of the sale proceeds when British Energy was bought by EDF. EDF does also make small annual payments to the NLF, amounting to £31m in 2011/12 (NLF 2012: 16). The current value of the fund exceeds the current discounted cost estimate of EDF's liabilities, but the NLF have publicly expressed the view that the fund may not in the end be large enough (NLF 2012: 2). This is mostly because the UK Government insists that the NLF fund is deposited almost entirely in the National Loans Fund, and earns a negligible annual rate of interest (well below the discount rate). This has the effect of reducing the overall UK Government's apparent public sector debt, but meanwhile it means that the NLF can earn negligible annual returns. The fund is on record as wanting to spread its investment portfolio more widely in order to earn higher rates of return, but even though it is an external segregated fund, it is clearly not independent of Government.

If or when there is nuclear new build a third financial system will operate. Potential operators of new nuclear power plants will be required to present a Funded Decommissioning Programme that needs Government approval before construction can start. This has two objectives. One is to cap the liability of operators for decommissioning and waste costs and the other is to aim to minimize the risk that taxpayers may have to contribute to the future costs of waste management.

These two objectives are in principle in conflict: a low cap will raise taxpayer risk while a high cap might provide a disincentive to developing new reactors. The official view is that the pogramme should ensure that operators pay "the full costs of decommissioning and their full share of waste management and disposal costs" (DECC 2011). The Programme will operate as an external segregated fund with regular scheduled payments by the operator, and an important part of this new scheme is the commitment by Government to take ownership of the wastes by a transfer pricing mechanism. The operators will need to pay Government a 'fixed unit price' per unit of waste, with the price designed to meet the total future costs of waste management. The intention is that there will be a substantial margin within the price for future cost escalations.

In this process, Government will be advised by a new body, the Nuclear Liabilities Financing Assurance Board.

3 Waste disposal concepts

3.1 Deep geological disposal

As outlined above, the UK Government has supported the idea that DGD was the ultimate aim for the management of all higher activity wastes since the early 1980s. An early programme of test drilling for DGD sites for higher activity was abandoned after vigorous local opposition at all sites. In 1982 Government decided that it was permissible to postpone decisions for the management of HLW for 50 years, and the new industry-owned body Nirex began a programme of exploring a wide range of sites for ILW disposal, strongly advocating the idea that a disposal site should include provisions for waste retrievability for up to 300 years. Local opposition continued but by the mid-1990s the choice had been narrowed down to a site very close to Sellafield and this was subject to a wide-ranging public inquiry. The Inspector at the Inquiry rejected the Nirex proposal for an Underground Rock Characterisation Facility (intended as a precursor to a repository) on multiple grounds.

When CoRWM was set up in 2003 it was instructed to start from a 'blank sheet of paper' – in other words to consider all long-term management options that had ever been seriously studied. While CoRWM followed these instructions, it rejected all options except two variants of DGD (early closure versus up to 300 years worth of retrievability) plus long-term storage, which it regarded as an interim not an eventual option (CoRWM 2006, Chapter 14). Its recommendation was to aim for the early closure of a DGD repository, while arguing that interim storage would be a priority for several decades. Government immediately endorsed the idea of DGD and the White Paper of 2008 (Defra 2008) set out a process for deciding on a site, using a voluntarist approach and embodying the right of a community to withdraw from the process up to a pre-specified point.

There has effectively been a 30-year continuity of official support for DGD, even though there was a hiatus between 1997 and 2006. The only major question within this support has been whether or not a DGD should deliberately design in retrievability of wastes, over a period of up to 300 years, or whether such a decision can be left to negotiations with potential host communities. Nirex, the body previously responsible for siting a repository, was a strong advocate of retrievability. Neither CoRWM nor Government is enthusiastic about delaying the closing of a DGD site though both have argued that there could be local negotiation on this once a clear decision to construct a DGD facility is made.

Two kinds of argument are advanced in the United Kingdom in favour of designed-in retrievability. The first is that it may be possible to discover a more effective way of managing wastes than currently available and so the wastes can be retrieved and managed better. The second is that some of the materials emplaced in the repository might be re-usable as fissile material in the future and could then be retrieved for this different purpose. Arguments in favour of early repository closure are that the long-term safety case is weakened by leaving a repository open and that (whether or not the safety case is worse) it is unethical to leave future generations to make decisions about closure and undertake the work to recover the material and/or close the repository (CoRWM 2006, especially chapter 15).

Finally, one variant of DGD has been debated in recent years. This is the idea that some wastes might be disposed in deep boreholes, up to several kilometers underground (CoRWM 2006). Because the capacity of individual boreholes is limited, the suggestion has been that boreholes might be used for wastes of small volume combined either with high political significance (e.g. plutonium) or materials unlikely ever to be useful if retrieved (e.g. vitrified HLW). This option is not currently being actively pursued, but might become significant as a variant of DGD for some wastes in the next one or two decades.

3.2 Storage

There has been very little debate about shorter-term storage policy in the United Kingdom, possibly because Government has assumed that giving attention to long-term storage might divert attention from its preferred option of DGD. Reprocessing spent fuel at Sellafield was the norm until the late 1990s and is still practiced for fuel from the older Magnox reactors, while spent fuel from AGR reactors still goes to Sellafield for storage. Given that the great bulk of all radioactivity from nuclear power is contained in spent fuel, this meant that Sellafield became a *de facto* national storage site for most wastes. The volumes and activity of wastes at almost all other sites was limited (except for Dounreay in Scotland).

However there are now changes in this picture. There is now a large volume of spent fuel stored indefinitely at the site of the United Kingdom's only PWR at Sizewell and it is clear that all future reactors will also need to store their spent fuel on site for many decades, given the absence of any operator interest in reprocessing. If significant amounts of waste are accumulated at dispersed reactor sites and the process of finding a DGD is protracted, this could lead to debate about the desirability of centralized or national storage over the next few decades.

3.3 Plutonium disposition

The UK-origin stockpile of plutonium is now around 96 tonnes and a further 24 tonnes of foreign-origin plutonium is also stored at Sellafield. This is by some distance the largest civilian stockpile of plutonium in the world. In practice demand for the return of foreign-owned plutonium is limited in the absence of fast reactors and because the cost of using plutonium in MOX fuel is high. As a result the United Kingdom may have to manage virtually all of the foreign-origin plutonium. An example of the process of ownership transfer is that over the last two years, some 7 tonnes of foreign-owned plutonium at Sellafield has been transferred via swaps into UK ownership (World Nuclear News 2013).

The UK Government has decided that continuing storage of plutonium is unacceptable and has been consulting on alternative disposition options. Plutonium has now been acknowledged as a waste rather than a 'zero-value' asset, as it was classified earlier (ONR 2013). The main alternatives currently under consideration are (DECC 2013a):

- Building a new mixed oxide fuel plant (MOX) to use up all the plutonium in future UK light water reactors and then manage spent MOX fuel as waste.
- Constructing new and untried reactors from overseas vendors (the Power Reactor Innovative Small Module (PRISM) fast reactor or a version of the CANada Deuterium Uranium (CANDU)) to consume the plutonium.
- Immobilising and disposing of the stockpile, probably in a ceramic waste-form, a process already being researched for the small fraction of plutonium that would be very difficult to use in MOX.

Government's stated current inclination is to build a new MOX facility, at an official net cost of some £ 3 bn, a figure which Government regards as potentially not necessarily cheaper – but less risky – than the alternatives (DECC 2013a). Its argument is based on the idea that MOX is the technologically most established management route. However there are serious difficulties with this argument. First the previous MOX plant built in the United Kingdom was a technological disaster, and closed having achieved scarcely 1% of its expected output. And a MOX plant currently being built for the US Government by Areva (the only credible supplier for a future UK plant) is suffering from chronic cost over-runs (Wald 2013) and an unwillingness of utilities to sign up to buy the output. This unwillingness extends to potential UK operators: both EDF and the Horizon project (the only two active developers in the UK at present) have unequivocally stated that they have no intention of using MOX (Broomby 2013). Even if UK utilities could be induced to accept MOX use ('buying' it is less likely due to the excess costs compared to using uranium-only fuel) reactors would have to be re-licensed and communities persuaded to accept use of plutonium-based fuels in their local power stations. Finally it is not clear whether export sales of MOX would be possible nor is it clear that positive revenue could be obtained.

The GE PRISM reactor is a revival of an old fast reactor design, never built to scale and abandoned in the United States in the 1990s. Details about the proposed version of the CANDU reactor are not yet clear. Concerns about both MOX and the untried reactor designs under consideration mean that immobilization is almost certainly the right way forward, even if further research is needed and some uncertainty exists. At present there is active research at the National Nuclear Laboratory into hot isostatic pressing of highly contaminated plutonium (about 5% of the current stockpile) – which will be treated as waste (National Nuclear Laboratory 2013). Once this is established, the remainder of the stockpile is 'purer' and less challenging to immobilize .

4 Legal framework

The earliest relevant legislation is the Nuclear Installations Act of 1965 establishing the safety regulatory framework and the role of the (then) Nuclear Installations Inspectorate, now superseded by the Office for Nuclear Regulation (ONR). The most important legislation from the perspective of the environment is the Radioactive Substances Act of 1993, later incorporated into the Environmental Permitting (England and Wales) Regulations of 2011.

In other domains, the Energy Act of 2004 established the NDA as a public body charged with owning and managing all public sector radioactive wastes. The act also set up the framework under which the NDA manages site licence companies (licensed by ONR), the senior management of which is open to periodic competition. The Justification of Practices Involving Ionising Radiation Regulations 2004 established the framework under which new radiological practices need a formal justification.

The Energy Act 2008 is wide-ranging legislation covering electricity, gas and energy regulation. In the present context it introduced new legislative provisions allowing Government to require future operators of nuclear power plants to set up the Funded Decommissioning Programmes referred to above.

5 The institutional framework

The UK institutional landscape is complex. The Department of Environment Food and Rural Affairs (Defra) had historic ministerial responsibility for waste management but this was transferred to the new Department of Energy and Climate Change (DECC) after its formation in 2008. DECC now has policy responsibility for wastes, via a relatively new Office for Nuclear Development within the Department. However this responsibility only covers England and Wales. In Scotland, the Scottish government, through the Cabinet Secretary for Rural Affairs and the Environment, has acquired devolved responsibility for waste policy.

Beneath DECC sits the NDA, with ownership and oversight of all public sector radioactive sites and their wastes. Nirex, the body that previously had responsibility for finding radioactive waste management routes for ILW (and originally owned by the industry and later Government) was abolished in 2005 and a high proportion of its employees became the core of the Radioactive Waste Management Directorate in the NDA. This directorate is focused on the issue of siting a DGD, while the NDA as a whole is focused on all radioactive waste management issues.

The NDA itself is also subject to scrutiny from the Shareholder Executive, a part of the Department for Business Innovation and Skills (BIS) which aims to manage in a broadly commercial way those assets held in the public sector but at arms' length from central Government. While the NDA owns the relevant public sector sites, it is an oversight body which contracts out management of sites to site licence companies (SLCs). These companies have a long-term existence as licence-holders, but their top management is periodically subject to competitions organized by the NDA and the winner is installed as the 'Parent Body Organisation' (PBO). The PBO owns the shares in the SLC for the duration of its contract, and it is supposed to provide strategic management of the SLC by bringing in secondees from its own organization to try and improve efficiency, introduce innovation and control costs. In principle there can be some risk transfer to the PBO and in return, PBOs can earn fees which vary according to performance.

A current example is Nuclear Management Partners (a consortium of three contracting companies: Amec, Arriva and URS), which temporarily owns the Sellafield SLC under a 17-year contract with break points for contract review.

The poorly characterized state of wastes at Sellafield mean that the current arrangements are of a cost-plus type (i.e all risks remain with the public sector). The first such break point in this contract was in 2013, at which point – despite serious reservations about the adequacy of NMP's performance (KPMG 2013) – the NDA agreed to allow NMP's contract to continue.

The NDA/SLC/PBO management arrangements, with periodic competitions for PBO status, are complex and unique. Their origin lies in the desire of Government to introduce competitive pressure into waste management and also to introduce a wide range of international experience into waste management. However it proved impossible to allow the PBOs to have full control over waste sites as the licensing process for site management companies takes up to two years. As a result it was decided that there should be ongoing site licence companies, with temporary ownership and senior management from PBOs at each round of competition. The relationship between the management of the ongoing SLCs and the management secondees from PBOs has, as at Sellafield, proved ambiguous and problematic.

There are three main regulators for waste sites. The Office for Nuclear Regulation (ONR) is responsible for issues of safety, security and transportation of nuclear materials. It issues site licences for waste sites and has a UK-wide remit. In England and Wales the Environment Agency (EA) has responsibility for regulating all site discharges and emissions, while for Scotland the same function is performed by the Scottish Environment Protection Agency (SEPA).

At sub-national level, local authorities at various levels – unitary authorities, county councils, councils, district councils and parish councils – are all potentially involved in negotiation processes for siting new radioactive waste management facilities.

6 Status of siting procedures and information policy

The main siting process on which the England and Wales authorities are currently engaged is for a DGD. If there is significant new reactor build, and – as currently intended – this means that spent fuel is stored on reactor sites for long periods into the future, pressure for a new, formal national storage site (which could be Sellafield) might grow.

Since the 2008 White Paper on DGD processes, Government has actively pursued its voluntarist agenda in seeking sites. This has involved issuing invitations for communities (as represented in the first instance by their local authorities) to join dialogue before expressing a formal declaration of willingness to negotiate.

When Government issued its invitations for communities (in practice local authorities) to declare a potential Expression of Interest in hosting a DGD, only the three relevant authorities around Sellafield (Cumbria County Council, and the two lower level district councils of Allerdale and Copeland within the Cumbria area) seriously pursued the possibility. All three entered into the West Cumbria MRWS Partnership process. Shepway District Council in Kent briefly considered making an Expression of Interest but did not follow this through before rejecting the idea of participating at all.

As expressed in the 2008 White Paper (Defra 2008) there are several stages in the process of volunteering. These are:

- Expression of Interest, a decision to enter into discussions with Government, but without commitment, which the Cumbrian authorities took.
- Decision to Participate, a point at which local authorities decide to enter into a commitment to participate in a repository siting process, but again without a commitment to host.
- Community Siting Partnerships – as outlined earlier the West Cumbrian MRWS Partnership was set up to deliberate on the issues.
- Right of Withdrawal – a right exercised by Cumbria County Council in 2013.
- Engagement Packages – resources made available to local communities to allow them to participate in negotiations without significant local costs.
- Benefits Packages – these are facilities that Government could offer local communities in exchange for their willingness to host a DGD. They might include such features as: improved local training, public services, transport infrastructure and healthcare.

The level of transparency displayed in the procedures undertaken in search of a specific site – primarily the Cumbria partnership process – has been high by historical standards and involved very wide-ranging civil society participation and deliberation, both on the part of stakeholder organizations and the wider public. In the West Cumbria MRWS Partnership, set up under the auspices of the NDA to discuss the Expression of Interest of local authorities and communities around Sellafield, a wide range of other stakeholders were actively involved in deliberation.

Members of the Partnership included local authorities, trades unions, chambers of commerce, religious groups, and tourist promotion organizations (DECC 2013b). Environmental groups were invited to join but declined. The partnership held a wide range of public events and two rounds of formal public and stakeholder consultations. High levels of information were provided at all stages and the NDA has a sophisticated system for managing issues that are raised by any stakeholder or members of the public (NDA 2013b).

The practice of more open governance dates back to the formation of CoRWM in 2003. Its processes included holding meetings in public, publishing all internal and submitted documents, and engaging in four rounds of comprehensive public and stakeholder engagement, involving meetings at nuclear sites, schools projects, national stakeholder forums, engaging with Citizens Panels and setting up deliberative forms of on-line public discussion.

A major difficulty for DGD policy emerged after Government became enthusiastic for new nuclear build from 2006 onwards. Before that time there was an emerging consensus that – with new build apparently moribund – it was necessary to find a 'least worst' solution to the problem of legacy waste. However once the prospect of new build became real, the issue for new build waste was different – on ethical, social and political grounds – from the need to manage legacy waste. Here it was possible to reject new nuclear build and instead consider the wider alternatives to new nuclear construction: there was no longer an inevitability about managing the resultant waste. However the UK Government has conflated the issues of legacy and new build waste, and the fact that new build is now probable has meant that constructing consensus around waste solutions has now become much more problematic.

7 Conclusions and lessons learned

There has been much change in the UK government's approach to nuclear waste management in the last decade, both in terms of substance and institutional structure.

First, serious attention has at last been given to the short- and medium-term management of legacy wastes, especially those representing high potential hazards at Sellafield. BNFL was wound up and a new institution – the NDA – was set up to focus explicitly on clean-up and waste management.

Second, Government has acknowledged that the stockpile of separated plutonium can no longer be regarded as an asset, must be managed as a waste, and should not be stored indefinitely. However, its currently proposed strategy of aiming to turn plutonium into MOX is highly problematic and immobilisation seems a more plausible outcome.

Third, policy to develop a long-term strategy for waste management has been radically overhauled after policy came to a halt in 1997 with the rejection of the Nirex proposal for an underground laboratory near Sellafield. This marked the end of two decades of a policy approach of decide-announce-defend (to which can be added 'abandon', or DADA). In this old process decisions were taken in secret by limited groups of 'insiders' – industry, Government and a few scientists – and there followed unsuccessful attempts to persuade local communities and wider publics that a particular site or sites would be suitable for deep disposal (MacKerron and Berkhout 2009).

The setting up of CoRWM in 2003 marked a radical departure in policy approach. CoRWM was given a 'blank sheet of paper' in relation to both technology and policy process, and consisted of a wide variety of members, including a founder member of Greenpeace UK, specialists in public engagement, social scientists as well as people from natural science and the nuclear industry. By conducting a public and transparent process, CoRWM's recommendations, broadly endorsed by Government, created some degree of public trust in the decision process (CoRWM 2007).

Government has continued the new tradition of an open and deliberative policy style in developing waste policy, but has yet to make progress in finding a site. Government's enthusiastic endorsement of new nuclear build since 2006 – before any clear long-term management route for wastes had been established – has been problematic for waste governance development. It has made the process of maintaining legitimacy in waste policy-making much more difficult.

References

Broomby, R. (2013, November 6). Hinkley Firm Denies Fuel Claim. *BBC News, Science & Environment*. http://www.bbc.com/news/science-environment-24834932.

Committee on Radioactive Waste Management (CoRWM) (2006): Managing our radioactive waste safely. CoRWM's recommendations to Government Doc 700, July.

CoRWM (2007). The CoRWM process – lessons learned. Doc 1896 September.

Defra (2008). Managing radioactive waste safely: a framework for implementing geological disposal Cm 7386, June.

Department of Energy and Climate Change (DECC) (2011). The Energy Act *2008*. Government response to the consultation on the amendment to the radioactive waste and decommissioning (Finance and Fees) regulations 2011, February.

DECC (2013a). Management of the UK's plutonium stocks: consultation responses, 24 February.

DECC (2013b). Managing radioactive waste safely: implementing geological disposal. Annual report April 2012–March 2013.

DECC (2013c). Review of the Siting Process for a Geological Disposal Facility, URN13D/250, September.

House of Commons Public Accounts Committee (2013). Nuclear Decommissioning Authority: Managing Risk at Sellafield HC 746 4 February.

KPMG (2013). The PBO Model at Sellafield: Performance to 31st May 2013. Final Report, September.

MacKerron, G. (2012). Evaluation of nuclear decommissioning and waste management. A report commissioned by the Department of Energy and Climate Change, March.

MacKerron, G., and Berkhout, F. (2009). Learning to listen: institutional change and legitimation in UK radioactive waste policy, *Journal of Risk Research*, 12 (7-8): 989-1008.

National Nuclear Laboratory (2013). 'Plutonium Disposition – Immobilisation of Pu residues' presented to IPFM, London, March.

Nuclear Decommissioning Authority (NDA) (2013a). Annual Report and Accounts 2012/2013 HC236 June.

NDA (2013b). Geological disposal: RWMD approach to issues management Report NDA/RWMD/081 Version 3.

NDA/ DECC (2010). Radioactive Waste in the UK: A Summary of the 2010 Inventory. http://www.nda.gov.uk/ukinventory/documents/Reports/upload/2010-UK-Radioactive-Waste-Inventory-Summary-of-the-2010-Inventory.pdf, last accessed 9 June 2014.

Nuclear Liabilities Fund (NLF) (2012). Nuclear Liabilities Fund Limited Annual Report for the Year ended 31 March 2012.

NLF (2013). http://www.nlf.uk.net/news.html, last access 9 June 2014.

Office for Nuclear Regulation (ONR) (2013). Annual figures for holdings of civil unirradiated plutonium 2013.

World Nuclear News (2013). UK swaps plutonium to balance MOX policies, 23 April 2013.

Wald, M. (2013, June 25). US moves to abandon costly reactor fuel plant. *The New York Times*. http://www.nytimes.com/2013/06/26/us/us-moves-to-abandon-costly-reactor-fuel-plant.html?pagewanted=all&_r=0, last access 9 June 2014.

Megaproject Underway
Governance of Nuclear Waste Management in France

Markku Lehtonen

1 Introduction

Given France's role as a nuclear energy "superpower", with 58 nuclear reactors supplying 75% of the electricity consumed in the country, the current French plans for the creation of a deep geological disposal repository for intermediate-level long-lived and high-level radioactive waste (ILW-LL and HLW) have been generating worldwide interest. The search for a site has been ongoing since the late 1970s, when the government declared deep geological disposal as the preferred option for radioactive waste management. However, like in many other countries where nuclear power is used, the site investigations conducted in the late 1980s generated intense local opposition, prompting the government to declare a one-year moratorium on the search of a site in 1990. After extensive consultation, the Parliament adopted a law on the management of radioactive waste in 1991; it was the country's first law on nuclear power. This "Bataille Law" reopened the search for solutions by stipulating a 15-year period of research on three options: a) geological disposal, which had until then been seen as the only option, b) long-term near-surface storage, and c) partitioning and transmutation.[1] The Law also marked the beginning of a more dialogic and participatory approach to waste management policy, and introduced the principle of reversibility. Since 1998, reversibility has indeed been the cornerstone of French radioactive waste management policy. It was confirmed in the 2006 "Planning Act", enacted after a public consultation process organised by the National Commission of Public Debate in 2005-2006.

In 1998, the government selected Bure, a small village on the border between the departments of Meuse and Haute-Marne in the north-east of the country, as the site for an underground research laboratory (URL) for deep geological disposal. Since then, the Bure site has been the primary candidate for hosting the final disposal facility. According to the timetable established in 2006,

1 The law was named after the parliamentarian Christian Bataille, appointed to lead the extensive stakeholder consultations in preparation of the law.

the construction of the repository was to begin in 2017 and be finished in 2025, when the installation would receive the first waste packages. However, the project, which is among the largest industrial projects ever to be undertaken in Europe, faces considerable challenges, partly because of its location in a sparsely populated area without a tradition of nuclear power. Following the difficulties encountered in the public consultation process in 2013, the timetable was subsequently revised, and an industrial pilot phase was added to the programme, to be launched when the installation is put into operation. The national agency for the radioactive waste management, Andra, plans to submit its application for approval by authorities in 2017, with a view of obtaining a construction licence around 2020.

2 The present policy

After the failure of the site investigations conducted in the late 1980s and the subsequent "reopening" of the radioactive waste policy process as stipulated in the Bataille Law 1991, the government reintroduced an initiative in 1993 that encouraged departments to volunteer as hosts for an underground research laboratory (URL). In 1994, the government launched site investigations in four of the thirty departments that expressed interest: Gard (clay formations), Vienne (granite), Meuse (clay), and Haute-Marne (clay). Local opposition arose rather quickly, although local politicians were predominantly in favour of the projects. In 1996, the sites of Meuse and Haute-Marne were merged to constitute the current site at Bure, and in 1997, public inquiries were conducted on the three sites regarding the creation of an URL. Both scientific and political (local opposition) reasons led to the elimination of Gard and Vienne from the selection process, leaving Bure in 1998 as the only candidate. The construction of the URL started in 2000. The site in Bure also hosts an exhibition centre displaying scientific and technological aspects related to the repository, monitoring stations operated by the permanent observatory of the environment, and an environmental databank ("écothèque") opened in 2013.

Gradually, over the course of the 2000s, Bure became the preferred location for construction of a deep geological repository. The absences of other alternatives and competing sites were among the most central points of criticism raised by local citizens and politicians in and around Bure. The transformation from an URL to a repository site was further consolidated in 2005, when Andra concluded that Bure was "perfectly appropriate" to host such a facility, and chose an area of 250 km^2 for the construction. [2]

In 2005-2006, at the request of Andra, the National Commission of Public Debate (CNDP) organised a national consultation on the country's waste management policy, whose conclusions in turn informed the parliamentary debate leading to the adoption of a new law on radioactive waste management in 2006. This "Planning Act" saw reversible geological disposal as the reference option, but stipulated that research should continue on two other options, i.e. long-term near-surface storage, and partitioning and transmutation. An Act on nuclear transparency and security was also enacted in 2006, notably creating an independent safety regulation agency (the National Agency for Nuclear Safety (ASN)).

In 2009, Andra submitted its proposal to create the geological disposal facility, called Cigéo (Centre industriel de stockage géologique). The initial area of 250 km^2 was then scaled down to about 100 km^2 in informal discussions with geologists[3], and finally – after consultations with local stakeholders – further narrowed down to 30 km^2.[4] In March 2010, the government approved the proposal, subsequent to its examination by the ASN, the National Assessment Board (CNE), and by international experts. In May 2013, the National Commission of Public Debate (CNDP) launched a mandatory public consultation process, which, however, had to be fundamentally restructured because of obstruction by local opponents of Cigéo. This included the cancellation of the planned public hearings, and the organisation of a consensus conference.

The fact that a nuclear "megaproject" of such a dimension should be carried out in a sparsely populated and economically declining region without a tradition of nuclear energy created specific socio-economic challenges (e.g. Lehtonen, forthcoming). The planning of the aboveground and underground facilities of Cigéo was partly guided by the desire to ensure that Cigéo would be equally beneficial to both departments. For example, the underground and surface installations are situated on different sides of the departmental border. Other measures aimed at assisting the local and departmental authorities to prepare for the possible arrival of the project included the economic support measures and the elaboration of an interdepartmental scheme for the development of the region (SIDT 2013).

Table 1 summarises the milestones of the French radioactive waste policy leading to the Cigéo decision.

2 This area is known as "zone de transposition".

3 Thibaud Labalette, Andra, personal communication, 18 November 2013.

4 "Zone d'intérêt pour la reconnaissance approfondie" (ZIRA).

Table 1: Milestones in French high-level radioactive waste management.

1978	Decision in favour of reprocessing: the planned sequence of RWM: vitrification – interim storage at La Hague – deep geological disposal.
1979	With the creation of Andra, geological disposal becomes institutionalised as the reference option.
1987-1990	Site investigations meet strong local opposition, and prompt the government to declare a one-year moratorium on site investigations.
1991	The country's first law on nuclear policy stipulates a 15-year period of research on three options : geological disposal, long-term near-surface storage, and transmutation and partitioning.
1994-1997	Site investigations initially at four sites.
1998	The government chooses Bure as the site for the Underground Research Laboratory. Government adopts *reversible* geological disposal as the reference option.
2005	Andra concludes on the suitability of Bure as a site for the repository.
2005-06	Public debate on the general options and orientations of the French radioactive waste management policy.
2006	A new law on radioactive waste management, the so-called Planning Act, mandates Andra to develop reversible geological disposal as the solution to HLW management.
2009	Proposal by Andra for the creation of a geological disposal facility at Bure, between two Departements (Meuse and Haute-Marne) and two regions (Lorraine and Champagne-Ardenne).
March 2010	Government validates the proposal, after consultation with the safety authority, the National Assessment Board, and local stakeholders.
May-Dec. 2013	Mandatory consultation, "public debate", on Cigéo. Opponents prevent two first public meetings from going ahead, and meetings are replaced by debate on the internet. Conclusions of the debate released on 15th February 2014.
Dec. 2013 - February 2014	Organisation of a consensus conference on the Cigéo project, in response to the problems encountered in the public debate.
May 2014	Andra publishes the actions it intends to undertake in reaction to the public debate. These include an "industrial pilot phase" and a slight adjustment of the timetable, with the planned date for construction licence in 2020.

The planned timetable for the realisation of Cigéo

Currently, about 90% of the volume of French radioactive waste is being placed in interim storage installations for low-level or short-lived radioactive waste in the department of Aube. The high- and medium-level wastes, which are to be disposed of in Cigéo, are stored in facilities in La Hague, Marcoule, and Cadarache. Andra plans to submit a final proposal for the creation of the disposal site by 2017. The application will then be assessed by the National Assessment Board (CNE), ASN, local and regional authorities, and the parliamentary office of science and technology (Opecst). The proposal must also describe the conditions for the practical implementation of the principle of reversibility, which will subsequently be stated in a governmental decree. Andra will then gather comments and feedback from evaluators and stakeholders and submit a revised proposal to a mandatory public inquiry, as the final step towards a government decree authorising the creation of Cigéo. Andra should start the construction of Cigéo by 2020, provided that the safety authority (ASN) grants the agency a licence. The operation of the facility would start with an industrial pilot phase in 2025. An assessment of the pilot phase would be conducted 5-10 years later. If the experience proves successful, the facility would be gradually made fully operational, and would be periodically checked for safety at ten-year intervals, conducted in consultation with the involved parties.

2.1 Categories and quantities of waste

The exploitation of nuclear power began in France as a "by-product" of nuclear weapons production. Hence, an estimated 11% of radioactive plutonium waste originates from military installations, yet this waste is subject to defence secrecy and therefore does not figure in the national radioactive waste inventory that Andra is required to submit every three years (Andra 2012a: 50). The majority (59% in 2010) of the waste comes from the 58 operating commercial nuclear reactors, while research activities produce about a quarter of the waste. France has opted for reprocessing spent nuclear fuel at the La Hague reprocessing plant, operated by Areva. [5] The reprocessed and recoverable materials (notably uranium and plutonium) are considered as resources, and therefore not included in Andra's waste inventory. The non-recoverable high-level waste is vitrified and conditioned in stainless-steel containers. The estimated 27 m^3 of spent fuel from the Brennilis heavy-water reactor is considered to have a limited potential for reutilisation, and is therefore destined for disposal at Cigéo. Unlike many other countries, France does not have a minimum radiation threshold which a waste must exceed to be classified as radioactive. Therefore, any waste originating from a designated installation must be disposed of at an Andra site. A specific disposal facility has therefore been created for very-low level waste originating mainly from decommissioning operations.

5 As part of the restructuring of the country's energy sector in response to market liberalisation, the government created Areva in 2000 as a new "national champion". It operates at virtually all stages of the nuclear fuel cycle: uranium mining, fuel production, reactor construction, and reprocessing. The state ownership in the company remained at 78%.

2.2 Volumes of existing waste

At the end of 2010, the total volume of existing radioactive waste was estimated at approximately 1,320,000 m^3. This volume was distributed across five waste categories, distinguished according to their level of radioactivity and half-life, as shown in Table 2.

Table 2: Volume of radioactive waste according to category in 2010 and expected volume by 2030 (in m^3)

Waste Category	**Volume (end of 2010, conditioned, in m^3 equivalent)**	**Expected volume by 2030**
High-level waste (HLW)	2,700	5,300
Intermediate-level long-lived waste (ILW-LL)	40,000	49,000
Low-level long-lived waste (LL-LL)	87,000	133,000
Low- and intermediate-level short-lived waste (LIL-SL)	830,000	1,200,000
Very low-level waste (VLL)	360,000	1,300,000
Non-classified waste (NCW)	3,600	
Total	**1,323,300**	**~ 2,700,000**

Source: PNGMDR (2013).

Some 60% of the intermediate-level and 30% of the high-level waste have already been generated, and are currently stored in facilities in La Hague, Marcoule and Cadarache. The medium- and high-level waste represents just over 3% of the total waste volume, but almost the totality of radioactivity contained in waste. The total waste volume is expected to reach about 1,900,000 m^3 by 2020, and 2,700,000 m^3 in 2030 (see Table 1). These estimates are based on the assumptions that the lifetime of a nuclear reactor will be extended to 50 years and that reprocessing will continue. The inventory takes into consideration the 58 existing reactors as well as the new EPR (European Pressurised Reactor) under construction in Flamanville.

The high- and medium-level radioactive waste is currently stored at the sites of La Hague, Marcoule, and Cadarache. In addition, Andra has three storage sites for short-lived low- and intermediate level waste – two in the department of

Aube and one in Digulleville, near the reprocessing plant of La Hague in the La Manche department in Normandy. Over half a million m^3 is disposed of at the La Manche facility, which was closed down in 1994, and has entered a post-closure monitoring period that will extend over several centuries. It is notable that the Cigéo project is situated in a relatively "virgin" region, i.e. far from other nuclear installations.

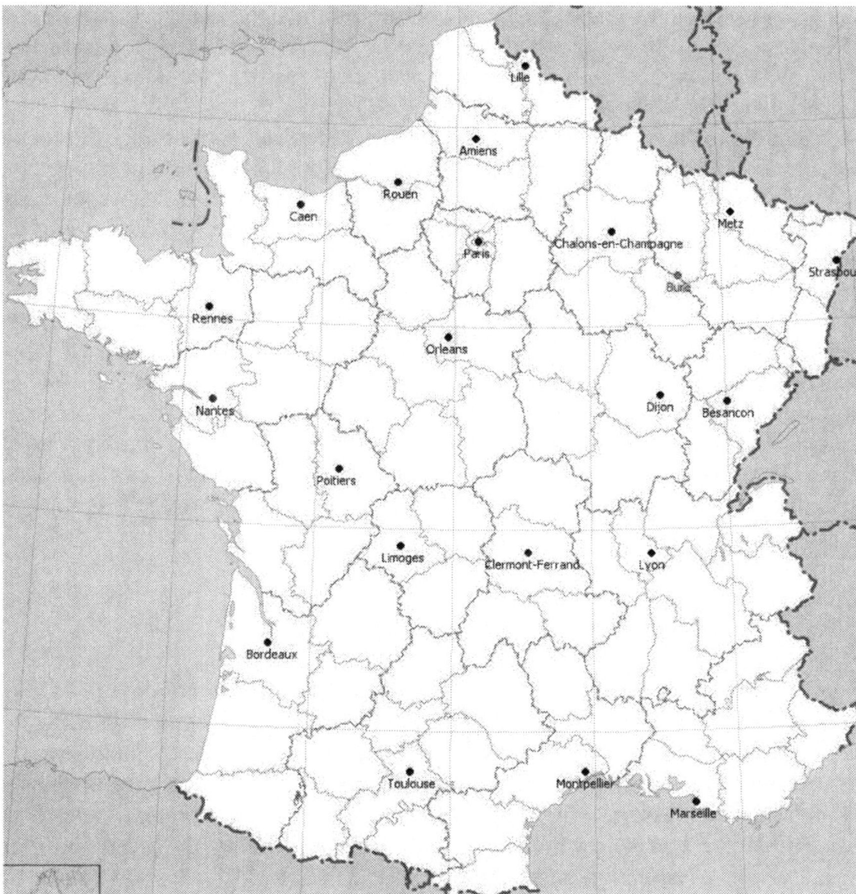

Figure 1: The localisation of Cigéo (Bure).

Source: http://www.cartesfrance.fr/carte-commune/55/55087/carte-administrative-lambert-regions-Bure.jpg, last accessed 28 October 2014.

2.3 Uncertainties concerning the waste inventory

The waste inventory is subject to numerous uncertainties, such as the amount of spent fuel reprocessed and the expected operational lifetime of reactors, as well as the timeable for and the amount of waste from reactor decommissioning. The waste inventory is affected by possible policy changes concerning reprocessing (reclassifying part or the total of spent fuel as waste would automatically increase the inventory), and decisions concerning the continuation vs. phase-out of nuclear energy in France. As a precaution, Andra is required to ensure that Cigéo will also be able to host the additional spent fuel that would be generated if France gave up reprocessing (Labalette 2013).

Significant amounts of the spent fuel classified as recoverable cannot at present be reprocessed at La Hague. About 1% of this fuel consists of plutonium, part of which is used for producing MOX (mixed oxide) fuel. However, only about a third of French reactors can at present use MOX fuel. The vast majority (95%) of the spent fuel is uranium, of which about 10% is in principle reusable; yet France does not currently possess the necessary technology to carry this out. Hence, the uranium from reprocessing is sent to Russia for enrichment, with the expectation that the so-called Generation IV reactors, which are capable of using enriched uranium fuel, will become commercially available. Estimates concerning the total amount of radioactive waste produced in France therefore depend decisively on the definitions of waste. Andra estimates the amount of radioactive waste produced annually at about 2 kg per inhabitant – a figure that, according to the anti-nuclear network "Réseau Sortir du Nuclear," should be multiplied by a factor of 50-100.[6]

2.4 The cost and financing of waste management

Financing radioactive waste management is based on the polluter pays principle, with the waste producers (essentially EDF, CEA, and Areva) being obligated to designate earmarked funds to cover the entire cost of waste management. Furthermore, the companies must secure these reserves through investments in secure financial assets, which are regulated by the government. Andra is responsible for estimating the total costs of Cigéo during the over 100-year period of the construction and operation of the facility. After having consulted the ASN and the waste producers, the Ministry of Energy sets the payments

6 In particular, to take into account the recoverable nuclear material, liquid and gaseous radioactive emissions, and the residues from uranium mining. http://fr.wikipedia.org/wiki/Gestion_des_d%C3%A9chets_radioactifs_en_France

required from the waste producers. This process of cost estimation is repeated every three years. Payments are determined in relation to the amount of waste produced: EDF is currently responsible for 78%, CEA for 17%, and Areva for 5% of the payments.

Research and studies on the management of high- and intermediate-level waste are financed by an additional tax on nuclear installations, which is deposited into a specific research fund managed by Andra. Between 2010 and 2012, the tax amounted to about € 118 m per year.

The latest, and the currently most "official", estimate of the total cost of Cigéo dates from 2005, when a committee consisting of representatives of Andra, the waste producers, ASN, and the energy ministry agreed upon a reference cost of € 13.5-16.5 bn over a period of the expected 100-plus year operational lifetime (including construction, operation, and closure) of the facility. Discounting is applied to determine the current value of expected future costs, and subsequently set the payments required by the waste producers. A new estimate is in preparation by a working group set up in 2009, made up of representatives from the same organisations.

2.5 Uncertainties concerning costs and financing

The cost estimates and the financing arrangements are subject to a number of uncertainties. Relatively minor changes in the applied discount rate can significantly affect the estimated total cost of Cigéo, and consequently the payments required from waste producers. It is likely that the discount rate will be revised and lowered in the next cost estimate, given the declining rate of return on the long-term state bonds over the past four years, which is used as a basis for determining the discount rate. Lowering the discount rate would automatically increase the payments required from the waste producers, thereby aggravating the already difficult financial positions of EDF and Areva. Furthermore, the total cost of the project is a source of repeated disputes between Andra and the waste producers; while Andra has an interest in ensuring that it has sufficient funds for an appropriate implementation of the project, the waste producers would benefit from minimising the cost estimates, and thereby their obligatory payments. In its highly-publicised 2012 report on the costs of nuclear energy in France, the National Audit Office (*Cour des Comptes*) suggested figures as high as €35bn as more realistic cost estimates, and predicted that in any case the new estimate is likely to be considerably higher than the 2005 figure of €13.5-16.5bn.[7]

7 The 2005 total cost figure was at the low end of the least costly estimates among the three scenarios produced by Andra, and the costs of inputs (labour, materials, equipment, etc.) were

3 Waste disposal concept(s)

Cigéo is designed to accommodate all high-level and long-lived intermediate-level radioactive waste generated by the current French nuclear fleet during its entire lifetime, assuming the continuation of reprocessing. The waste is to be buried in a 160 million year-old Callovo-Oxfordian clay formation located at an average depth of about 500 metres (Andra 2012b: 4), and this solution must be reversible. The option of burying the waste in granitic formations was studied in the 80s and 90s, but was subsequently abandoned because of geological and political reasons (i.e. lack of support from the local communities). The facility is expected to host about 10,000 m^3 of high-level waste (about 60,000 waste packages) and 70,000 m^3 of intermediate-level waste (about 180,000 packages). Depending on the nature of the waste, different means of stabilisation will be applied: vitrification (for the high-level waste), cementation, or asphalting.

The government adopted the principle of reversible geological disposal as its reference option in 1998, and this choice was approved in the 2006 Planning Act.[8] The project must be reversible for a minimum of 100 years (Labalette 2013). The Act also rules that Andra should submit its proposal on the detailed definition of reversibility, together with its application for a licence in 2015, for government approval.[9] A new Act will then validate the conditions of reversibility. Although no specific regulation has been put into place, the dualcharacter of reversibility as both a technical and a political concept has been accepted as the basic principle. The technical retrievability of the waste packages is accompanied by "decisional" reversibility, i.e. the ability to return to the previous decision-making phase at any time in the project´s development.[10] Numerous observers have expressed scepticism concerning the true value of reversibility, given that Cigéo is intended to advance in a stepwise manner towards closure, making waste retrieval extremely difficult and costly.[11]

---

based on 2002 figures, which certainly have evolved since then (Cour des Comptes 2012: 143). For example, the prices of steel and of concrete have increased by about 50–100% since 2005 (Labalette 2013).

8 While the details of reversibility are to be defined in a specific decree, Andra (2010: 33) has noted that reversibility "is not the direct result of a technical or scientific need, but of social demand and of a political choice, which engineers and scientists are duty-bound to accept".

9 However, this original timetable was revised recently (see section 2 above).

10 Loi n° 2006-739 du 28 juin 2006 de programme relative à la gestion durable des matières et déchets radioactifs: http://legifrance.gouv.fr/affichTexte.do?cidTexte=JORFTEXT000000240700. Planning Act No. 2006-739 of 28 June 2006 Concerning the Sustainable Management of Radioactive Materials and Waste: http://www.andra.fr/download/andra-internationalen/document/editions/305va.pdf.

11 Barthe and Meyer (2012: 9) distinguish between a "restricted" and "extended" notion of reversibility. The former would not necessarily imply a change in the hierarchy of solutions

However, Andra underlines the importance of checking on the project every 10 years as an essential component of reversibility. The current approach put forward by Andra includes a "flexible closure schedule", i.e. society should have the power to decide on when and how it wishes to close down the facility (Labalette 2013).

4 Legal framework

The general principles of radioactive waste management were first laid out in the 1991 Waste Act, the "Bataille Law". The Act was then modified in 2006 by the Planning Act on the sustainable management of radioactive materials and waste, which instructed Andra to look into the implementation of a reversible geological disposal and submit an application for a construction licence in 2015. In response to the outcome of the public consultation in 2013-2014, Andra plans to submit the first elements of the licence application (concerning safety measures and reversibility) in 2015, followed by the final application in 2017.

The Bataille Law and the Planning Act laid down three key principles:

- sustainable management of radioactive materials and waste, protection of personnel health, safety, and the environment;
- the search for a definite and safe manner of managing radioactive waste, in order to prevent or limit the burden on future generations;
- producer responsibility, i.e. the duty of the producers of spent fuel and radioactive waste to safely manage these substances, within the limits of the general responsibilities that these parties have as operators of nuclear installations.

Nuclear waste management policy is governed, firstly, by the National Management Plan for Radioactive Materials and Waste (Plan national de gestion des matières et des déchets radioactifs (PNGMDR)), guided by the national waste inventory produced by Andra. When producing a revised version of the PNGMDR every three years, the energy ministry and the ASN draw on the advice from a multi-stakeholder committee. The plan is completed by the

proposed so far, while the latter would constitute a fundamental change of "doctrine" that should lead to change this hierarchy.

government and then passed on to Parliament. The latest plan, for 2013-2015, was made public in September 2013.[12]

The construction and operation of Cigéo will be governed by a number of legislative acts, including the decree on reversibility, and the law authorising the construction of the repository (enactment planned for 2020).

Legislation concerning public participation also influences radioactive waste management. The Arhus Convention on public participation and access to information, as well as the creation of the National Commission of Public Debate (CNDP), have been major drivers towards greater public participation in a hitherto highly closed system of decision-making on radioactive waste management (e.g. Barthe and Meyer 2012).

5 Institutional framework

5.1 National-level decision-making and management

The ministry responsible for energy, more specifically its Directorate General for Energy and Climate (DGEC),[13] elaborates national radioactive waste management policy, liaising, and arbitrating between the various stakeholders and state organisms. The policies and safety concerning military waste are the responsibility of a specific authority, DSND,[14] which reports to the Ministries of Industry and Defence. The civil nuclear safety authority, ASN, is independent of the government and the stakeholders, and is responsible for ensuring the safety of civil nuclear installations; the Autorité de Sûreté Nucléaire Défense (ASND) fulfils the same functions concerning military waste. ASN is assisted by a technical support organisation, IRSN, which is responsible for radiation and nuclear safety.

Andra, the agency that manages radioactive waste, was created in 1979 as part of the Ministry of Energy, with the specific task of designing an irreversible geological disposal system. The "Bataille Law" 1991 transformed Andra into an industrial and commercial agency, making it independent of the waste producers. The agency today is also responsible for the publication of the national waste

12 http://www.french-nuclear-safety.fr/Information/Publications/Others-ASN-reports/French-National-Plan-for-the-Management-of-Radioactive-Materials-and-Waste-2013-2015, last accessed 26 May 2014.

13 Currently a part of the ministry for the environment, energy and sustainable development.

14 Délégué à la sûreté nucléaire et à la radioprotection pour les activités et installations intéressant la défense.

inventory; conception, exploitation, and surveillance of the waste storage sites; coordination of R&D on geological disposal and interim storage; collection of waste from the producers; cleaning up abandoned contaminated sites; and providing public information.

5.2 Evaluative and consultative bodies

The parliamentary office of science and technology Opecst (Office parlementaire d'évaluation des choix scientifiques et technologiques) assesses strategic choices concerning a number of questions relating to science and technology, including radioactive waste management.

The Planning Act of 2006 ruled that a Commission (CNEF) should be set up to assess matters related to the financing of radioactive waste management. The Commission was then established in 2009, but only managed to submit its first report in 2012, allegedly because of turf battles between the various involved parties.

The National Assessment Board (CNE) was created in keeping with the Bataille Law 1991 as an independent organisation set up to evaluate research and studies annually, notably those conducted by Andra.

The High Level Committee (CHN), chaired by the Minister of Energy, has the main responsibility for the advancement and monitoring of the local economic support measures associated with the Underground Research Laboratory and Cigéo.

The multi-stakeholder High Commission for Transparency and Information on Nuclear Security (HCTISN) organises periodic consultations and debates on the topic of radioactive waste management.

The National Audit Office (Cour des Comptes) oversees the legality and efficiency of the use of public money. In January 2012, it published a highly "mediatised" report on the costs of nuclear energy in France. The report was the first independent assessment conducted by a public authority on the economics of nuclear energy in France, and also included an assessment of the costs and financing of Cigéo.

The authorisation of Cigéo is not only based on an assessment by ASN, consultation with the local and regional authorities, and the enactment of a decree on reversibility. It must also be preceded by a public consultation (see the section on information and citizen participation below).

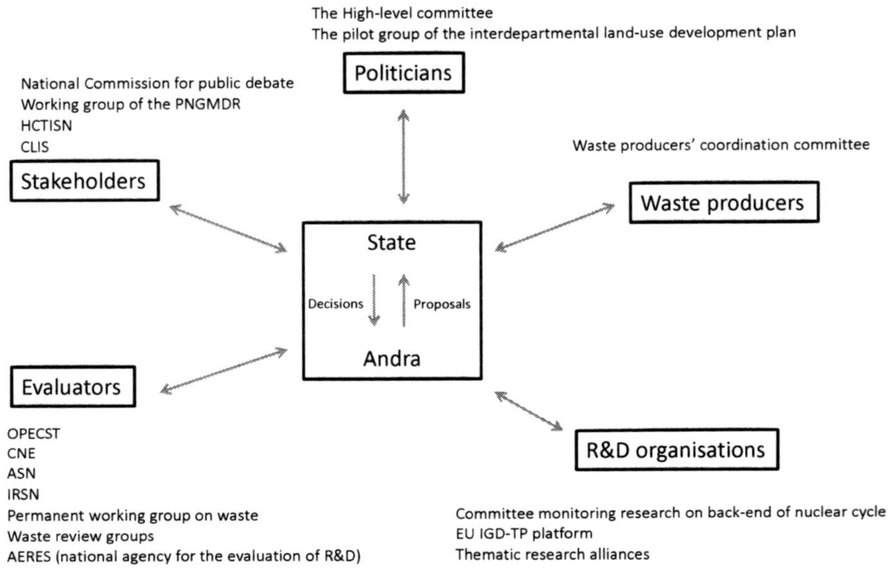

Figure 2: The governance of Cigéo.

Source: Adapted from Andra (2013: 74-76).

5.4 Local and regional level

The planned underground and surface installations of Cigéo would be situated at the border between two departments, Meuse and Haute-Marne, and two regions, Lorraine and Champagne-Ardenne. Currently the prefecture of Meuse is representing the central government in piloting the project at the local level. However, many local actors are calling for a stronger involvement by the government, arguing that they need more precise information about the planned execution of Cigéo (timetables, needed workforce, etc.) in order to plan their own policies (housing, transportation, health services, etc.). The departmental authorities and the local municipalities in proximity to Cigéo are the state actors' main counterparts in the planning process, especially in socio-economic matters related to the project. The regional-level authorities have so far been only loosely involved in the governance of Cigéo. A committee (GIP – Groupement d'intérêt public) has been established in each department to manage the economic support packages (see the next section).

The Local Information and Oversight Committee (CLIS), which includes representatives of the state, Andra, local politicians, and local business and civil society organisations, was set up in Bure in November 1999, after the government had chosen the village as the site for the URL. Its tasks are to inform the public, to enable dialogue between stakeholders, and to monitor the activities of URL and Cigéo. The Aarhus Convention helped to strengthen the role of CLIS and other local information and monitoring committees (CLIs) set up at localities hosting nuclear installations.[15]

5.5 Economic support mechanisms

The idea of compensation payments to the host municipalities was first introduced in the Bataille Law 1991. Since the start of the construction of the URL at Bure, the neighbouring municipalities have received economic support financed through a specific tax levied on the waste producers. The money is distributed via the two GIPs (*Groupement d'intérêt public*), the organisms specifically established to manage these funds in Meuse and Haute-Marne. Each GIP is led by the chairman of the departmental council, whose membership consists of local and regional authorities; Andra; the waste producers; the chambers of commerce, agriculture and craft trades; and the state (represented by the prefecture). The amount of support has increased progressively over the years, and now amounts to €30m per year for each department. The Planning Act of 2006 gave priority in funding allocations to the municipalities within a ten-kilometre radius of the URL. These municipalities can freely use approximately 10% of the GIP funding allocated on a pro rata basis in accordance with the population size of the municipality. However, the bulk of the funding is attributed to specific projects, which must fall within designated priority areas of activity. Each GIP decides by vote on the allocation of project funding – a procedure that essentially grants the departmental authorities and the prefecture the decision-making power, while leaving local politicians with very little say.

Officially, the economic support channelled through the GIPs is not described as compensation, but it is primarily designed to ensure that the local communities possess the skills, capacities, and infrastructure necessary to host the facility. In principle, funding should be granted to projects that create true value added and lasting socio-economic development. Yet in practice, part of the funding has been used to finance the statutory duties of the local and departmental authorities. Furthermore, at least in some cases, municipalities have constructed what could be qualified as "luxury facilities" (e.g. swimming halls,

15 The Bataille Law (1991) made local information and monitoring committees mandatory at each site hosting a nuclear installation in France.

community festival halls, etc.) that remain underutilised because of a lack of sufficient local population base. GIP support has also been criticised for creating competition and envy amongst the often extremely small municipalities (e.g. Bure has less than 100 inhabitants), and for being merely a form of bribery, designed to "buy" the acceptance of the local communities. However, both the opponents and the defenders of Cigéo acknowledge that the opposition against the project was a crucial trigger for the introduction of economic support.

In addition to the support by the GIPs – which is based on a legal mandate – the waste producers provide voluntary direct support aimed at strengthening the skills and capacities in the region and enhancing the acceptability of Cigéo.[16] Many of these projects focus on the environment and sustainability, thus triggering allegations that the support represents "greenwashing".

The future of the economic support is uncertain, as the system is likely to undergo a major reform once the current financing period expires in 2014. The waste producers want to have a greater say in the allocation of the support and demand that the GIPs be abolished. In their view, GIPs will be unnecessary if Cigéo is implemented, given that the project should generate its own tax revenue for the communities.

5.6 Estimated socioeconomic impacts and requirements for the installation of Cigéo

A central feature of Cigéo is its location in a sparsely populated, largely rural – and previously also industrial – area faced with decades of population decline and loss of industrial activity. The local population frequently refers to the fate of the region as a "sacrificed land," a buffer zone against Germany during the war hostilities in the 19th and 20th centuries, and virtually abandoned by the government. Many see Cigéo as yet another manifestation of this region being sacrificed in the interest of the nation. The opinions of local people are often marked by a certain ambiguity, combining huge expectations concerning the socio-economic benefits of Cigéo and serious concerns about its potentially harmful socio-economic and environmental impacts.

The interdepartmental land-use development plan prepared for the region in 2011-2013 estimated that Cigéo would generate 1,300-2,300 direct jobs during the construction of the facility, between 2019 and 2025, and 600-1,000 during its

16 The forms of support include the establishment of a spare parts distribution centre by the EDF, the archives of EDF and Areva, and a "second generation" biofuel project by the CEA. Efforts are also made to help local firms developing the specialist skills and expertise needed to collaborate with nuclear operators.

100 years of operation (SIDT 2013). At earlier stages during the preparation of the scheme, the generation of around 6,000-7,000 indirect and induced jobs was promised, but the version of the report released for the public debate in 2013 settled with the more conservative estimate of 2,000-4,000. In a sparsely populated region with a declining industry and an ageing population with a relatively low level of training and education, such perspectives trigger great expectations. At the same time, various local stakeholders, citizens, and politicians evoke fears and concerns, including:

- excessive dependence on a single industry, and the creation of a boom-and-bust cycle following the completion of Cigéo;
- a rise in the cost of housing, transports and services;
- loss of "freedom" associated with rural lifestyles (freedom to hunt, low transport volumes, etc.) triggered by the greater control needed to ensure the safety and security of waste transports;
- the fear that jobs would only or primarily benefit other regions, due to insufficient capacities and resources among local businesses to respond to project calls from Cigéo;
- weakening of local businesses, as Cigéo would absorb the most skilled and capable workforce;
- inability of the region to attract skilled workers and their families to a region lacking sufficient attractiveness;
- degradation of the image of the region that might be perceived as a "nuclear waste bin";
- loss of support to Cigéo, if the promised socio-economic benefits fail to materialise, and the greatly inflated expectations remain unfulfilled;
- concerns about radiation risks; and
- congestion, air pollution, and accidents resulting from increasing transport volumes.[17]

6 Information policy and forms of participation including civil society

Until the late 1980s, planning and decision-making on radioactive waste management, as well as nuclear power in general, was highly expert-led in France, with the "state engineers" (Hecht 2009) of the "Corps des mines" holding key positions while the formal decision-making remained in the hands of the

17 CEA, EDF and Areva have estimated that 700-900 waste packages per year could be transported in 2030-2040.

government. The failed site investigations in the late 1980s triggered an interest in more dialogue and greater citizen engagement. Barthe (2006) described the change as "reversibilisation," as the far-advanced process towards irreversible geological disposal was suddenly "reopened" to the participation of various groups of society on one hand, and to various waste management options on the other. The new approach involved broad consultations at different levels of policymaking, led by the parliamentarian Christian Bataille. Despite the more open and participatory site investigation process launched in 1994, local opposition was quickly organised in all three candidate communities. The government finally chose Bure, partly because it represented the "least-resistance" option. More intensive opposition against the project grew only once the construction of the URL began in 2000.

The local information and monitoring committee, CLIS, currently has 91 members representing the state, the departments, local politicians, business community, and NGOs. Given its pluralistic composition, including both advocates and opponents of geological disposal, CLIS could constitute a forum of inquiry, discussion, and dissemination of information from various perspectives. In practice, the opponents often describe CLIS as a "paper tiger" or even as a body serving to legitimise Cigéo, while the advocates of Cigéo sometimes accuse CLIS of providing opponents an open forum for disseminating their "propaganda". However, most critics acknowledge that CLIS has gradually evolved towards a truly independent platform for discussion.[18]

The public consultations ("public debates") organised by the National Commission for Public Debate (CNDP), constitute the main vehicle for participation at the national and regional levels. CNDP was created in 1995 and given the status of an independent administrative authority in 2002. At the request of the developer, CNDP starts a consultation by nominating a specific commission, CPDP (*Commission particulière du débat public*), consisting typically of six independent experts representing a range of expertise to run the debate. The developer then submits to the CPDP a background document describing the project, and interested parties are invited to give written feedback. The CPDP then compiles and summarises this feedback. To complement the material, CPDP often invites new participants to provide counter-appraisals. At the end of the debate, the commission then drafts a concluding report providing a summary of the discussions at the public meetings organised during the four-month debate. Within the three months following the debate, the developer must provide explanations on how the public debate is to be taken into account in

18 Such a change is manifested, for instance, in the fact that the committee is no longer headed by a representative of the state, but by a local or national politician.

project development. The CNDP does not have any decision-making power, nor does it make recommendations.

In 2005-2006, a national debate was organised on the general orientation of French radioactive waste management policy.[19] This was designed to instruct the parliamentary debates in preparation for the 2006 Planning Act. Many opponents of the disposal project deemed the debate relatively successful – a step forward compared with the 2003 "national debate on energy," which was considered by many observers as a mere legitimisation exercise orchestrated by the government.[20] The perception of relative success prevailed, despite the fact that one of the main conclusions of the consultation – the examination of near-surface interim storage as an option on an equal footing with geological disposal – was not included in the new Act (Lehtonen 2010a).

The Planning Act of 2006 stipulated a public debate organised by CNDP as a prerequisite for the authorisation of a disposal facility. Andra submitted its proposal, a CPDP was nominated, and the debate was started in May 2013, with a break scheduled during the summer vacations.[21] However, the opponents of Cigéo prevented the two first public meetings from taking place, prompting the CPDP to cancel the remaining planned public meetings. Instead, the duration of the consultation was extended by two months, and exchanges continued on the Internet, including a series of thematic debates confronting experts with varying opinions on the project. Many observers – and not only the opponents of Cigéo – remained highly sceptical about the value and true objectives of the public debate.[22] To compensate for the failures of the public debate, a consensus conference was organised between December 2013 and February 2014. For the CNDP, the Cigéo debate represented once again a challenge and an opportunity for renewal, forcing the organisation to innovate and seek new forms of citizen engagement.

Other forums and instruments of public engagement include the visitors' centre at the URL in Bure, which receives over 10,000 visitors annually, from

19 The CNDP organised in parallel two other nuclear debates – one on the new EPR reactor in Flamanville and another on the transmission line needed to connect the new plant to the grid.

20 This debate was not organised by the CNDP, but by the government.

21 The starting date of the debate was, for quite some time, an object of controversy, given that a national debate on energy transition was underway. This debate was launched by the president, and not organised by the CNDP. After pressure and criticism from many experts and stakeholders, the start date of the waste debate was postponed sufficiently to allow the first conclusions of the transition debate to be available.

22 The internet debates have not been spared from boycotts; the expert invited to discuss the costs of Cigéo as a representative of the critics of the project left the table after only a few minutes in the debate, claiming that the background documentation provided by the developer and the CPDP lacked the sufficient detail necessary for an informed discussion.

France and abroad. Public inquiry is a mandatory process designed mainly to give landowners the potential to be heard, but it does not represent a major instrument of public participation. Environmental Impact Assessment is likewise mandatory, but tends to be rather technically oriented, and therefore provides little opportunity for citizen involvement.

7 Conclusions

If the government and regulatory authorities indeed greenlight the construction of a French disposal facility, this will be among the largest industrial projects ever implemented in Europe. Apart from the usual problems associated with complex "megaprojects" , e.g.construction delays, budget overruns, failure to deliver the promised benefits) the integration of the project in the local territory will present major challenges, largely because the project is located in a declining, sparsely populated, and "non-nuclearised" region (Flyvbjerg et al. 2003; Lehtonen 2014). Such a siting choice may have been tempting, given the relative enthusiasm of local politicians and the weak opposition to the project. However, the choice also entails high risks, especially if the great expectations regarding the socio-economic benefits of the project do not materialise, and the prevailing atmosphere of mistrust towards the state persists or even intensifies. The widespread local perceptionthat the region will once again be sacrificed in the interests of the nation further accentuates the challenge. In such a context, the ambitious timetable, with the first waste packages scheduled to go into the repository in 2025, was slightly revised in May 2014 as a consequence of the public consultation under the auspices of CNDP.

The governance structures of the French disposal project are particularly complex, largely due to the complex nature of the country's political and administrative system. The governance of Cigéo involves a broad range of public, semi-public, and private actors at multiple levels of decision-making, with partially overlapping competences and evolving responsibilities. Tensions are clear, not only between "the state" and "the locals", but also between the different state actors and the largely state-owned nuclear industry, between Andra's headquarters in Paris and its local office in Bure, and between the numerous small municipalities that compete for the economic support and expected benefits from the project. These tensions are further amplified by the numerous technical and political uncertainties, as well as by the politically sensitive nature of the matter, which has discouraged key actors from taking responsibility for the unavoidably difficult decisions. However, the uncertainties related to the project and the complexity of its governance structures may well be also an asset, to the extent that in a context of multiple responsibilities, no single actor has complete command over the project. Instead, the actors are subject to multiple lines of mutual control and accountability. The requirement that the governance system of Cigéo must at any time allow a return to earlier phases of decision-making has become a potential major trigger for innovative and flexible governance solutions, as well as for greater openness in decision-making to different perspectives and participants. However, the virtues of reversibility are not automatic, since the concept also provides an opportunity for strategic manipulation and legitimisation, while keeping largely intact the old concept of irreversible disposal (e.g. Lehtonen 2010b; Barthe and Meyer 2012).

In the practice of participatory and deliberative democracy in France, the radioactive waste management policy has been somewhat of a forerunner and a test case, having undoubtedly contributed to the gradual opening up of the traditionally closed and non-transparent French "nucleocracy". The "opening up" of this process, subsequent to the dead-end reached in the late 1980s, entailed replacing the hitherto exclusively technical treatment of the problem with the consideration of a range of waste management alternatives, clearer separation of responsibilities between key actors, enhanced independence of safety authorities and experts, and the inclusion of new participants in the process. However, the deep mistrust of experts and state actors in nuclear matters still persists, while attitudes are characterised by polarisation and ambiguities, such as those between state-led authoritarianism (participation and governance must be "public" – "from the state and by the state" to be legitimate) and "grassroots romanticism" (radical grassroots activism), as well as between simultaneous calls for stronger central government leadership and criticism of excessive state intrusion in local and regional matters.[23]

23 While recent opinion polls among citizens have given Andra relatively high acceptance ratings, with 64% of respondents saying they trusted Andra in general (Andra 2012c), mistrust and ambiguities were clearly manifest in the interviews conducted in 2013 as a part of a research project involving field interviews in the communities surrounding the disposal project.

References

Agence nationale pour la gestion des déchets radioactifs (Andra) (2010). Making nuclear waste governable: Deep underground disposal and the challenge of reversibility. Châtenay-Malabry.

Andra (2012a). Rapport de synthèse: inventaire national des matières et déchets.

Andra (2012b). Etude de contexte: Le territoire du projet Cigéo.

Andra (2012c). Le journal de l'Andra; Edition Meuse/Haute-Marne, automne 2012. No 12.

Andra (2013). Projet Cigéo – Centre industriel de stockage reversible profond de déchets radioactifs en Meuse/Haute-Marne. Le dossier du maître d'ouvrage. Débat public du 15 mai au 15 octobre 2013.

Barthe, Y. (2006). *Le pouvoir d'indécision: La mise en politique des déchets nucléaires*. Paris: Economica.

Barthe, Y., and Meyer, M. (2012). Identifying remaining socio-technical challenges at the national level: France. InSOTEC Project working paper (WP 1 – MS 6). 21 May 2012.

Cour des Comptes (2012). Les coûts de la filière électronucléaire. Rapport public thématique. Paris. Janvier 2012.

Flyvbjerg, B, Bruzelius N., and Rothengatter, W. (2003) Megaprojects and Risk: An anatomy of Ambition. Cambridge: Cambridge University Press.

Hecht, G. (2009). The Radiance of France: Nuclear Power and National Identity after World War II. Cambridge, MA: The MIT Press. (first published in French in 1998)

Labalette, T. (2013). Cigéo as a megaproject? Presentation at the workshop "Dealing with uncertainties in socioeconomic evaluation of megaprojects". Andra, Chatenay-Malabry, 24 June 2013.

Lehtonen, M. (2010a). Deliberative decision-making on radioactive waste management in Finland, France and the UK: Influence of mixed forms of deliberation in the macro discursive context. *Journal of Integrative Environmental Sciences* 7(3): 175-196.

Lehtonen, M. (2010b). Opening up or Closing Down Radioactive Waste Management Policy? Debates on Reversibility and Retrievability in Finland, France, and the United Kingdom. *Risk, Hazards & Crisis in Public Policy* 1(4): 139-179.

Lehtonen, M. (2014). Evaluating megaprojects: from the "iron triangle" to network mapping. *Evaluation* 20(3): 278-295.

Lehtonen, M. (forthcoming). Evaluation of 'the social' in megaprojects: tensions, dichotomies, and ambiguities. *International Journal of Architecture, Engineering and Construction*.

PNGMDR (2013). French National Plan for the Management of Radioactive Materials and Waste 2013-2015. Ministry for Ecology, Sustainable Development and Energy and French nuclear safety authority (ASN). Paris.

SIDT (2013). Schéma Interdépartemental de Développement du Territoire Meuse – Haute-Marne. Projet de schéma, document pour le débat public. Préfecture de la Meuse; version 2.6, 21 mars 2013, Comité de Haut Niveau du 4 février 2013.

IV. Countries With Direct Disposal of Spent Nuclear Fuel (Including Cases with Additional Disposal of Radioactive Waste from Reprocessing)

Advanced Research, Lagging Policy
Nuclear Waste Governance in Belgium

Jantine Schröder, Anne Bergmans and Erik Laes

1 Introduction[1]

Belgium developed a rather extensive nuclear research and development (R&D) programme quite early due to the ready supply of uranium from the former colony of Congo and its contribution to the Manhattan project. Belgium, once had the national ambition of developing a full nuclear fuel cycle. Nuclear power provides 52% of the national electricity supply in Belgium in 2014.[2] Belgium's seven reactors – four in the Flemish municipality of Doel, three in the Walloon municipality of Tihange – became operational between 1975 and 1985. The history of the Belgian nuclear programme is for a large part one of 'fait accompli' politics and has been characterized by a general lack of transparent decision making (Laes et al. 2007). Strategic investment decisions were made in the mid-1960s without much attention to issues such as nuclear power plant (NPP) siting and the nuclear waste problem. Starting in the mid-1970s, a protracted and polarised debate between nuclear proponents and opponents eventually resulted in a nuclear phase-out law, which was adopted in 2003.[3] According to this law, all commercial Belgian NPPs should be closed down after a 40-year operational lifetime (i.e. in the period 2015-2025) and no new reactors for nuclear energy production should be constructed. However that law does not apply to nuclear research reactors. In addition a 'force majeure' clause related to the security of electricity supply was included. After the approval of the law in 2003, gloomy pictures were painted of future energy shortages and some efforts

1 This article is for a large part an adaptation of the report "Identifying remaining socio-technical challenges at the national level: Belgium", a working paper produced within the European FP7 project InSOTEC, coordinated by University of Antwerp (see Schröder and Bergmans 2012). The descriptive sections in this chapter have been taken directly from this report, which is part of a research project funded through the European Atomic Energy Community's Seventh Framework Programme (FP7/2007-2011) under grant agreement number 269906.

2 Eurostat – Electricity production and supply statistics, http://epp.eurostat.ec.europa.eu/statistics_explained/index.php/Electricity_production_and_supply_statistics.

3 Law of 31 January 2003 on the Gradual Phase-out of Nuclear Energy for the Purposes of Industrial Electricity Production.

were made to bring nuclear back on the agenda, as was done by the Nuclear Forum.[4] In this case, various political parties argued for at least postponing the shut-down of the three oldest NPPs. Today, the political compromise is to shut down the two oldest units in Doel (Doel 1 and Doel 2) as laid out in the phase-out law (2015), while keeping the oldest unit in Tihange (Tihange 1) operational for 10 additional years (until 2025).

Summarized, Belgium has a relatively large nuclear research and nuclear power programme and a relatively large amount and variety of radioactive waste. Its low- and intermediate-level radioactive waste management programme is quite advanced, but the status of the high-level / long-lived radioactive waste management programme is rather ambiguous. Although dedicated research on geological disposal has been taking place since the 1970s, there still is no formal policy in place. Throughout this article we will elaborate on the overall Belgian radioactive waste management situation and discuss some of the challenges faced (notably related to high-level, long-lived waste) in this particular setting.

2 The present policy for radioactive waste storage

In the early years of nuclear power in Belgium, reprocessing of spent fuel was the official policy. This played a role in postponing the high-level waste issue to a somewhat more distant future. Other waste streams were initially managed by the Belgian Nuclear Research Centre, SCK•CEN. Only in 1981, was a dedicated waste management agency, NIRAS-ONDRAF (N/O), established. This agency is responsible for all radioactive materials on Belgian territory, once they have been declared as waste. Radioactive waste is legally defined as "any substance for which no possible use is foreseen and which contains radionuclides in a higher concentration than the values considered acceptable by the competent authority to be used or disposed of without supervision".[5] After the compliance with acceptance criteria has been established, payment for its long-term management by the radioactive waste (RW) producers has been made, and after its transfer to N/O's storage facilities, N/O becomes the owner of the waste. At present this is not the case for spent fuel, which is stored on site at the NPPs under responsibility of the utility operators, because a moratorium (no definitive legal ban) on reprocessing was put in place in 1993 (Bergmans et al. 2006). It is however the case for the vitrified high-level waste from previous reprocessing contracts, which over the last decennium has been returned from AREVA in La Hague. N/O is also in charge of the 'nuclear passiva', including high-level waste

4 See http://forumnucleaire.be/nl?noredirect=1.

5 Royal Decree of 30 March 1981, article 1, own translation.

from former fuel cycle and research activities in Belgium, for which no provisions were made so the responsibility was transferred to the state.

2.1 Waste types and amounts

In line with the international classification developed by the IAEA, N/O makes a distinction between:[6]

- Low- and intermediate-level (LILW), short-lived waste (category A);
- Low- and intermediate-level, long-lived waste (category B);
- High-level (HLW), short- or long-lived waste (category C).

According to an inventory made in 2008 (calculated on the assumption of a 40-year lifetime of the Belgian NPPs), the total amounts of conditioned waste estimated to exist by 2075 are as follows:

- Category A: 69,900 m^3;
- Category B: 11,100 m^3 in the case of reprocessing, 10,430 m^3 without re-processing;
- Category C: 600 m^3 in the case of reprocessing, 4,500 m^3 without repro-cessing.[7]

2.2 Waste management programme(s)

Starting as a technocratic organisation, N/O began to change its organisational culture at the end of the 1990s, at least with regards to its programme for the management of category A waste. After Belgium's late ratification of The London Convention on the Prevention of Marine Pollution by Dumping of Wastes and Other Matter, N/O invited strong public opposition because of its top-down siting policy for a LILW disposal facility. Over 90 municipalities were listed as potentially suited, but without exception, all refused to cooperate in further investigations. Consequently N/O was forced by the government to organise a more bottom-up site selection procedure. Voluntary, participatory approaches were developed with the municipalities of Mol, Dessel, and Fleurus

6 See http://www.niras.be/content/classificatie.

7 See http://www.niras.be/content/volumes.

and Farciennes (Bergmans 2005; Bergmans et al. 2006). Finally, in 2006 the government selected the Dessel site for surface disposal of LILW.

Research on geological disposal for category B and C waste started early in Belgium and an inventory of potential geological formations was already made in the 1970s. In 1974 detailed investigations started concentrating on the Boom clay layer, conveniently available under the nuclear research centre (SCK•CEN) in Mol. In 1980 the construction of an underground research laboratory named HADES was started in the Boom clay layer, the first of its kind worldwide. Since 1995 HADES has become a joint venture between SCK•CEN and N/O, and is now known as EURIDICE. Although geological disposal in Boom clay has been the reference scenario from the start of Belgian radioactive waste management (RWM), research and dedicated research has steadily continued, up until now no formal political decision has been made on a long term management policy for category B and C waste.

In a state-of-the-art report on category B and C management research published in 2001, N/O explicitly asked the federal government to provide a framework for a transparent and legitimate decision-making process, with the intention of launching a public dialogue on the management of this waste (NIRAS/ ONDRAF 2001). In 2004, the government turned the tables on N/O and ordered them to do exactly this. In line with this governmental request and the foresight of a European Directive, N/O drew up a strategic waste management plan between 2009 and 2011.[8] This 'Waste Plan' was accompanied by a mandatory strategic environmental impact assessment (SEA). Different options and alternatives were discussed and public consultations were organised (we elaborate on this later in the section dedicated to public information and participation). In fact what happened in the framework of the Waste Plan and accompanying SEA, represented the first time that category B and C waste was explicitly and deliberately put on the public agenda. In September 2011 N/O handed over the final version of the Waste Plan to the federal government, which now has to decide on the further strategic direction of category B and C waste management. Except for the approval of the Waste Plan by N/O's board of directors and a ministerial letter telling N/O to continue developing its waste management strategy, over two years later a formal decision still had not been made (December 2013).

8 EURATOM, Council Directive 2011/70/EURATOM of 19 July 2011 establishing a Community framework for the responsible and safe management of spent fuel and radioactive waste, which requires Member States to put in place national policies by 2015.

2.3 Costs and financing

All N/O's costs are financed by the waste producers (including the Federal State, which is as responsible for legacy waste), based on the polluter pays principle. Two main financing mechanisms have been put in place for the long term management of radioactive wastes (NIRAS/ ONDRAF 2011a: 61-66). The first serves for the long-term management of the waste, mainly to be understood as final disposal. This 'Fund for the Long-Term management of all radioactive wastes' (FLT) has been operational since 1999 and is dedicated to the financing of temporary storage and eventually final disposal of the radioactive wastes under the ownership of N/O. Anyone who passes on radioactive waste to N/O needs to deposit a certain amount into this fund (based mainly on the quantity and radiological character of the waste, and the (periodically reviewed) estimated costs of future disposal). The FLT thus accumulates gradually, proportionate to the batches of waste being transferred to N/O. It is structured as a capitalisation fund, managed by N/O in keeping with the investment strategy stipulated by its board of directors. In 2012 this fund amounted to about €106m (NIRAS/ ONDRAF 2012: 52).

The second mechanism provides for the future funding of the decommissioning of nuclear power plants and for the management of spent fuel (which, as explained before, is currently not being reprocessed, but also not officially considered waste). Synatom, a subsidiary of the nuclear power producer GdF-Suez/Electrabel, looks after these provisions. This second mechanism was introduced after the liberalisation of the energy market and the approval of the phase out law. At the end of 2012 about €7bn were allocated for this fund (Synatom 2012: 9). There are however questions about Synatom´s transparency and the way in which it manages its funds (e.g. Laes et al. 2007: 242-243).

With the repository project for LILW heading towards the implementation phase, a third financing mechanism has recently been introduced. Whereas the FLT is dedicated to the technical aspects of final disposal, the 'Fund for the Middle-Term' (FMT) was specifically set up to comply with "the complementary conditions" of the host community (NIRAS/ ONDRAF 2011a: 65). The law authorizing this new fund provides for the general principles and management structure (similar to the FLT fund), so that it can be used for all types of future radioactive waste disposal facilities. The amount for the FMT in the framework of the surface disposal of category A waste has been set at €130m.[9] For every new concrete project a separate royal decree will be issued to determine how much additional provisions should go into the fund and on which basis.

9 Law of 29 December 2010 concerning various regulations.

Bearing in mind the magnitude of remaining uncertainties (notably for geological disposal, e.g. related to the waste inventory, location, depth, operational safety issues, retrievability demands, ...) and the general relativity of long-term cost estimates of any complex project, the total cost for the technical development and implementation of a surface disposal for category A waste has been estimated between €734m and €878m (NIRAS/ ONDRAF 2010: 126) and at around €3bn for a geological disposal for category B and C waste (NIRAS/ ONDRAF 2009).

3 Waste storage concept(s)

At present, all waste types are kept in temporary surface storage facilities. For LILW, the long term management concept which is currently undergoing a licencing process is that of a surface repository, which is reversible (cf. section 2), but considered to be final. As far as HLW management (potentially including spent fuel) is concerned, different alternatives were discussed in the context of N/O's Waste Plan and the SEA. The Waste Plan explains that N/O strives for an integrated "transdisciplinary" solution covering four dimensions: a technological, financial and economic, environmental and safety, and a societal and ethical dimension. 'Sustainability' is presented as a guiding principle in the search for a balance between the four dimensions. 'Precaution' and 'participation' are to be taken into account and the 'polluter pays' (or should pay) principle should be applied. In the Waste Plan recommendations, N/O among others requires a refinement of the institutional framework to guarantee the implementation of these overarching principles (NIRAS/ ONDRAF 2011a).

As ordered by law, the SEA included a comparison of various potential final options for the long-term management of category B and C waste. Some options were rejected straight away, because they are in violation of international treaties or conventions which Belgium signed (e.g. sea dumping and disposal in ice sheets), and/or of the Belgian legal and regulatory framework (e.g. disposal by injecting liquid waste in the deep underground), and/or judged not to provide adequate safety guarantees (e.g. surface disposal). The remaining options, listed as eternal storage, geological disposal, disposal in deep boreholes, long-term interim storage ('awaiting something else') and the option of continuing the current situation ('status quo') were compared by means of a cross-disciplinary assessment. In the end, two options were withheld: disposal in an appropriate geological formation and long-term interim storage (NIRAS/ ONDRAF 2011b: 8-9).

Prolonging storage for 100 to 300 years before deciding on disposal is rejected by N/O on the basis of the agency's interpretation of the precautionary principle,[10] as well as for reasons of security concerns (proliferation risks), maintenance cost and funding uncertainty. In its Waste Plan, N/O asks for a governmental decision in principle to continue developing the reference solution of deep geological disposal in a clay layer in Belgium, in a 'stepwise', 'flexible' decision-making process. During the operational phase of putting the waste into the disposal site, this flexible process would allow for reversibility (defined as the possibility of taking back the waste by the same means as the ones used to place the waste in the repository (NIRAS/ ONDRAF 2011a: 135)). A period of retrievability (defined as the possibility of taking back the waste after full or partial closure of the repository, when artificial barriers are still expected to be intact) could be an option for N/O, as long as it does not compromise the safety of the repository.

4 Legal framework

Belgium is a federal state with three regions (the Flemish, Walloon and Brussels Capital Region) and three communities (the Dutch, French and German speaking Community). All issues related to the development, deployment, applications and consequences of nuclear technology and radioactivity are governed at the federal, national level. The fact that other potentially connected domains, such as environmental legislation, spatial planning, employment, certain aspects of emergency management, etc. are in the hands of the regional and local regulatory levels, adds to the overall complexity of RWM in Belgium (Bergmans et al. 2006).

The basic regulations for the safety of nuclear activities was first laid down in the law of 29 March 1958 pertaining to the Protection of the Population against the Hazards of Ionizing Radiation, accompanied by the royal decree of 28 February 1963. It governed the licensing of nuclear facilities, the inspection and control regime, radiological protection, medical applications of ionising radiation, import, transit, distribution and transport of radioactive substances, and RWM. With regard to RWM, these initial legal documents merely state that each actor is responsible for the safe management of the radioactive waste it produces. This law of 1958 was repealed and replaced by that of 15 April 1994, which also constituted the legal basis for the Federal Agency for Nuclear Control (FANC/

10 The meaning of the precautionary principle is susceptible to interpretation. One could also read it in favour of prolonged interim storage, as is to a certain extent the case in for example the Netherlands (e.g. Schröder 2012).

AFCN) to be the Belgian nuclear regulatory body. In fact the installation of both a "national regulatory agency for all nuclear activities" and of a "governmental agency for the management of radioactive waste and spent fuel" was already foreseen in 1980, in a law on budgetary proposals.[11] The National Institution for the Management of Radioactive Waste and Enriched Fissile Materials (NIRAS/ ONDRAF) was founded and its responsibilities specified by a royal decree in 1981.

Furthermore, the Euratom directives on radiological protection and on basic nuclear safety standards have been implemented and the International Treaties on nuclear safety and on the safe management of nuclear waste have been ratified. The licensing process for nuclear installations is amongst other criteria based on the IAEA safety fundamentals and safety series.

In 2010 FANC/AFCN drafted a proposal for a new legislation on the licensing procedure for final disposal. An amendment of the licensing legislation for nuclear installations of 1994 was proposed to account for the specific character of installations for final disposal, notably their long life time and the fact that they are not planned to be dismantled in the future. A royal decree was drafted, but it has not yet been formally published (December 2013).

With regard to public participation in RWM, the government assigned N/O the responsibility to develop a societal or dialogue process for the long-term management of LILW waste in 1998 and for the long term management of HLW in 2004 (see also the section on public information and participation). Federal legislation is based on international conventions and directives.

5 Institutional framework

Belgium has a rather large diversity of public, semi-public and private nuclear actors. Throughout this section we shall briefly describe in our opinion the most prominent ones and their main responsibilities and activities with regard to radioactive waste (management).

The Société belge des Combustibles Nucléaires (Synatom) is a private company that owns the nuclear fuel and is responsible for the management of the Belgian nuclear fuel chain, from the supply of uranium to the handling of spent fuel. Synatom is responsible for collecting and managing the provisions for the decommissioning of the NPPs and the management of spent fuel. This task includes establishing decommissioning strategies and scenarios for spent fuel, as well as providing cost estimates and securing and managing the actual financial

11 Law of August 8, 1980 concerning budgetary proposals 1979-1980.

provisions. Synatom is a 100% daughter of GdF-Suez/Electrabel, the power company which owns both NPPs in Doel and Tihange.

The National Agency for Radioactive Waste and Enriched Fissile Materials carries all formal responsibility and authority for both the short term as well as the long term management of all RW on Belgian territory. N/O is a public agency under the tutelage of the Federal Minister(s) responsible for Energy and Economy. It is, however, financed by the waste producers (including the Federal State, as responsible for legacy waste), based on the polluter pays principle. After compliance with acceptance criteria and payment by the RW producers, N/O becomes the owner of the RW. In addition to the identification of the acceptance criteria, the responsibilities of N/O include: transport, treatment, conditioning, storage and disposal of all RW in Belgium, as well as keeping an inventory of all nuclear installations and sites (including obsolete ones) and the radioactive substances they generate.

For the execution of many of these responsibilities, N/O subcontracts third parties (e.g. Transubel to deal with nuclear transport and Tractebel-engineering to carry out infrastructural and other work). The same goes for research, which is done by SCK•CEN (cf. section 2.2), universities and other research institutes, private (e.g. engineering) companies, and in international cooperation.

All substances that have been declared as waste and of which the ownership is transferred to N/O are stored at the premises of Belgoprocess (BP), including the returning vitrified and compacted waste from past reprocessing contracts. The location choice for this interim storage was never the subject of a public deliberation process, but followed automatically from the location of the two first large waste producers, namely the nuclear research centre SCK•CEN and the former reprocessing plant Eurochemic. BP is the industrial daughter company of N/O. It is in charge of the industrial management of the processing and storage of radioactive waste, whereas N/O is responsible for the administrative management and research. BP is currently processing, conditioning and storing all categories of radioactive waste resulting from both the nuclear fuel cycle activities and from the production and uses of isotopes in medicine, agriculture and industry.

The FANC/AFCN is the Belgian regulator, responsible for the protection of the population and the environment against the dangers of ionising radiation. It is in charge of both the licencing of activities involving ionising radiation (including the transport of radioactive material) as well as the supervision of these activities. FANC/AFCN defines the rules, licenses the installations to which these rules apply, and supervises compliance with the rules. With regard to the big nuclear installations (the so called 'class I installations', including future waste disposal installations), a licence is granted by royal decree, issued by the federal Minister of Interior Affairs, FANC's tutelage minister, on the latter's advice. The actual inspection of compliance with the license and the license conditions of these installations is carried out by FANC's subsidiary, Bel V.

The Belgian Nuclear Research Centre (SCK•CEN) is a Foundation of Public Utility under tutelage of the Federal Ministry of Energy. It was founded in 1952 to promote the peaceful application of nuclear energy. The centre created its own 'waste department' in the early days, and gradually also started to look after other parties' wastes with similar characteristics. But as the amount and diversity of RW expanded, notably due to the start of nuclear energy production, the functionality of the waste department of SCK•CEN became too limited. Furthermore the waste procedures used lacked any legal basis and the need for an independent waste management organisation became apparent (Bergmans et al. 2006; Laes et al. 2007: 105). After the establishment of N/O, close collaboration with SCK•CEN continued, and the centre initially retained the responsibility for the treatment of certain waste types. A scandal involving some SCK•CEN collaborators as well as the need for additional treatment and storage facilities also brought an end to these activities, limiting the centre's RWM role to supportive and fundamental research.

Environmental movements that regularly intervene in the debate on nuclear energy and RWM are most notably Greenpeace Belgium and Interenvironnement Wallonie. To complete the picture, the local partnerships, STORA and MONA,[12] involving citizens of the municipalities of respectively Mol and Dessel, should be mentioned. They play a pertinent role with regard to the long term management of LILW. But they also follow up on evolution regarding HLW waste and nuclear issues in general, as the nuclear site of Mol-Dessel-Geel hosts a diversity of related activities.

6 Siting procedures

An official siting decision for LILW waste was made in 2006, and a licencing application is currently being processed. As indicated before, with the help of social scientists, this siting procedure was elaborated into an approach in which the implementer N/O and the volunteering communities formed partnerships to investigate the technical feasibility and social desirability of becoming a host for a repository for LILW. The municipalities formulated among others the creation of a local fund as a condition of acceptance. The total amount of this fund will be between €90m and €110m and will be jointly managed by the partnership of

12 See http://www.stora.org/nl; http://www.monavzw.be.

Dessel (the host community) and Mol (the neighbouring community that also volunteered). It is meant to last for several generations and to be invested in projects that create socio-economic added value.

No formal political decision has been made regarding geological disposal as proposed by N/O to be the way forward for HLW. The question of siting has not been addressed as such, and currently no specific procedure (other than what is required in terms of general environmental impact assessment regulations) has been developed. Nevertheless, by letting the R&D programme which is dedicated to national geological disposal carry on for almost 40 years, implicitly, choices have been made that not only relate to the preferred technology (geological disposal) but also to the preferred host rock. Since stable geological formations such as salt or granite are not nationally available in a convenient manner, nuclear researchers judged poorly indurated clay as appropriate already in the 1970s (Laes et al. 2007: 150). This left three national options: Boom clay, Ypresian clay or schist formations. The reason Boom clay became the reference formation is simple and coincidental: it was available underneath the nuclear research centre SCK•CEN in Mol. In response to the waste manager's 2011 Waste Plan and related SEA procedure, the regulator, FANC/AFCN, pointed out the need to investigate the potential alternative of the Ypresian clay layer. Previously, this had also been recommended in international peer reviews of 10-yearly state-of-the-art research progress reports. N/O will follow up this recommendation, firstly by investigating to which degree the knowledge of Boom clay can be transferred to Ypresian clay (Minon 2012). Although it does consider Ypresian clay as a potential alternative, N/O clearly states that it is financially impossible to do the same extensive in situ research (i.e. 30 years of dedicated underground research laboratory experimentation, €360m spent (Eggermont and Hugé 2011: 30)) on another host rock than Boom clay. At this stage the latter thus remains to be the 'reference formation'.

Without a decision in principle (be it on geological disposal, or something else) and a clear view on any procedures following such a decision, it is difficult to anticipate how the Belgian HLW management programme will further unfold. What is clear today, is that a legacy of more than 30 years of R&D put into a particular management strategy in a particular host rock type has left its mark, pre-empting in many peoples' minds the identification of potential host sites. Throughout the public consultations on the Waste Plan for instance, concerns were expressed about the Boom clay layer not being deep, thick nor homogeneous enough for geological disposal (Minon 2012), often by actors in the concerned regions (cf. e.g. Weyns 2010).

7 Policy regarding public information and participation

With regard to public participation in RWM, we have already mentioned that the government assigned N/O to develop a societal process for the long term management of nuclear waste. Below we focus on public participation processes tied to B and C waste.

Current formal, federal legislation on information and participation is among others based on translations of international conventions and directives (the Aarhus Convention on Access to Information, Public Participation in Decision-making and Access to Justice in Environmental Matters and the European Directives on Environmental Impact Assessment and Strategic Environmental Assessment). [13] In carrying out these laws, a number of consultations had to be organised before N/O's Waste Plan for HLW could be finalised. Within this framework, the general public is invited to post written comments for a certain period of time (officially 60 days) after an announcement is made in the Law Gazette, the federal portal website and on the initiator's website (in this case N/O). Consultation with a number of advisory councils, such as the Federal Council on Sustainable Development, is mandatory. As initiator, N/O is also obliged to show how it has taken into account the comments received throughout these consultations in the final version of the Waste Plan presented to the government. Clearly such a legally mandatory consultation does not automatically lead to a broad societal dialogue. Only well-informed individuals and groups, having the necessary resources (time, expertise) to get involved in this complex issue, will raise their voice in the process. N/O was aware of these limitations, and therefore organised so-called 'NIRAS dialogues' (April-May 2009) – a series of consultation days open to participants from civil society – and an 'interdisciplinary conference' (30 April 2009) where experts on the topic were invited to share and discuss their opinions. The principal aim of these consultations was to draw up a 'map' of public concerns; the instruction for the facilitators was to avoid polarised debates and to look for consent wherever possible (Laes et al. 2009). N/O experts were present at these meetings, but did not participate.

13 The Law of 17 December 2002 (BS 24 April 2003, 22128), which transposes the Aarhus Convention into Belgian legislation, the Law of 5 August 2006 on public access to environmental information (BS 28 August 2006, 42538), which transposes the European Directives 2003/4/EC, the Law of 13 February 2006 on the assessment of the environmental impact of plans and programs and on the participation of the general public in the development of plans and programs related to the environment (BS 10 March 2006, 14491), which is a transposition of the European Directives 2001/42/EC on Strategic Environmental Assessment (SEA) and 2003/35/EC on public participation in decision making in environmental matters.

Furthermore, responding to earlier criticism about its role as organiser of the dialogue (Bombaerts and Eggermont 2009), N/O asked an independent body, the King Baudouin Foundation (KBS) (a well-respected centre for philanthropy, fostering projects on justice, democracy and diversity) to organise another public consultation, which it did by means of a citizens' conference. Belgian citizens with diverse backgrounds came together for three weekends to study and discuss, among themselves and with experts, the long-term management of category B&C waste and to deliver a report with recommendations.

In spite of the additional efforts made, the process of public consultations received quite some critique. The public dialogues, which had a poor turnout, were reproached for being organized in a hidden place in Brussels without sincere means of prior information (about the event itself and its topic). In addition, in light of RWM being a clear case of unequally distributed costs and benefits, the citizens' conference organised by the KBS was criticised because the technique focuses on the opinion of the 'average citizen' and ignores 'the potential local stakeholder'. N/O agrees that the actual dialogue still needs to start up, and that integrating the participative dynamics into the B and C programme had been largely ignored up until a few years ago (NIRAS/ONDRAF 2011b: 20). Besides the additional public consultations organised in the framework of the Waste Plan, at present there is no clear view on what participation on category B and C waste management should and could mean in practice, especially outside a concrete siting process.

8 Conclusions

One could say that for high-level and long-lived radioactive waste, NIRAS-ONDRAF, Belgium's nuclear waste management organisation, is at present confronted with the limits of 'implicit' or 'de facto' decision-making. The Belgian case of category B and C waste governance displays an ambiguous combination of highly specialised advanced and focused research with rather impassive, fragmented and belated policy making. For over 30 years, radioactive waste management has been in the research phase. The focus has been on scientific advancement with regard to geological disposal in the Boom clay layer, aided by the experiences gained from the underground research laboratory located in the municipality of Mol. The lab's advanced technical knowledge and know how is recognised by the international research community, yet remains rather isolated from society and politics. The public debate on category B and C waste management in the framework of the publication of N/O's Waste Plan remains limited. The 'momentum' it created appears to have died again (the

Waste Plan has been in limbo for over two years now, and there is little chance that the current government will make a decision on this matter in the run up to the 2014 elections). The challenge now being faced is the development of a process that will outline how a complex, long-term project like geological disposal can be implemented, taking into account a mixture of demands such as transparency, safety, flexibility, controllability, justice, and accountability. In such a context, participation cannot be limited to a few consultation moments within the framework of an SEA procedure. The local partnership approach adopted for category A waste may be something that could be strived for in a more advanced siting stage, but will require a broader framework and a dedicated revaluation with regard to its applicability to category B and C waste and geological disposal. The design and practical organisation of a robust participatory process should furthermore not be delegated to the waste manager alone. The time has come to initiate a 'dialogue on the dialogue', to discuss the shape and content of a decision-making process spanning decades, and to do this in a transparent way, taking into account the viewpoints of experts on participative methodologies and the stakeholders who could possibly influence this process or stand to lose or gain from it.

References

Bergmans, A. (2005). *Van 'de burger als beleidssubject' naar 'de burger als partner': de Belgische queeste naar een lange-termijnoplossing voor het beheer van het laagradioactief en kortlevend afval*. Doctoral thesis, University of Antwerp, Faculty of Political and Social Sciences, Antwerp.

Bergmans, A., Van Steenberge, A., and Verjans, G. (2006). CARL Country Report – Belgium.

Bombaerts, G., and Eggermont, G. (2009). Afval beheren en controle loslaten. Over participatie bij berging van nucleair afval. *Oikos* 48 (1), 28 – 38.

Eggermont, G., and Hugé, J. (2011). Nuclear Energy Governance. Deliverable 4.1, SEPIA project, Belgian Science Policy (Research Programme Science for a Sustainable Development), Brussels.

Laes, E., Chayapathi, L., Meskens, G., and Eggermont, G. (2007). *Kernenergie (on)besproken*. Acco, Leuven.

Laes, E., Eggermont, G., and Bombaerts, G. (2009). A risk governance approach for high-level waste in Belgium: a process appraisal. Paper presented at the "Managing Radioactive Waste" Conference in Göteborg, Sweden.

Minon, J. P. (2012). *Reacties op het Ontwerp van het Afvalplan van NIRAS/ ONDRAF*. Presented at MONA, 06/03/12, Mol.

NIRAS/ ONDRAF (2001). SAFIR 2 – Safety and Feasibility Interim Report 2. NIROND 2001-06 E, N/O, Brussels.

NIRAS/ ONDRAF (2009). Cost Evaluation of Geological Disposal of Category B&C Waste for the Long Term Fund (Revision of 2009). Report NIROND-TR 2009-15 E, 2009.

NIRAS/ ONDRAF (2011a). Afvalplan voor het langetermijnbeheer van geconditioneerd hoogradioactief en / of langlevend afval en overzicht van verwante vragen. NI-ROND 2011-02, N/O, Brussels.

NIRAS/ ONDRAF (2011b). Executive Summary of the Waste Plan for the long-term management of conditioned high-level and/or long-lived radioactive waste and overview of related issues. NIROND 2011-04E, N/O, Brussels.

NIRAS/ ONDRAF (2012). Annual report 2012. http://www.ondraf.be/sites/default/files/Jaarverslag_2012_NL.indd/HTML/files/assets/downloads/publication.pdf, last accessed 26 March 2014.

Schröder, J., and Bergmans, A. (2012). Identifying remaining challenges at the national level: Belgium. InSOTEC working paper, University of Antwerp, Antwerp.

Schröder, J. (2012). Identifying remaining socio-technical challenges at the national level: the Netherlands. InSOTEC working paper, University of Antwerp, Antwerp.

Synatom (2012). Annual report. http://www.synatom.be/files/Jaarverslag/Jaarverslag%20 2012_Web.pdf, last accessed 26.03.2014.

Weyns, W. (2010). Definitieve berging van hoogradioactief en / of langlevend kernafval in een kleiformatie in de Belgisch-Nederlandse grensstreek. Written for Greenpeace Belgium and Greenpeace the Netherlands, Michel Genet, Brussels.

Participation under Tricky Conditions
The Swiss Nuclear Waste Strategy Based on the Sectoral Plan

Peter Hocke and Sophie Kuppler

1 Introduction[1]

There has been a considerable delay in constructing civil repositories for high active waste considering former time schedules established by industrial countries with developed nuclear sectors. Only a few countries like Sweden and Finland have been making good progress. Others are far behind their original schedule, but have recently made some progress. This progress in decision-making is happening in Switzerland, where an elaborated site selection procedure with a strong promise of substantial participation started in 2008 (SFOE 2008).

In this chapter, we will give an overview of the Swiss case which could be instructive for a number of siting projects in the field of radioactive waste management. As other countries, Switzerland started with ambitious nuclear programs in the 1960s and at the end of that decade the first nuclear reactors were brought on line (Minhans and Kallenbach-Herbert 2012: 5). Much confidence was put in nuclear energy production leading to its robust development. This also led to an increase in the annual percentages of nuclear power in electricity production (Brunner 2009: 389). This changed when the accidents in Three Mile Island (1979), Chernobyl (1986) and Fukushima (2011) mobilized anti-nuclear social movements and political change. Phase-out scenarios were discussed in different countries and over time nuclear waste policy in Switzerland became an ever more complicated national task. Civil society established a social movement sector, which was successful in mobilizing important groups from the interested public against the civil use of

---

1 This chapter is based on the notes paper prepared for our presentation for the 18th REFORM Group Meeting (Salzburg, August 26th – 30th, 2013). The work presented here is part of the ITAS work package within the Project ENTRIA funded by the German Federal Ministry of Education and Research. ITAS is the Institute of Technology Assessment and Systems Analysis at Karlsruhe Institute of Technology (KIT). ENTRIA is the acronym for the German language equivalent of "Disposal options for radioactive residues: Interdisciplinary analyses and development of evaluation principles" (www.entria.de). The number for funding is 02S9082D.

nuclear power and the industry's plans to install repositories for radioactive waste without substantial participation of the public in the affected areas. Citizens' expectations to be involved in decision-making processes on contested technologies became more and more important, especially in most western democratic countries. "Decide-Announce-Defend" strategies led to less progress than expected and an enormous delay in realizing repositories became fact. Established collective actors in different policy areas reacted by adopting approaches of New Governance developed by think tanks and governmental organizations (e.g. Eberlein and Kerwer 2004). Also in the social scientific debate questions of participation and deliberative democracy gained importance (see e.g. Benz and Dose 2010). Around this time, experts of the nuclear community became familiar with some of these ideas. In Switzerland, a decision to adopt central ideas from this discussion was taken in the early 2000s. In the reconceptualization of the search process for Swiss nuclear disposal sites, which has been on-going since 2008, aims from international discussions were adopted (see e.g. NEA 2004 or AkEnd 2002). Our thesis is that the Swiss responsible collective actors were successful in "modernizing" the radioactive waste management concept before it could get blocked by central actors from politics, industry and civil society.

We show in this chapter that the Swiss government and responsible organizations turned the difficult situation in the beginning of the 2000s by introducing an alternative planning strategy and a number of procedural changes, which gave them a new range of flexibility. They adopted a limited number of political innovations within the setting of an already existing planning tool at the national level, the so-called Sectoral Plan. Through these changes they got a new chance for carrying through a potentially successful siting process. The outcome of this setting is open, but it is an innovative modernization track for Swiss national nuclear waste policy.

2 Governance as a conceptual frame

Some scientific literature about the Swiss case of radioactive waste management is available. The Wellenberg siting approach, where the implementer National Cooperative for the Disposal of Radioactive Waste (Nagra) and the Swiss authorities were stopped two times by a cantonal veto, was the issue of analyses made in the late 1990s and over the last decade (e.g. Krütli et al. 2010a; Scholz et al. 2007). The same research group at the ETH Zürich also published some theoretical work on adequate decision-making processes (Krütli et al. 2010b; Flüeler 2006). From our point of view it is remarkable that the current process is not in the focus of social science research although there has been a small increase of publications over the last years (see Solomon et al. 2010 and Strandberg and Andrén 2011). Studies focusing on sociotechnical and social science aspects in general are very limited. We know from conferences and professional contacts that there is also a "grey literature" for Switzerland (like SVA s.a.).

Radioactive waste management and the related social conflict classify as doubly complex and "wicked" problems. They are doubly complex as the modes of clarifying and deliberating on sociotechnical options and solutions for managing the technical and the social sides of the waste problem under conditions of uncertainty bring up extremely specialized and complex research questions (Kuppler 2012). Following Brunnengräber et al. (2012) nuclear waste can be seen as a "wicked problem" where uncertainties are accompanied by strong conflicts about values and preferences and a number of strong veto-players are involved in decision making. Altogether, no solution to the radioactive waste problem can fulfil all the expectations of stakeholders and society given contradictory interests and different perceptions of what "the problem actually is". Our research focus is on the description of the "context structure" of the societal conflict in Switzerland, the dissent between and the collective action of experts, and the modes of governance and effects of political and societal deliberation.

Not only in the Swiss case is radioactive waste management a state-driven activity. The Swiss mode of approaching its waste management duty can be seen as a "new governance" issue. New Governance is an approach which was developed and strengthened parallel to the debate about ongoing globalization. It is based on the observation that classical top-down approaches lost their steering capacity – also in concrete local processes like siting procedures. The World Bank and European institutions were early promoters of this approach and different disciplines in the social sciences engaged in this discussion.[2] New Governance studies focus on the design of decision-making processes and institutionalized structures and designs that influence how problems are picked up and solved. (Grande 2012; Hoppe 2011; Chhotray and Stoker 2009).

Two basic ideas which are highlighted in those concepts are (1) the promotion of networking and (2) the coordination of stakeholders (Haus 2010, Benz and Dose 2010). The coordination of stakeholders in issue specific debates on technology is an ambitious challenge, as they are more or less autonomous collective actors with widely differing expectations related to their participation

2 See World Bank (1992), Gramberger (2001) and Scharpf (2001). Various scientific perspectives especially that of political sciences, globalization research, innovation studies, science and technology studies (STS) and technology assessment are important in this debate.

in the preparation of collectively binding decisions and their role in concrete deliberations of decisions. In governance arrangements, not only established lobby-groups, responsible authorities and industrial associations are involved in decision-making, but also non-established stakeholders (like environmental associations or citizens' initiatives). This will most probably lead to struggle and conflicts as this includes groups such as the extra-parliamentary opposition and groups which are critical of representative democracy. Modes of deliberation and bargaining gain importance and processes that take place in the "shadow of hierarchy" stay dependent on resources like time, money and the ability to have influence on formal decision makers.[3]

In the mid-2000s, the official Swiss nuclear waste policy was to restart nuclear waste management with a stepwise approach to site selection, integrating the interested public, non-governmental organisations (NGOs), authorities, industries and a wide range of (socio-)political actors in their "Sectoral Plan for Deep Geological Repositories" (SFOE 2008). The early participation of many stakeholders and public participation, especially at the local level, were encouraged. This was done with the expectation of reducing complexity while buffering the potential for an increase of contingencies during the overall process of deliberation and decision-making.[4]

Such expectations are difficult to fulfil in complex processes of opinion formation and deliberation. It is an ambitious aim to expect to be able to close debates like these within the foreseen timeframes with any stepwise approach. In the next section the characteristics of this attempt are discussed before a problem-oriented interpretation gives some hints as to which lessons could be learnt from the Swiss case.

3 The failure and the restart

Historically, electricity generated in Switzerland came exclusively from hydropower without any recourse to fossil fuels (DETEC 2011: 5). Fossil fuels were not available as a natural resource and so the promise of cheap and unlimited energy from the civil use of nuclear power in the 1950s and 1960s had a special attractiveness for this central European state. The Atomic Act was put into force in 1959 and during the 1960s several projects for nuclear power plants

---

3 On the strong influence of hierarchy see Börzel (2008) and Marschall (2008); for a more sceptical view on a more conceptual level see Offe (2008) and Mayntz (2009).

4 On the increase of contingency and processes to reduce contingency on the same track see Haus (2010: 11).

(NPP) were initiated (DETEC 2011: 5). Four of them were realized, but some were rejected in part due to ecological protests (e.g. Kriesi 1982).

Currently, five reactors at four sites are working (Beznau 1 and 2/ Canton Aargau, Gösgen/ Canton Solothurn, Leibstadt/ Canton Aargau und Mühleberg/ Canton Bern); these were commissioned between 1969 and 1984.[5] Further important nuclear facilities are the three research reactors and interim storage facilities at the NPPs and the two research labs focused on nuclear waste at Grimsel (Canton Bern) and Mont Terri (Canton Jura). In the past few years, Switzerland belonged to a group of a very limited number of countries, which had a plan to increase the number of NPPs without public money. But in May 2011, following the nuclear accident in Fukushima, the Federal Council announced its intention to abandon plans to build new NPPs.[6] The current plan is to close down the last NPP in 2034. The first closure of an NPP is planned for May 2019.

In Switzerland nuclear waste is divided into three categories: (a) high-level waste (HLW): (a.1) spent fuel not destined for further use; (a.2) vitrified fission product solutions from reprocessing of spent fuel; (b) alpha-toxic waste, with a content of alpha emitters that exceeds 20,000 Becquerel/g of conditioned waste; (c) low- and intermediate-level waste (L/ILW): all other radioactive waste (SFOE 2008: 12, and Art. 51 Nuclear Energy Ordinance). In total 7,300 m^3 of high-level waste are expected based on a reactor life-time of 50 years. Further 60,000 m^3 of low- and intermediate level waste are expected from operation and decommissioning of the nuclear power plants. Each NPP has facilities for the conditioning and the interim storage of radioactive waste from its operations. For the dry storage of spent fuel elements and the vitrified high level waste there is an additional facility at the Beznau site (ZWIBEZ); its operation started in 2008. For the wet storage of spent fuel elements a separate building was installed in Gösgen NNP in the same year (s. DETEC 2011). Beside these decentralised installations there is a further central interim storage facility for all kinds of nuclear waste located in the northern Swiss town of Würenlingen (ZWILAG).

5 The waste inventories for each of the four NPP sites is documented in the Joint Convention Report (as per 2010) (DETEC 2011: 27).

6 The Federal Council (Bundesrat) is composed of seven ministers of equal rank and acts as the federal government. It is linked with the chambers of parliament (the National Council and the Council of States, which represents the 46 very differently structured Swiss Cantons). Linder (2005) classifies the Swiss democracy as a "half-direct" democracy, as decisions in the Swiss democratic system are not taken solely by the public, as in direct democracies, but also not solely by the elected representatives, as in representative democracies. Instead, each "group", the public, the parliament and the government, plays a specific role in the decision-making process.

Financial resources for radioactive waste management are paid into a special fund (*Entsorgungsfond*). The waste producers are under a legal obligation to dispose of the waste at their own cost. The waste management costs arising during operations (e.g. for reprocessing of spent fuel, Nagra's investigations, construction of interim storage facilities) are met on an ongoing basis.[7] The total expected costs are 15.97bn Swiss Francs (2011 estimate). The costs during the operation phase will be paid by the operator itself. Until 2011, these were around 5.0 billion Swiss Francs (SFOE 2012). Until completion of the phase-out, they will amount to 7.52bn. The fund will cover the missing 8.44bn. At the end of 2011, the total fund capital was 2.83bn (SFOE 2012).

The Swiss decision to favour waste repositories in deep geological underground facilities, and there in clay formations, is central to their waste management concept, just as is the decision to realize the project on national territory (Rodríguez 2009: 59). A pilot site is planned as part of the repository (see Kuppler and Hocke 2012: 46). The pilot site will allow for monitoring activities over the whole operation period as well as for a certain monitoring period after closure. The option to retrieve the waste should the site not develop as planned is part of the plan.

This robust infrastructure is the result of a complex process of planning nuclear waste management. This process was not without problems and setbacks. After the failure at the Wellenberg site, where siting plans in this rock formation were stopped by two cantonal votes (1995 and 2002), and the production of the required proof of safe management and disposal of HLW (*Entsorgungsnachweis*) for the Benken site in 2006, the responsible authorities prepared an alternative strategy (plan B). They did not start an underground exploration at Benken for HLW but instead presented a new strategy starting with a nationwide process of identifying suitable regions and integrating stakeholders and the interested public.

For those plans, legal frameworks were prepared. The operators of nuclear facilities and actors in decommissioning as well as other producers of nuclear waste are responsible for managing radioactive waste. The Swiss Nuclear Energy Act was modified in 2003 and came into force in 2005 together with the Nuclear Energy Ordinance.[8] The Nuclear Energy Act covers all aspects of nuclear waste management from preparation of the waste for disposal to all aspects related to the disposal facilities (siting, construction and operation) (Streffer et al. 2011:

7 Decommissioning costs and waste management costs arising after the shutdown of the plants are secured by payments made by the waste producers in the mentioned waste management fund and a second one: the "decommissioning fund" (SFOE 2008: 7).

8 For the important legislative and regulatory framework see Minhans and Kallenbach-Herbert (2012: 4-5).

396). With the passing of the Nuclear Energy Act, the cantonal veto right on facility construction was abandoned and replaced by a right to be heard in all decisions regarding facility siting, but only if this does not unnecessarily prolong the process. Further, the right for an optional national veto on the general licence was introduced. The general licence is issued by parliament at the end of the siting procedure. The Nuclear Energy Ordinance determined that the siting procedure has to follow the general outline of a sectoral plan.

As Minhans and Kallenbach-Herbert (2012) underline, Swiss sectoral plans are well established instruments in the context of spatial planning for projects of national relevance. The Sectoral Plan for Deep Geological Repositories is a modified version of this instrument – prepared on the long hand and affiliated with intensive expert and public debate in very early stages of its development (Minhans and Kallenbach-Herbert 2012: 7). Central principles and ideas developed by the AkEnd, a German expert group (1999-2002), were reflected and modified for the Swiss case (AkEnd 2002). The Conceptual Part of the Swiss Sectoral Plan emphasizes the following aspects: criteria-based, stepwise approach, safety first!, transparent decision-making and integration of local public. Some necessary differentiations and more detailed specifications were developed for the Sectoral Plan. By implementing the approach of the Sectoral Plan, three steps of site selection for two deep geological repositories were fixed and adapted to the different types of waste – one branch for low- and intermediate level waste and one for high-level waste – within the same site selection process. The possibility to combine the repositories for both types of waste at one site is not excluded, but neither is it the preferred option.

The decision between the separated or the unified solution is part of the site selection process, which will not be completed until 2023. The end of the construction phase and start of the operation of the L/ILW underground repository will be around 2030, for HLW close to 2040 (Rodríguez 2009). The three central steps defined by the Sectoral Plan are:

- Step 1: Selection of geological siting areas for L/ILW and HLW (started in April 2008).
- Step 2: Selection of at least two sites each for L/ILW and HLW (started in the end of 2011).
- Step 3: Site selection and general licence procedure for L/ILW and HLW.[9]

9 For the steps see also Streffer et al. (2011: 402).

At the end of each step, the Federal Council has to decide whether to continue with the next step. The general licence is granted by parliament and can be subjugated to a national referendum.

In the literature, the Sectoral Plan in the context of the Nuclear Energy Regulation is qualified by the characteristics of containing "basic requirements of a substantive and procedural nature for the selection process that are binding to all authorities" (Streffer et al. 2011: 401). This means that the technocratic approach, i.e. allowing experts to decide where siting should happen, was integrated in forms of new governance. Important are the statutory requirements and a number of programmatic positions, which are fixed in this document. "Long-term nuclear safety and the lasting protection of human health and the environment are regarded the primary substantive process; spatial planning and the economic concerns as well as the attitudes of society have to be considered but are given a lower weight" (Streffer et al. 2011: 401-402; see also SFOE 2008). Basic criteria for the selection of candidate regions for siting are:

- sufficient extent of suitable host rock,
- favourable hydrological circumstances,
- long-term geological stability.

More site specific criteria for relevant sites were developed and linked with a conceptual frame, in which those sites are considered to be equivalent, where – due to geological characteristics – it can be ruled out that doses of more than 10 μSv per year will be reached. Based on this, planning and economic concerns can be considered (Streffer et al. 2011: 402).

4 The actors in Swiss radioactive waste management

The central actors responsible for radioactive waste management are the National Cooperative for the Disposal of Radioactive Waste (Nagra), the Swiss Federal Office of Energy (SFOE) and the Swiss Federal Nuclear Safety Inspectorate (ENSI) (Jost 2012: 139). As in most Western countries the delay and the societal debate about nuclear waste and nuclear energy in general led to a setting of concerned collective actors, who feel responsible for economic and political decision-making in this field.[10] Collective actors from politics, the economy, and culture/science/civil society have been involved in the struggle for a nuclear waste solution.

10 For the struggle in three of this Western countries between society and politics see Rucht (1994).

The Nagra acts as an implementer for the planned repositories. It was created by the waste producers and the Swiss Confederation in 1972. In the process of the Sectoral Plan for deep geological repositories, its role is to elaborate the necessary geological information required in the search for sites and to propose the potential siting regions and later the specific sites. In the end of the site selection process, Nagra has to apply for the license of construction. During the site selection process it is responsible for suggesting areas and sites, but has no decision power. Nagra is organized as a cooperative with members of the Swiss Confederation (represented by the Department of Home Affairs) and the owners of the nuclear power plants. It supports the Sectoral Plan in a strong way. Members of the cooperative are the BKW FMB AG, Kernkraftwerk Gösgen-Däniken AG, Kernkraftwerk Leibstadt AG, Axpo AG, and Alpiq Suisse SA.

On the political side the Federal Council, Federal Department of the Environment, Transport, Energy and Communications (DETEC) and SFOE as federal authorities are the central collective actors. The Federal Council (Bundesrat), which in the Swiss model of democracy has the same role as a cabinet of federal ministers in western political systems, is the most influential political actor. Seats in the Federal Council are assigned to members of all important political parties. Each member also acts as a head of ministry. The established mode of the consensus democracy in Switzerland requires them to come to decisions that are supported by all parties represented in this council as well as by representatives of affected stakeholders, such as labour unions or associations. The DETEC is responsible for radioactive waste management.[11] The final approval of the license has to be given by the Swiss Parliament. The parliamentary decision for license approval can be subjected to a nation-wide referendum if sufficient support for such a referendum can be mobilized.

The SFOE works on energy-related questions within DETEC's field of responsibility. It has the lead in the Sectoral Plan for Deep Geological Repositories and the general licensing process and is responsible for the whole process of regional participation (initiation, control and closing). It keeps the public informed about the status of work. As leader of the Sectoral Plan, SFOE controls the procedural progress and in a practical sense it is checking the quality of decision-making.

Guaranteeing nuclear safety is the duty of the ENSI as the national regulatory body that cares for the technical security aspects, complementary to the procedural overview guaranteed by the SFOE. ENSI also reviews the license application of the deep geological repository before the Federal Council and the

11 As a Federal Councillor Doris Leuthard currently (2014) is the head of the Federal Department of the Environment, Transport, Energy and Communications. She is a member of the Christian Democratic People's Party.

Parliament make their final decision. It is responsible for evaluating Nagra's proposals in terms of safety aspects. As an institution it is independent from DETEC and controlled by a committee.

Further authorities relevant in the current siting procedure are the ARE, FOEN and FOPH. The Swiss Federal Office for Spatial Development (ARE) examines and assesses spatial planning aspects and supports the DETEC. The examination and assessment of environmental aspects in support of DETEC is managed by the Swiss Federal Office for Environment (FOEN). Other Federal Offices like the Federal Office of Public Health (FOPH) or the Paul Scherrer Institute (PSI) support the SFOE in specific questions. Specific scientific expertise is centred at three official institutions. The first one is the Expert Group on Nuclear Wste Disposal (EGT). It was established in 2011 and took over the role of the former KNE (Commission for Nuclear Waste Disposal). This expert commission is set up by the DETEC and the Federal Department of Defence, Civil Protection and Sport (DDPS). It provides expert advice to ENSI in questions related to geo-scientific, long-term safety and construction issues. It is responsible for reviewing sites and carries out its own studies on specific safety and construction issues. To work on fundamental issues related to nuclear safety is the duty of the Swiss Federal Nuclear Safety Commission (NSC); it is an advisory board of the Federal Council, the DETEC and the ENSI. They are accompanied by an independent national advisory board, which is called *Beirat Entsorgung* (Waste Management Advisory Board) and is appointed by the DETEC. Its responsibility is to identify conflicts and risks at an early stage and propose solutions. It has the aim to promote dialogue among all actors.

As in most western societies intermediary organizations placed in the sector of civil society and culture play an important role in Switzerland. In a functional sense they are all interfaces for communication about nuclear waste policy. But they differ in their linkages to governmental organizations, authorities and industry. Their central role in nuclear waste management comes from the social processes within "interest aggregation". For national politics they "organize" the contact between affected siting regions, interest groups and stakeholders as well as between different levels of governmental organizations and authorities. Under certain conditions they can be very influential, as they are used as "interfaces" in communication and decision-making between central actors in multi-level governance.

For the Swiss case the Cantonal Commission (*Ausschuss der Kantone*) is a collective actor very close to official politics. According to the Sectoral Plan and given their traditionally strong influence in the Swiss political system, the cantons play a central role in the siting process – though their veto power was abolished by the revision of the Nuclear Law in 2003/04. The cantons accompany the whole process both professionally and politically. They coordinate the collaboration with the affected communities and give statements to the Federal Government. Several committees working at the political level as well as specific working groups (such as "Information and Communication" and "Spatial Planning") have been set up to ensure qualified and transparent decision-making on the cantonal level. Those organizations, including the decisions they take, are very closely observed by the Swiss national collective actors responsible for the processing of the Sectoral Plan.

Furthermore, a "Technical Forum Safety" (TFS) has been established with special duties in communication. The TFS discusses and answers technical and scientific questions of the interested public, communes, siting regions, organizations, cantons, and authorities in neighbouring states. The forum consists of experts from the SFOE, from other bodies with supervisory or supportive roles (e.g. ENSI, Swiss Federal Office of Topography [swisstopo]), from commissions (like KNE), from Nagra, from the cantons, as well as one representative from each of the siting regions.

NGOs and citizenship initiatives also intervene in the debate on nuclear waste and participate to a certain amount with official actors. The *Schweizerische Energie-Stiftung* and *KLAR!-Schweiz/Deutschland* and a number of local protest groups and national environmental interest groups (like Greenpeace) play an active role in this field.[12]

5 Current status and forms of participation

After the restart of the siting process in 2008, the new tools of governance organized within and around the Sectoral Plan became dominant in Swiss radioactive waste management. In total their impact can be interpreted as an explicit shift away from the decide-announce-defend approach. Switzerland was successful in establishing a modernized form of an integrative search for deep geological repositories. Further, a new level of transparency with regard to arguments used in decision-making is established by demanding a high number of documents and research reports to be published for general access. These documents explain the ongoing developments, clarify abstract rules from the conceptual part of the Sectoral Plan and transform general rules into a number of guidelines.

12 The results of social movements research on anti-nuclear opposition and current anti-nuclear waste initiatives are not well developed. Only more general research on this topic is available from the 1990s (see e.g. Wisler and Kriesi 1997).

Information and, within certain boundaries, participation, play an important role in nuclear waste management in Switzerland. This already was introduced with the decision to restart the siting process for final disposal. The whole process of conceptualization of plans, steering ideas and documentation of decision-making was accompanied by professional communication and argumentation by authorities like ENSI and SFOE and also Nagra. The draft of the Sectoral Plan for Deep Geological Repositories was discussed with stakeholders and the interested public as of 2005/06 using different formats such as focus groups, information events and a hearing. The final version was modified based on the comments received and put into force by the Federal Council in 2007 (SFOE 2008).

Since 2008, some dozens of conferences, workshops and meetings with the (inter)national public and concerned citizens were organized each year and documented in an easily accessible form.[13] These activities, which were carried through during step one and two of its implementation, were prepared involving stakeholders, industry and politics in a wide range of open or closed events, meetings and workshops. Only after these consultations, the Federal Council decided about the conclusion of step 1. The media and the interested public were at all times provided with detailed information in a remarkably open way.

Since 2010/11 six Regional Conferences were established in the selected potential siting areas according to the rules of the Sectoral Plan.[14] They are committees built of different regional stakeholders (especially representatives of communities, but also NGOs and selected citizens). Each regional conference has about 100 members. They have the mandate to represent regional interests. They comment on Nagra's reports and develop a strategy for regional development. Within the regional conferences three smaller working groups prepare comments and decisions on specific topics. One of those working groups, the group "Safety" was not part of the original setup but established upon explicit request from local stakeholders in the Benken area. The regional conferences have relatively large freedom in how they organize themselves. They receive financial support from the Nagra and can invite experts they trust to their meetings. True participation in the sense of "having a say in something" is restricted to decisions on above-ground facilities. In all other areas, they have the right to be heard, i.e. to ask questions and receive answers to them. The Technical Forum Safety mentioned above is an important instrument for the

13 See e.g. SFOE (2009), AGNEB (2013), SFOE (2011a) and SFOE (2011).

14 For more details on the five regional conferences and the "Plattform Wellenberg" with a special self-conception which is more critical than the other five conferences (see Jost 2012: 146-149; PLANVAL 2014; and Canton Nidwalden 2013: 9).

discussion of technical and scientific topics between members of the public, the responsible authorities and the Nagra.

The impact of all these activities was that suitable geological features (regions) were identified, within the reviewing of these results no serious contradictions were uncovered, which would hint at the selection process not being criteria-based. Further, no arguments were brought up against the proposed fact sheets, which included the description of affected municipalities. All aspects of step 1 were widely accepted.

Also German communities, which are very close to four of the six candidate regions (see figure 1), were integrated in the preparation of the regional conferences. After some debate German communities within a certain distance from the potential siting areas were allowed to send representatives as members of the regional conferences. They make up around 13% to 17% of all members.

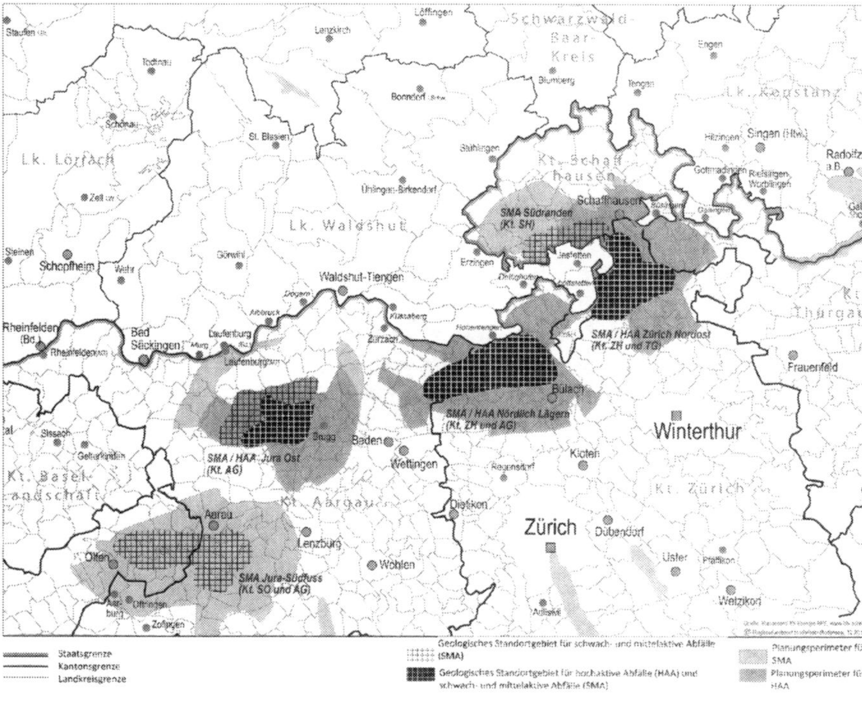

Figure 1: Potential siting areas Switzerland (five of six)*

* In this map the Wellenberg site is not included. It is a site where HLW should not be stored.
Source: Deutsche Koordinationsstelle Schweizer Tiefenlager (Waldshut/ Dez. 13).

During the implementation of step 2 of the Sectoral Plan, in which at least two sites for HLW and two for L/ILW should be identified for an in-depth analysis, the first serious tensions emerged. This put the stakeholders' and interested public's general openness to a test. Further, some regional conferences did not acknowledge the set time schedule and called for an interim stop until their points of doubt were clarified.

6 Tensions are becoming stronger

Given stakeholders' differing interests tensions between Swiss collective actors are complicated. Also if the procedure of the Sectoral Plan is accepted as rational, it is not easy for many collective actors to play a constructive role as the debate about nuclear waste and the civil use of nuclear power has not yet been concluded. Political culture and the level of trust and distrust in responsible governmental organizations play an important role in interaction and cooperation. Their relevance can be seen in the following central characteristics: (1) Civil society in Switzerland on the one hand knows that referenda within the Swiss "half-direct democracy" (Linder 2005) offer different chances to intervene. But this is restricted to late steps within the approach for selection and siting. (2) Throughout the whole process elaborated modes of transparency are installed, but only limited rights for substantial involvement are given to the affected local public and the national interested public. In general, those rights have to be classified as "rights to be heard" and not as egalitarian structures of decision-making. (3) Openness and willingness for broad discussion and documentation of positions and necessities of optimizing the procedure is a central part of the structured stepwise approach of the Sectoral Plan during all phases (including preparation of the plan). In this way "voice" is given to the public and the stakeholders (see Hirschman 1990).

As the number of actors in total is not very high, the debates and conflicts could be interpreted as being a problem of "friends and enemies" or better put, supporters and opponents of the Sectoral Plan. Such an oversimplified classification of individuals in a debate can polarize a debate in such a way that deliberation is no longer possible. The argumentative approach of the Sectoral Plan would not work in such a case. When reflecting on the linguistic turn in social sciences and research on knowledge politics, a more reflexive description of the current challenges can be offered. Late modern societies, like the Swiss one, are familiar with more or less open processes in the beginning of planning for large-scale technical projects (like central railway stations, international airports or nuclear facilities). Every responsible actor, independent of whether the actor is an implementer or political regulator, is faced with a growing list of public expectations on how interested and affected people, organizations and stakeholders from industry, politics and civil society should be "respected" in decision-making on a nuclear waste repository. The trust in the officials of western nuclear institutions is not very high, as events like the Fukushima nuclear disaster (2011) and the Chernobyl accident (1986) are discussed in a way that suggests that they were not prepared and did not expect accidents like those were possible.[15] The processes in which society classifies an institution as reliable and extends its trust to it are complex. The way in which the responsible institutions reacted to serious problems in repositories like the Asse Mine in Germany, or similar scandals that have occurred over the last years, indicate that their understanding of those processes is not sufficiently differentiated. In consequence society trusts them even less. Serious accidents become part of public "knowledge" and a reminder "not to trust" (Wynne 1996). For those reasons, finding a surface installation for a nuclear underground repository is more than a technical challenge.

There are multidimensional expectations among inhabitants of potentially affected regions about safety and the correct order of decisions to be taken in a stepwise approach. The selection of sites for the surface installations of a repository is one central part of local participation in the six Swiss candidate regions and was supposed to take place before the exact location of a potential underground repository was decided upon. As could be expected, stakeholders argued that it was not acceptable to decide about the best place for a surface installation without knowing what kind of underground installation would be built. In the 2012/13 dicussions, it was a communicative challenge to convince the members of the regional conferences that this question would be answered later, as this decision was only to be taken in the end of step 2 of the Sectoral Plan. The "knowledge problem" became a line of conflict within the Swiss debate. In step 2, one prominent and highly qualified expert addressed this "societal" problem by going in conflict with his colleagues in official positions. In the end, he stepped down from his official role in the Sectoral Plan. He is, however, still actively participating in the debate and he has been invited for several public events. These events can be interpreted as a form of rational societal discourse, but also as part of an antinuclear counter-mobilization against the Sectoral Plan.

The institutional setting responsible for "rationality" within the process of Swiss site selection is a second central challenge. As the start of the operation phase of a future underground repository is still more than a decade away (and

15 See Kallenbach-Herbert et al. (2011: 48-49) and Bauchmüller (2010).

for highly active waste even more than two decades) a "master plan" is needed, which guarantees the allocation of the time and resources needed for cooperation between stakeholders and the integration of the interested public in decision-making. Without this, institutional and social robustness of decisions cannot be expected. The financing of the siting process and the implementation of the facilities with operation and closure is one side of this challenge. The other less discussed side is the institutionalization of safety and control and the unclear public attention over time. Responsible (governmental) organizations have to take into consideration situations in which public attention declines. In such situations, control over the technical process of filling the repository still has to continue without any loss in professionalism. The quality of the process of burying the canisters must be monitored, a process of checks and balances must exist, and safety standards need to be guaranteed. At the moment, planning for optimizing the procedure is still in the beginning phase and the concept for long-term activities is very open (Kuppler and Hocke 2012). The necessity for long-term checks and balances leads to the second consideration: the social necessity of institutionalization. Which types of institutions can organize "long-term stewardship" and which basic organizational elements are needed to realize sufficient safety levels? Stewardship is generally recognized as the assignment of responsibility to shepherd and safeguard the valuables of others. This term was first used in U.S. activities around military nuclear installations and military waste in the beginning of the new century (Probst and McGovern 1998; La Porte 2004; Tonn 2001). Some of the problems and approaches discussed in the debate on stewardship with regard to military waste are also applicable to civil waste. Stewardship is more than guardianship. The discourse about stewardship has shown that the waste problem is characterized by "extraordinary complexity" and that the idea that stewardship is an answer to long-term responsibilty is "ambiguous". At the same time the idea of stewardship has manifested itself in a way that it is "mobilizing technology, money, labour, and consent" (Taylor 2007: 219, 203). The stakeholders' discourse according to Taylor et al. (2007) became a site of struggle between groups arguing over who has what responsibility, for how long, concerning the management of persistent hazards (Taylor 2007: 219).

7 Conclusions

Switzerland is right in the middle of the implementation of a very innovative approach for finding one or several repository sites. The Swiss have taken an established instrument, the Sectoral Plan, and "upgraded" it by including provisions for transparency and public participation. In this process they have picked up debates that have been had within the social sciences and the international nuclear community on public participation and deliberation. At the same time, the Sectoral Plan and its legal context have limited the public's right to actually have a say in the decisions to be taken by abandoning the cantonal veto and granting the right to be heard regarding most issues at stake.

Even though some problems have occurred during implementation and more will certainly come up in the future, the process can so far be called a success with regard to the inclusion of a wide range of actors in consultations. Still, some challenges lie ahead if the process is to be concluded successfully. This holds particularly true for the questions of long-term stewardship. If such debates on the long-term institutionalization of radioactive waste management concepts are to be carried out, there is a chance that nations will find agreement on who is responsible, for how long they are willing to take responsibility and which actors and institutions will care for the final closing of a repository after five or eight decades. Still, the actors have to find an innovative model of independent checks, balances and forms of democracy which protects against unintended side-effects and guarantees professional precaution and intervention if it is necessary. Whether political aims of transparency and western democracy are prepared for such debates is an open question – not only in Switzerland.

Note: For the support in finishing this paper we would like to thank Svenja Nollau (ITAS at KIT) and the editors for their instructive comments. One of the authors of this chapter, Peter Hocke, is a member of the the pluralistically structured German expert group "ESchT", which is structured pluralistic and is working for the Federal Ministry, the state Baden-Württemberg and the German South-West as scientific support (see www.escht.de). Some of his observations and knowledge of protocols, internal information and observation have influence the structuring of this chapter.

References

Arbeitsgruppe des Bundes für die nukleare Entsorgung (AGNEB) (2013). Jahresbericht 2012. Bern. http://www.bfe.admin.ch/radioaktiveabfaelle/01275/02612/index.html?lang=de&dossier_id=00752, last accessed 17 March 2014.

AkEnd (2002). Site Selection Procedure for Repository Sites. Recommendations of the AkEnd (Commitee on a Site Selection Procedure for Repository Sites), Ms., 248 pages.

Bauchmüller, M. (2010, 20 March). Verstrahlt, vergraben, vergeigt – das Desaster des Atomlagers Asse. *Süddeutsche Zeitung*, 8-9.

Benz, A., and Dose, N. (2010). *Governance – Regieren in komplexen Regelsystemen.* Springer, Wiesbaden.

Börzel, T. A. (2008). Der „Schatten der Hierarchie" – Ein Governance-Paradox? In Zürn, M., Schuppert, G. F. (Eds.) (2008). *Governance in einer sich wandelnden Welt.* VS-Verlag, Wiesbaden, 118-131.

Brunnengräber, A., Mez, L., Di Nucci, M. R., and Schreurs, M. (2012). Nukleare Entsorgung: Ein "wicked" und höchst konfliktbehaftetes Gesellschaftsproblem. *Technikfolgenabschätzung Theorie und Praxis*, 21(3), 59-65.

Brunner, C. U. (2009). Switzerland – What's left from the Glaciers in the Alps? In Mez, L., Schneider, M., and Thomas, S. (Eds.) (2009). *International Perspectives on Energy Policy and the Role of Nuclear Power.* Multi Science Publishing, Brentwood Essex, 387 – 403.

Chhotray, V., and Stoker, G. (2009). *Governance Theory and Practice*, Palgrave Macmillan, Houndmills (NY).

Federal Department for the Environment, Transport, Energy and Communications (DETEC) (2011). Implementation of the Obligations of the Joint Convention on the Safety Spent Fuel Management on the Safety of Radioactive Waste Management – Fourth National Report of Switzerland, Brugg.

Eberlein, B., and Kerwer, D. (2004). New Governance in the European Union: A Theoretical Perspective. *Journal of Common Market Studies*, 42(1), 121-142.

Flüeler, T. (2006). *Decision Making for Complex Socio-Technical Systems. Robustness from Lessons Learned in Long-term Radioactive Waste Governance*, Springer, Dordrecht.

Gramberger, M. (2001). Citizens as Partners. OECD Handbook on Information, Consultation and Public Participation in Policy-Making. Paris Cedex.

Grande, E. (2012). Governance-Forschung in der Governance-Falle? Eine kritische Bestandsaufnahme. *Politische Vierteljahresschrift*, 53(4), 565-592.

Haus, M. (2010): *Transformation des Regierens und Herausforderungen der Institutionenpolitik*, Nomos, Baden-Baden.

Hirschman, A. O. (1990). *Exit, Voice and Loyality: Responses to Decline in Firms, Organizations and States*, Harvard University Press, Boston.

Hoppe, R. (2011). *The Governance of Problems: Puzzling, Powering and Participation*, Policy Press, Portland.

Jost, M. (2012). Entsorgung radioaktiver Abfälle: Akteure und Aufgabenverteilung in der Schweiz. In Müller, M. (Ed.) (2012). Endlagersuche: Auf ein Neues? Der Weg zu einem gerechten und durchführbaren Verfahren. *Loccumer Protokolle 25/12*, Rehburg-Loccum, 139-153.

Kallenbach-Herbert, B. (2011). Germany (with contributions by Minhans, A., and Pistner, C.). In Feiveson, H., Mian, Z., Ramana, M.V., and von Hippel, F. (Eds.) (2011). *Managing Spent Fuel from Nuclear Power Reactors. Experience and Lessons from Around the World*, IMPF, Princeton, 43-51.

Kanton Nidwalden (2013). Brennpunkt Wellenberg. Der Kanton Nidwalden im Sachplanverfahren geologisches Tiefenlager, Stans, Schweiz.
http://www.nw.ch/dl.php/de/525cd88a09231/ Brennpunkt_Wellenberg.pdf /, last accessed 17 March 2014.

Kriesi, H. (1982). *AKW-Gegner in der Schweiz. Eine Fallstudie zum Aufbau des Widerstands gegen das geplante AKW in Graben*, Ruegger, Diessenhoffen.

Krütli, P., Flüeler, T., Stauffacher, M., Wiek, A., and Scholz, R. W. (2010a). Technical Safety vs. Public Involvement? A Case Study on the Unrealized Project for the Disposal of Nuclear Waste at Wellenberg. *Journal of Integrative Environmental Sciences*, Vol. 7, 229-244.

Krütli, P., Stauffacher, M., Flüeler, T., and Scholz, R. W. (2010b). Functional-Dynamic Public Participation in Technological Decision-Making: Site Selection Processes of Nuclear Waste Repositories. *Journal of Risk Research*, Vol. 13, 861-875.

Kuppler, S. (2012). From Government to Governance? (Non-) Effects of Deliberation on Decision-Making Structures for Nuclear Waste Management in Germany and Switzerland. *Journal of Integrative Environmental Sciences*, Vol. 9, Issue 2, 103-122.

Kuppler, S., and Hocke, P. (2012). Monitoring in einem Pilotlager. Kontrollierte Deponierung von Nuklearabfällen im Konzept eines Schweizer Tiefenlagers. *Technikfolgenabschätzung –Theorie und Praxis*, 21(3), 43-51.

La Porte, T. R. (2004). Elements for Long Term Institutional Stewardship in a Hazardous Age. Views from a "Stewardee" Session on Institutional Challenges for Long-Term Stewardship of Contaminated Sites, (Seattle, WA) Association for the Advancement of Science.

Linder, W. (2005): *Schweizerische Demokratie*, second ed. Haupt Verlag, Bern/ Stuttgart/Wien.

Marschall, S. (2008). Kommunikation und Entscheidungsfindung im Parlament. In Sarcinelli U., Tenscher, J. (Eds.) (2008). *Politikherstellung und Politikdarstellung*, Herbert von Halem Verlag, Köln, 44-62.

Mayntz, R. (2009). *Über Governance. Institutionen und Prozesse politischer Regelung*, Campus, Frankfurt am Main.

Minhans, A., and Kallenbach-Herbert, B. (2012). Identifying Remaining Socio-Technical Challenges at the National Level: Switzerland. InSOTEC Working Paper WP 1 - MS7.

Nuclear Energy Agency (2004). Stepwise Approach to Decision Making for Long-Term Radioactive Waste Management. Experience, Issues and Guiding Principles. OECD NEA / Radioactive Waste Management: NEA 4429, Paris.

Offe, C. (2008). Governance – „Empty Signifier" oder sozialwissenschaftliches Forschungs-programm? In Zürn, M., Schuppert, G. F. (Eds.) (2008). *Governance in einer sich wandelnden Welt*. VS-Verlag, Wiesbaden, 61-76.

PLANVAL (2014). Aufbau der regionalen Partizipation im Sachplanverfahren zur Standortsuche von geologischen Tiefenlagern. Umsetzung und Erfahrungen. Bern.

Probst, K. N., and McGovern, M. H. (1998). *Long-Term Stewardship and the Nuclear Weapons Complex: The Challenge Ahead (Resources for the Future)*, Johns Hopkins University Press, Washington.

Rodríguez, J. (2009). Tiefenlagersuche in der Schweiz. In Müller, M. (Ed.) (2009). Endlager und kein Ende? Wege aus einer unendlichen Geschichte. *Loccumer Protokolle 29/9*, Rehburg-Loccum, 59-66.

Rucht, D. (1994). *Modernisierung und neue soziale Bewegungen*, Campus, Frankfurt am Main.

Scharpf, F. W. (2001). European Governance: Common Concerns vs. the Challenge of Diversity. MPIfG Working Paper, No. 01/6, Köln.

Scholz, R. W., Stauffacher, M., Bösch, S., Krütli, P., and Wiek, A. (Eds.) (2007). Entscheidungsprozesse Wellenberg – Lagerung radioaktiver Abfälle in der Schweiz. ETH-UNS Fallstudie 2006, Zürich.

Solomon, B. D., Andrén, M., and Strandberg, U. (2010). Three Decades of Social Science Research on High-Level Nuclear Waste. *Risk, Hazards & Crisis in Public Policy*, Vol. 1, 13-47.

Strandberg, U., and Andrén, M. (2011). Editorial. In Strandberg, U., and Andrén, M. (Eds.) *Nuclear Waste Management in a Globalized World*, Routledge, Abingdon.

Streffer, C., Gethmann, C. F., Kamp, G., Kröger, W., Rehbinder, E., Renn, O., and Röhlig, K.-J. (Eds.) (2011). *Radioactive Waste. Technical and Normative Aspects of its Disposal*, Springer, Berlin/ Heidelberg.

Swiss Federal Office of Energy (SFOE) (2008). Sectoral Plan for Deep Geological Repositories. Conceptual Part, Bern. http://www.bfe.admin.ch/radioaktiveabfaelle/01375/04389/index.html?lang=en., last accessed 17 March 2014.

Swiss Federal Office of Energy (SFOE) (2009). Leitfaden Aufbau regionale Partizipation. Bern.

SFOE (2011). Konzept regionale Partizipation: Grundlagen und Umsetzung in Etappe 1 und 2. Bern.

SFOE (2011a). Sachplan geologische Tiefenlager. Bericht über die Ergebnisse der Anhörung zu Etappe 1. Bern.

SFOE (2012). Höhere Beiträge für Stilllegung der Kernkraftwerke und Entsorgung der radioaktiven Abfälle bestätigt. Bern. http://www.bfe.admin.ch/energie/00588/00589/00644/index.html?lang=de&msg-id=46787, last accessed 17 March 2014.

Schweizerische Vereinigung für Atomenergie (SVA): Basisdokumentation – Kernenergie in der Schweiz. http://www.nuklearforum.ch/sites/default/files/media/Chronik_dt_fr_bis_1999.pdf, last accessed 17 March 2014.

Taylor, B. C. (2007). (Forever) At Work in the Fields of the Bomb. Images of Long-Term Stewardship in Post-Cold War Nuclear Discourse. In Taylor, B. C., Kinsella, W., Depoe, S., and Metzler, M.S. (Eds.), *Nuclear Legacies*. Lexington Books, Lanham, 199-234.

Taylor, B. C., Kinsella, W., Depoe, S., and Metzler, M.S. (Eds.) (2007). *Nuclear Legacies*. Lexington Books, Lanham.

Tonn, B. (2001). Institutional Design for Long-Term Stewardship of Nuclear and Hazardous Waste Sites. *Technological Forecasting and Social Change*, Vol. 68, 255–273.

Wisler, D., and Kriesi, H. (1997). The Policing of Mass Demonstration in Contemporary Democracies. RSC No 97/5 EUI Working Papers, Florence.

World Bank (1992). Governance and Development. Washington, D.C.

Wynne, B. (1996). May the Sheep Safely Graze? A Reflexive View of the Expert-Lay Knowledge Divide. In Lash, S., Szerszynski, B., and Wynne, B. (Eds.), *Risk, Environment and Modernity: Towards a New Ecology*. Sage, London, 44-83.

Always the Same Old Story?
Nuclear Waste Governance in Germany

Peter Hocke and Beate Kallenbach-Herbert[1]

1 Introduction

German nuclear waste management policy envisages geological disposal for all kinds of nuclear waste. The Konrad repository near Salzgitter in northern Germany which is licensed and designated to deal with "non-heat-generating" waste (roughly comparable to the categories of low- and intermediate-level waste according to the IAEA classification) is currently under construction.[2] Earlier the Asse research mine (from 1967 to 1978) and the Morsleben repository "ERAM" (from 1971 to 1991 and from 1994 to 1998) situated in the former German Democratic Republic were also used to dispose of low- and intermediate level waste. However, a disposal solution for "heat-generating" waste, which is mainly spent fuel from nuclear power plants and vitrified high-level waste from reprocessing, is still pending.

Siting activities for a geological repository for heat-generating waste in Germany have been going on for more than 30 years and have been highly controversial in society and politics. Since 1978 the Gorleben salt dome in the state of Lower Saxony has been the only candidate site and there have been underground explorations since 1986. Since neither an expert nor a political agreement could be reached on the geological suitability of the Gorleben salt dome, several attempts have been made to reorganize the site selection process. A political agreement on a new siting procedure was first reached in 2013. It comprises a stepwise, criteria based, comparative selection process which is

---

1 For the support in finishing this paper we would like to thank Svenja Nollau and Elske Bechthold (ITAS at KIT) and the comments of the editors. The work presented here is closely connected with the Project ENTRIA funded by the German Federal Ministry of Education and Research and an ITAS work package with a special cooperation with the editors of this volume. ITAS is the Institute of Technology Assessment and Systems Analysis at the Karlsruhe Institute of Technology (KIT). ENTRIA is the acronym for the German language equivalent of: "Disposal options for radioactive residues: Interdisciplinary analyses and development of evaluation principles" (www.entria.de). The number for funding is 02S9082D.

2 For the criteria used to define and categorize radioactive waste in Germany in comparison to the classification of the International Atomic Energy Agency (IAEA) see BMU (2012: 50-53, esp. 53).

described in the general terms of the "Repository Site Selection Act" passed in July 2013 by the two chambers of the German Parliament. Due to the federal elections, the implementation of the Act was postponed. As a first step a Commission for the Disposal of High-active Waste was appointed in April 2014 and, as laid out in the Act, it discusses basic questions pertaining to nuclear waste management and evaluates the Site Selection Act.

This Site Selection Act marks – in our perception – a new "milestone" in the long lasting process of siting a repository for heat-generating waste and represents an important step in finding a political compromise on the framework and on the siting procedure. This article describes the ongoing process over the years – especially developments during the last 15 years. A major focus is put on the intertwined political and societal expectations and on the way they influenced the governance concept for heat-generating nuclear waste over time. Information on relevant sites and waste streams in reference to the category of heat-generating wastes, the historical background, the German concept for the management of heat-generating waste and the legal and institutional framework are given as background information at the beginning of the article.

2 Overview of the waste inventory and waste management sites

The largest amount of heat-generating waste results from nuclear power reactors either as spent fuel foreseen for direct disposal or as vitrified high-level waste from the reprocessing of spent fuel. Further small amounts stem from spent fuel produced by research and prototype reactors (see Table 1). Following the Fukushima accidents in March 2011, the German government shut down eight nuclear power plants immediately and introduced a phase-out plan for the remaining nine plants by shutting them down in stages by 2022 in keeping with an amendment of the atomic law. The nine NPPs still under operation produce about 600 billion kWh per year (2011: 608.9bn kWh; 2012: 617.6bn kWh) (Deutsches Atomforum 2013).

From the beginning of 2012 until the end of their operational time, the remaining nine NPPs will have produced about 2,760 tonnes of heavy metal (t_{HM}) spent fuel or 16% of the total amount of spent fuel from German nuclear reactors (see Table 1).

Table 1: Amount of spent nuclear fuel in Germany per NPP and waste management path (in t_{HM}, estimated figures including 2011)

NPP	Spent Fuel Elements Production in t_{HM}		Repro-cessing	Direct Dis-posal (without reprocessing)
	until 31 Dec. 2011	during total operating time		
NPPs in Operation				
Brokdorf (KBR)	634	974	198	776
Grohnde (KWG)	715	1.049	288	761
Emsland (KKE)	646	997	113	884
Philippsburg (KKP 2)	700	951	208	743
Neckarwestheim (GKN 2)	567	888	0	888
Gundremmingen (KBR-B)	733	1.002	195	807
Gundremmingen (KBR-C)	704	1.070	186	884
Isar (KKI 2)	599	944	179	765
Grafenrheinfeld (KKG)	773	952	391	561
NPPs Shut Down				
Brunsbüttel (KKB)	464	464	296	168
Krümmel (KKK)	694	694	324	370
Unterweser (KKU)	922	922	536	386
Biblis (KWB-A)	897	897	427	470
Biblis (KWB-B)	976	976	420	556
Philippsburg (KKP 1)	646	646	391	255
Neckarwestheim (GKN 1)	655	655	446	209
Isar (KKI 1)	723	723	339	384
NPPs under Decommissioning				
Stade (KKS)	539	539	539	0
Obrigheim (KWO)	352	352	243	100[1)]
Mülheim-Kärlich (KMK)	96	96	96	0
Würgassen (KWW)	346	346	346	0
Gundremmingen (KRB-A)	125	125	117	0[2)]
Lingen (KWL)	66	66	66	0
Greifswald+Rheinsberg	893	893	283	583[3)]
Jülich (AVR), Hamm (THTR)	8.7	8.7	0	8.7
Test/Prototype Reactors	171	171	163.8	0.54[4)]
Research Reactors	?	14.8 (by 2025)	4.5	8.1
Total	**14,645**[5)]	**17,416**	**6,795**	**10,567**

Source: Neumann (2013: 24). Table 3 added and translated.

[1)] 9 t_{HM} shipped to Sweden for permanent disposition. [2)] 8 t_{HM} shipped to Sweden for permanent disposition. [3)] 27 t_{HM} shipped to Hungary for further use. [4)] 6.9 t_{HM} shipped to Sweden for permanent disposition. [5)] This is the calculated sum of the figures in the column above. Neumann (2013: 24) counts 15,538 t_{HM}.

The German nuclear power plants (NPPs) are concentrated in two larger "clouds", one in the south of Germany, the other in the north west (see Figure 1). The "Big Four" of nuclear power production (E.On, RWE, EnBW and Vattenfall) are the owners of the NPPs and are responsible for the operation of their remaining plants as well as for their stepwise closure according to the phase-out regulations and for the decommissioning of all these and the former operated reactors (Stahl and Strub 2012).

Furthermore, they are responsible for the management of nuclear wastes generated during operation and decommissioning until it is taken over for disposal by the responsible public authority. This final step is carried out by the federal government (BMU 2012: 109-13).

Figure 1 also shows an on-site interim storage facility at almost every reactor site in Germany. The construction and operation of these facilities go back to an agreement in 1999 between the federal coalition government (which constituted the Social Democratic Party (SPD) and the Green Party) and the "Big Four" to abandon reprocessing spent fuel abroad from July 2005 on. Since then, on-site interim storage has been the only option of predisposal management of spent fuel from commercial nuclear power plants in the German atomic law. Further capacities for storage of spent fuel exist in the off-site interim storage facilities at Gorleben, Greifswald, Jülich and Ahaus.

The four sites mentioned as repository or research mines are all located very closely together; three are in the state of Lower Saxony and the fourth one is only about a dozen kilometers away in the neighboring state Saxony Anhalt.[3] They are all at different stages. The Morsleben site in Saxony Anhalt and the Asse mine are both in the process of being closed. The future of the Asse mine is being discussed after severe problems with brine ingress, which has caused growing public concern since the second half of the last decade. The Konrad repository for non-heat-generating nuclear waste (near the German town Salzgitter) is licensed for construction and operation and a test operation should start there around 2022/23 mainly because several delays during construction have postponed the expected completion several times. The fourth site is the most important in the context of heat-generating waste: the Gorleben salt dome which is labeled as a research or exploration mine (BMWi 2008). This site is the main focus of the societal and political nuclear waste management conflict in Germany (Kallenbach-Herbert 2011, Roose 2010) in which all forms of contradictions and protest have taken place on an impressive level for decades.

3 The term "research mine" is used here for the Asse and Gorleben sites. Asse 2 was a former salt mine which was reused for research and disposal of 126,000 barrels of low- and intermediate-level waste. The Gorleben mine was installed for exploration of the salt dome with regard to its suitability for radioactive waste disposal.

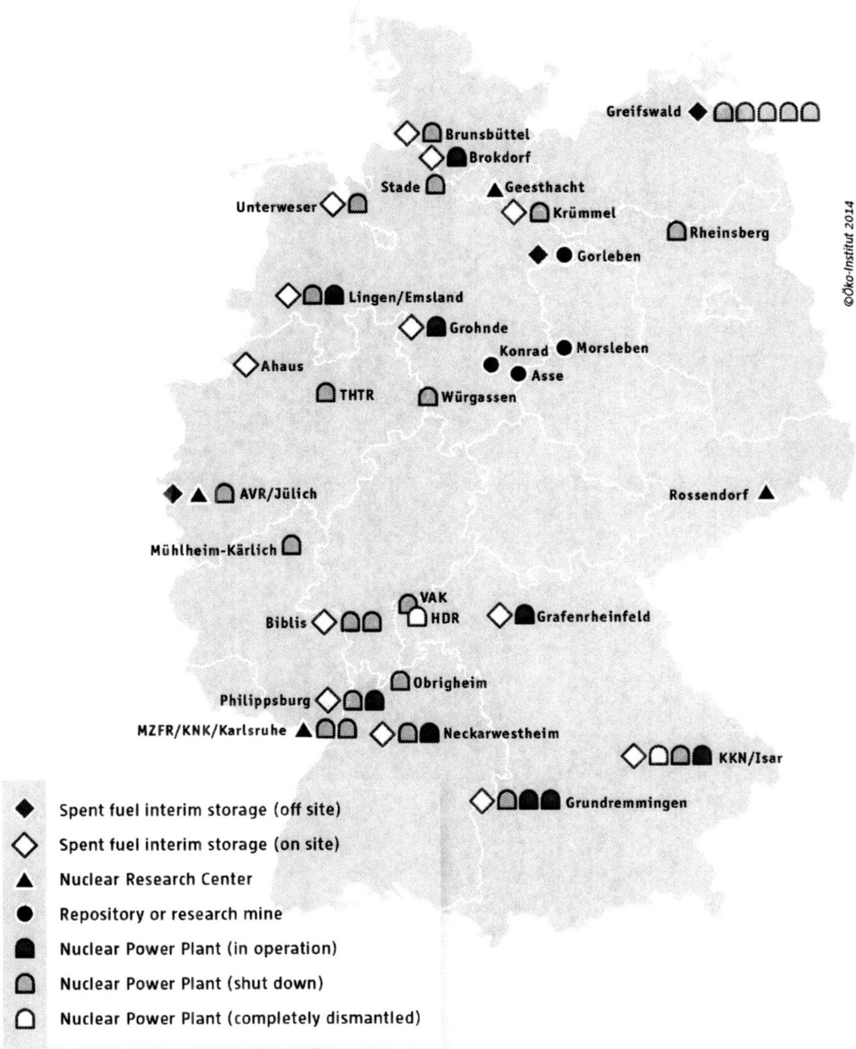

Figure 1: Locations of Germany's nuclear power plants, spent fuel interim storage facilities, nuclear research centers and sites of disposal projects.
Source: Oeko-Institut (2014)/ authors' compilation.

3 Historical context

The German research and development in the field of civil use of nuclear energy began in 1955, after the federal government officially renounced the development and possession of nuclear weapons. In the same year the German government decided to become a member of EURATOM and of the Nuclear Energy Agency (NEA) of the Organisation for Economic Cooperation and Development (OECD) (BMU 2012: 35). With the aid of U.S. manufacturers, and highly politically and financially supported by the German government, the power plant companies began to develop commercial NPPs.[4] In 1960 the Federal Republic of Germany (FRG)'s first NPP, the prototype reactor VAK at Kahl in Bavaria began operating. All further reactors went on line between the late 1960s and the early 1980s.[5]

The history of nuclear waste management in the FRG is closely linked to the history of the civil use of nuclear power. The public debate, the extra-parliamentary opposition and the anti-nuclear power movement with its clear positions found gradually more and more support in society and from a number of political actors (see Rucht 2007: 254-57). These groups opposed national and state decisions and planning. As a result, a long lasting and thorny social conflict came into life (Hocke and Renn 2011: 43-50).

Starting in the early 1960s, the Federal Institute for Geosciences and Natural Resources (BGR) – influenced by similar movements in the United States – undertook research on the suitability of rock salt for nuclear waste. The idea behind it was the complete confinement of waste (canisters) in dry conditions. In 1965, the Asse mine, where salt exploitation had taken place until 1964, was bought by the federal government and converted into a research mine for radioactive waste disposal. Although comprehensive conclusions on necessary and favourable conditions for geological disposal could not be reached during the following years (Comel 1982), the salt concept was a basic element of the siting process for a nuclear waste management center (Nukleares Entsorgungszentrum, NEZ) which started in 1973. It became the central hotspot of the federal government in the context of nuclear waste management.

The idea at that time was to establish a "closed fuel cycle" in which (fissile) plutonium is produced by breeder reactors from (non-fissile) uranium-238, separated in reprocessing plants and recycled as fuel. It was planned to concentrate all necessary facilities at a single site, including an industrial

4 Esp. Siemens/Westinghouse and AEG/General Electric were involved.

5 Minhans and Kallenbach-Herbert (2012: 2); see the list of German NPPs in Table 1. – No detailed research about the nuclear power and waste governance in the German Democratic Republic (GDR) is available.

reprocessing plant, conditioning plants for waste treatment, storage facilities as well as an underground repository for all types of nuclear waste from Germany. Research on the disposal of highly radioactive waste was focused on rock salt as a host formation. The first experience in Germany was gathered in the Asse research mine, which was established as a test site in 1965.[6] Thus the siting process for the NEZ concentrated on locations above massive salt formations in the northern German lowlands and its border regions. The state of Lower Saxony became the main target area for siting the nuclear waste management center.

The site selection process started in 1972. The final selection of the NEZ site was to be based on a comparative review of three sites which resulted from a broader, desk study review (Appel 2006: 57-8). The start of the first investigations, however, caused strong local opposition; the inhabitants of these regions and particularly the landowners were outraged, and the investigations were stopped in 1976. Some weeks later the government of Lower Saxony asked the federal government to stall the ongoing site investigation until they had designated a site of their own (Tiggemann 2004: 387). Differences between the federal and the state government aggravated the conflict (Rucht 1980: 72-3; Tiggemann 2006: 95), and in February 1977, the Prime Minister of Lower Saxony, Ernst Albrecht, designated the Gorleben salt dome as the most suitable potential NEZ site.[7] The federal government finally accepted this solution in spite of their earlier concerns regarding the proximity of the site to the German-German border, and the GDR accepted this. However the strategy of concentrating protests in one region did not work. Within a few months the public resistance against the site in Gorleben became a countrywide subject of media reports. The implementation of the "Gorleben-Hearing" in March/April 1979 was an important result of the protests. Supported by the prime minister of Lower Saxony, two groups of "pro" and "contra" international experts discussed for six days the safety of reprocessing and other general aspects of geological disposal. The hearing demonstrated that even the proponents of nuclear energy had some doubts about the safety issues of reprocessing.

As a result, the prime minister of Lower Saxony concluded that the technical feasibility of a reprocessing plant in Gorleben had found sufficient support from the members of the hearing, but that such a plant would not be politically feasible. The concept of an integrated nuclear waste management center was replaced by the plan to use the site as a final repository with attached interim storage and conditioning facilities.

6 The backfilling of the mine with salt to close unstable parts started in 1980 (BfS 2005, cited by Hocke and Renn 2011).

7 For the reasons see Hocke and Renn (2011: 46-7). – Parallel to this a comparative site selection process was on-going in the United States (see Solomon 2009: 1010-11; Fischer 2006).

In 1984, a shaft was installed and interim storage facilities were also constructed on the surface. Underground exploration of the Gorleben site began in 1986, but was accompanied by controversy. Criticism from mainly anti-nuclear experts and activists was raised about the (in their view) insufficient consideration of negative influences of the Gorleben quaternary channels existing in the overlying rock.

The next 12 years were characterized by intermezzos of investigations, court orders to stop any construction work and higher court orders allowing further construction. Ruetter and Partner (2005: 113) have characterized the period until 1996 as a 'stop-and-go policy'. Although there were periods of latency a local and national social movement against NPPs and the siting of nuclear facilities was stabilized and could again successfully mobilize the public for their campaigns (Rucht 2008, Fischer and Boehnke 2004, Kolb 1997).[8] In essence, the investigations for a geological repository at Gorleben took far longer than planned. The whole process could be described as a "messy muddling through", void of any coherent action plan (Hocke and Renn 2011).

Continuous civil disobedience, the ongoing transports of casks with spent fuel and vitrified highly active waste from reprocessing for surface storage in the Gorleben interim storage facility and – in this context – protests and even riots where the police had to guard the transports, brought the unsolved societal conflict surrounding nuclear waste back onto the national agenda.[9]

The German debate on nuclear issues was significantly influenced by the Chernobyl accident (1986), the shift to nuclear-skeptical positions within the Social Democrat Party and the end of the conservative government under conservative Chancellor Helmut Kohl (1999). As an attempt to restart the siting procedure for a national repository for heat-generating waste the following federal government coalition of the Social Democratic Party and the Green Party installed the "AkEnd" ("Committee on a Site Selection Procedure for Repository Sites") as a multidisciplinary high-level expert group which supported different positions pro and contra nuclear energy. The AkEnd was asked to develop a proposal for a transparent, criteria based, and convincing approach for a comparative site selection process, which should also include effective forms of public participation (see Hocke-Bergler and Gloede 2006: 91). So, they suggested that the Parliaments at the federal and state levels and the major

8 For the concept of "latency" in the context of social movement theory see Melucci 1998.

9 Shipments to the Gorleben interim storage facility started in 1995 with one spent fuel cask from the Philippsburg NPP. Three casks with vitrified highly active waste from reprocessing in La Hague and four casks with spent fuel from the Gundremmingen und Neckarwestheim NPPs followed in 1996 and 1997. From 2001 and 2011, a further 105 casks from reprocessing in La Hague were shipped to the Gorleben interim storage facility.

political parties in Germany (Christian Democratic Union (CDU), Christian Social Union (CSU), SPD, Free Democratic Party (FDP), and the Greens) should buy into the plan of a staged siting process starting with a "white map" because such a process would not be successful without the approval of the main political actors. In addition, they recommended that the process should be accompanied by a Supervisory Board from the very beginning.[10] Unfortunately, the AkEnd group and its supporters were not successful in convincing the key German political actors to follow their recommendations.[11] From today's perspective, one of the main reasons for AkEnd's failure was the lack of consent in the political arena and among societal groups about the need to restart the siting process for a high-level radioactive waste repository. In spite of a 10-year moratorium on explorations in Gorleben and a decision in 2002 to phase-out nuclear power, no breakthrough could be reached with regards to finding a political and societal consensus to the way forward (Kallenbach-Herbert 2011: 44). The "Gorleben moratorium" was terminated by the government of the CDU and FDP that won the national elections in 2009. However, their policy of "phasing-out the phase-out" and attempts to boost meaningful results regarding the suitability of the Gorleben salt dome came to a sudden end in March 2011.

The Fukushima accident was the next major turning point in German nuclear policy. Eight of the 17 operating NPPs were shut down immediately and a phase-out plan in stages for the remaining nine reactors terminating in 2022 was decided on. Along with the broad political and societal support for the nuclear phase-out decision, a new openness evolved towards solving the waste management problem. In June 2011, several state government leaders announced to support a country-wide site selection process (DAPD 2011). A working group of the federal and state governments was set up to find a compromise between the political parties and the federal and state interests regarding the future policy for geological disposal of heat-generating radioactive waste in Germany. In July 2013 the Repository Site Selection Act was enforced after it had been passed by the two houses of parliament (Bundestag and Bundesrat).[12]

10 In spite of the pluralistic composition the AkEnd members passed their recommendations unanimously. They are documented in detail (AkEnd 2002).

11 In the international community they won high reputation with their proposal (see Radkau and Hahn 2013: 354-5, and AkEnd 2002).

12 In German the term is „Gesetz zur Suche und Auswahl eines Standortes für ein Endlager für Wärme entwickelnde radioaktive Abfälle" (Standortauswahlgesetz – StandAG), 23 July 2013 (BGBl. I S. 2553).

4 The German concept of nuclear waste management

Germany's spent fuel policy was changed in two steps. The first step was the decision not to invest in one central national nuclear waste management center. The struggle for a new reprocessing site in Germany at the Wackersdorf site (Bavaria) was terminated by a disinterested industry and the political decision to stop this project in 1989 (Kretschmer 1988). Instead, Germany's nuclear operators invested in France's reprocessing facility at La Hague. Until 2005, they sent about half of their spent fuel to France and the UK for reprocessing and placed the other half in domestic interim storage mainly in a wet or dry storage at reactor sites, for direct disposal. The Nuclear Phase-out law of April 2002 ordered the termination of spent fuel shipments to reprocessing facilities abroad starting in June 2005.[13] The main reasons adduced were the safety risks and the costs associated with such shipments as well as a move towards abandoning the plutonium economy and minimizing spent fuel and high-level waste transports. Since then the only option has been the interim storage of spent fuel at the reactor site where it was produced and the subsequent geological disposal (Kallenbach-Herbert 2011: 43). The returning of vitrified reprocessing waste, however, has also continued after 2005 according to agreements of the German government with France and the UK. The interim storage facility at Gorleben which is the only destination in Germany for these waste streams has been criticized by political and societal representatives from Lower Saxony and the Gorleben region in the last years. The Repository Site Selection Act of 2013 therefore stipulates that free capacities of the on-site interim storage facilities should be used to store 21 casks with vitrified highly active waste from reprocessing in Sellafield (UK). Current negotiations also include comparable solutions for five casks with vitrified intermediate level waste from reprocessing in La Hague (France). Shipments are expected to start in 2015.

Besides the question of reprocessing or direct disposal of spent fuel the decision to use rock salt domes as a host geology (see section 3) for the disposal of heat-generating waste has significantly influenced the German waste management concept.

During the past 30 years different options of packaging spent fuel for disposal in salt rock were discussed. The so called "Pollux" concept provides that spent fuel in "transport and storage casks" is transferred to a national conditioning facility close to the repository site and repacked into smaller casks constructed for the conditions in deep geological formations. These "Pollux" casks should then be disposed of in tunnels or chambers built into the salt

13 For detailed references see Kallenbach-Herbert (2011: 44-6).

formation at a depth of several hundred meters. Another option might be the disposal of spent fuel in the transport and storage casks without unloading or repackaging, if it is technically feasible.

For vitrified high level waste the "borehole" concept is being discussed whereby the canisters containing the waste matrix are unloaded from the transport and storage cask and disposed of, without any further packaging, in a borehole, which is constructed from a drift at several hundred meters depth. An adapted borehole concept might also be an option for the disposal of spent fuel.

Although the suitability of the Gorleben salt dome could neither be proved nor confuted the possibility of opening the siting process to other host formations was not seriously taken into account before AkEnd was set up. In 1994, the Bundesanstalt für Geowissenschaften und Rohstoffe (BGR) published a study on crystalline formations followed by studies on salt and clay formations in 1995 and 2007 and then a summarizing report in 2007 (BGR 2007). In addition to broadening the view on potential host rock formations, AkEnd also evaluated different technical options of geological disposal including caverns and deep boreholes constructed from above ground. The question of retrievability was also tackled. These analyses confirmed the concept of maintenance-free deep geological disposal in a mere repository mine: The "disposal in deep geological formations (…) has decisive advantages in comparison with other disposal options … [R]epair and long-term monitoring are principally not required after sealing and backfilling of a repository..." (AkEnd 2002: 24-5).

The Asse research mine which is having problems for more than 20 years with brine ingress and stability that are affecting the safety of disposal, influenced the debate about suitable concepts considerably. The option of retrievability vis-à-vis a maintenance-free disposal concept became more interesting. This debate is still ongoing; however the possibility of recovering waste packages from the repository for a period of 500 years is stipulated in the latest German safety regulations for waste disposal (BMU 2010). The Asse case also raised doubts in society and politics about the suitability of rock salt for waste disposal. These doubts were increased by the fact that clay and crystalline are the favored options in other countries. On this background the debate on the most promising host rock was taken up again. However, different host rocks need different disposal concepts so that the technological solution as a whole has to be considered. For each potential host rock detailed explanations are necessary.[14] These discussions and international experience have also influenced the development of the Repository Site Selection Act (StandAG). As a result, salt, clay and crystalline are all considered as potential host rocks (§ 4).

14 The latest publication for salt in this context is Bollingerfehr et al. (2013).

5 The legal framework including the Repository Site Selection Act

The legislative and regulatory system which governs nuclear waste management in Germany is rather complex. A detailed account is provided in Germany's report to the Joint Convention on the Safety of Spent Fuel Management and on the Safety of Radioactive Waste Management (BMU 2012: 109-11), but in the following only the main features will be summarized.

As Germany is a federal state, the responsibilities for law-making and law enforcement are assigned to the organs of the federation and the states (Länder) according to their respective regulatory duties. There is a hierarchy of national regulation, and some specific regulations are adopted by different authorities or institutions and have different degrees of obligation: Basic Law ("Grundgesetz", adopted by parliamentary council and state parliaments), the "Atomic Energy Law" and the "Repository Site Selection Act" (adopted by parliament) and "Ordinances" (adopted by the federal government, federal council) are all compulsory. General administrative provisions are compulsory for all authorities (adopted by the federal government, federal council). Furthermore, there is a list of official documents which are binding by specification in the licence or by supervisory measures in the individual case (e.g. BMU guidelines and recommendations, KTA safety standards).[15]

The Basic Law (Grundgesetz (GG)) contains provisions on the legislative and administrative competencies of the federation and the states (Länder) regarding the use of nuclear energy. In addition, there are fundamental principles that also apply to the nuclear law. The basic legislation governing nuclear activities is set out in the 1959 Atomic Energy Act (AtG). This lays down the general national regulations for protective and precautionary measures including radiation protection and the management of radioactive waste and spent fuel and constitutes the basis for the associated ordinances.[16] For further detailing of the legal regulations, the Atomic Energy Act stipulates the disclosure of statutory ordinances by the federal government or by the competent federal ministry. Such ordinances need to be authorized by the Bundesrat. With regard to the disposal of radioactive waste the most important ordinances pertain to:

- radiation protection: "Radiation Protection Ordinance",
- advance payments for the construction of radioactive waste disposal facilities: "Repository Prepayment Ordinance",

15 Nuclear Safety Standards Commission („Kerntechnischer Ausschuss").

16 See BMU (2012: 107-40).

- prevention of any actions that could potentially influence underground explorations for a geological repository in the Gorleben salt dome: "Gorleben Development Freeze Ordinance".

Safety provisions for the disposal of heat-generating waste are laid down in the "Safety Requirements Governing the Final Disposal of Heat-Generating Radioactive Waste" (BMU 2010). Such safety requirements or guidelines are issued by the Federal Ministry for the Environment, Nature Conservation and Nuclear Safety (now BMUB, in former times BMU) following consultations with the states and generally by a consensus. These guidelines are designed to provide detailed specifications of selected technical and administrative issues.

For an approval, the implementer has to assess the long-term safety and has to give other evidence of safety for the operation and post-closure phase concerning radiological assessments as well as mining aspects or the protection of groundwater. Moreover, an Environmental Impact Assessment has to be conducted. Still, precise binding regulations that stipulate how safety assessments for disposal of heat-generating waste are to be performed do not exist yet.

The new German Repository Site Selection Act represents a political consensus about a stepwise approach for selecting a site for an underground repository for heat-generating waste and aims at proposing principles for developing improved site selection criteria.[17] This process starts with a "white map" of Germany – including under certain conditions the very controversially discussed Gorleben site.[18] This act mainly regulates the following issues:

- Appointment of a "Commission for the Disposal of High-active Waste"[19] which will prepare the siting process by December 2015.
- A three staged, criteria based site selection procedure considering different host rock formations (rock-salt, clay, crystalline) and the following schedule: The official site selection process should start in 2016, decision on the sites for surface investigations by the national parliament (year not speci-

17 This political consensus is articulated by the big political parties including the Greens and state governments, but not by all non-governmental organizations and interest groups. For this discussion see e.g. Smeddinck and Roßegger (2013); Gallego Carrera (2013), and Ausgestrahlt (2014).

18 For the struggle about the Gorleben site see Kallenbach-Herbert (2011: 49-50), and Schönberger (2013: 143-6 and 237-8).

19 This Commission should comprise 33 members representing the following groups: 1 chair person, eight members of the Federal Parliament, eight representatives of the federal states (these 16 members from the political arena have no voting rights), eight scientists, eight representatives of societal groups (trade unions, churches, industry and environmental organizations).

fied), decision on sites for underground investigations (2023), decision on the selected site (2031). Each step is concluded by a federal law as a binding decision for the next steps.

- Setting up the Federal Office for Nuclear Waste Management (Bundesamt für kerntechnische Entsorgung (BfKE)) as a new national authority acting as the regulator in the context of nuclear waste disposal.
- Involvement of concerned authorities and the public.

The 2013 law does not specify the concrete steps and the mode of evaluation in detail. But it fills the gap in the German legislative system that did not include regulations for the siting process before. It is one of the central tasks of the waste disposal commission to develop details for the criteria and the procedure for the site selection. A major challenge will be to show that lessons from the former conflicts were learnt and substantial participation and pluralistic views of experts are going to be integrated in the processes of decision making.

6 Institutional framework and financing

The Federal Ministry for the Environment, Nature Conservation, Building and Nuclear Safety (BMUB) is the regulatory authority, which is responsible for the development and implementation of the waste management policy and has a supervisory function with respect to the Federal Office for Radiation Protection (BfS) and the nuclear licensing authorities of the states. They are closely interwoven with two other federal ministries. On the one hand there are links with the Federal Ministry for Economic Affairs and Energy (BMWi) which is responsible for implementation related basic research on the disposal of radioactive waste and for the remediation of the former uranium mining sites of the former GDR. On the other hand an interface exists with the Federal Ministry of Education and Research (BMBF), which is responsible for the basic research in the nuclear waste management area.

Licensing is generally administered by the respective state where an installation is situated.[20] For example the environmental ministry of Lower Saxony is responsible for licensing the closure of the Asse mine and for the plan approval of the Konrad repository.

The BfS and BGR are central technical authorities. The BfS acts as the implementer for radioactive waste disposal. It is a subordinate authority under the BMUB and will be supplemented by the Federal Office for Nuclear Waste

20 As an exception to this the BfS is responsible for licensing of interim storage facilities for nuclear fuel bearing material.

Management (BfKE) in the near future. In fulfilling its tasks related to the construction and operation of repositories for radioactive waste, the BfS currently employs the services of the German Service Company for the Construction and Operation of Waste Repositories (Deutsche Gesellschaft zum Bau und Betrieb von Endlagern für Abfallstoffe (DBE)) as administrative aid. The BGR is the central geo-scientific authority and advises the German federal government on all geo-related questions. It is subordinate to BMWi.

Several commissions like the Nuclear Waste Management Commission (ESK), the Commission on Radiological Protection (SSK) and the Reactor Safety Commission (RSK) are advisers to the BMUB. The ESK, which is made up of national and international experts who represent a wide range of scientific and technical views, advises the BMUB on matters relating to nuclear waste management. The SSK advises the BMUB on issues involving the protection against the dangers of ionising and non-ionising radiation. It is a group of several national experts and one IAEA science and technology expert. The RSK advises the same ministry in safety-related matters to do with nuclear installations and until 2008 on radioactive waste management. With the establishment of the ESK in 2008, radioactive waste management became a topic of consultation of the ESK.[21]

These central institutions with official roles and decision-making powers are embedded in a highly politicized debate, in which a wide range of stakeholders, non-governmental organizations, citizens' initiatives on concrete nuclear sites, the churches and all types of media have become important collective actors in agenda setting. These also influence the debates between scientists and experts. Multiple interactions of these collective actors affect consensus building on specific topics, like with the new site selection act. They furthermore influence and establish the ongoing dissent among science, politics and society on certain points.[22]

Financing nuclear waste disposal is an interesting topic, which creates a fair share of conflict.[23] Industry has always underlined the position, that there are no problems in managing the financing of nuclear waste disposal including the special conditions of heat-generating waste. According to the Atomic law, the federal government is required to establish radioactive waste repositories (BMU 2012: 132). To cover the necessary expenses for federal facilities, the BfS collects – as explained in Germany's report to the Joint Convention – advance payments for cost-covering contributions from the future users of a repository

21 For further details see Minhans and Kallenbach-Herbert (2012: 3).

22 For a sociological approach to analyse the internal logics of interaction in nuclear waste management see Hocke 2006.

23 For this see the different positions of Müller-Dehn (2008) and Meyer (2012).

(BMU 2012: 138).[24] The amount of accrued liability ("Rückstellungen") and the mode of administrating itself seem to be complex, as the pressure for organizing a robust solution, which gives possibilities for clearance and reliability is enforced by the federal government. Meyer estimates that around €33 to €34m are available (2012). Since regulations for the publication of NPP-operating companies' specific annual financial statements (*Betreibergesellschaften*) and standards for reporting in a transparent and traceable way are not available, actors like the German Federal Court of Auditors (*Bundesrechnungshof*) have asked for independent verification (Meyer 2012: 47; Flauger and Fockbrock 2011).

The accumulation of internal funds under the responsibility of the private energy companies is furthermore associated with the question of assuring the availability of accrued liability in case of bankruptcy of the respective companies. Currently there is no financial security system for nuclear waste management in case of such circumstances. Meyer (2012) and others therefore demand an external fund solution which should be controlled by the public sector. So far this option has not found enough support in the federal government.

As the responsibility for disposal of nuclear waste rests with the states, the mode of financing the activities of the responsible governmental organizations is not easy to trace. Exact figures of the amounts allocated to nuclear waste management in the annual budget are not available. But it is documented that the financial means are fixed by the federal government (BMU 2012: 138). The BMUB, the BMWi, the BGR and the BfS have updated budgets for radioactive waste management duties. BMUB and BfS have an estimated radioactive waste management (RWM) budget of around €6 or €7m per year.[25]

7 Status of the siting procedure

After a period of investigations at the Gorleben site from the mid-1980s to 1998 which can be described as the stop-and-go policy (Ruetter and Partner 2005: 113) the implementation of the AkEnd as an interdisciplinary working group opened the discussion to a broad range of new issues.[26] Such issues comprise the

24 These cost-covering payments have to be paid in accordance with the Atomic Energy Act according to the Repository Prepayment Ordinance (see the act's chapter entitled, "The legal framework including the Repository Site Selection Act").

25 Authors' estimation based on BMU data (2012: 138).

26 The progress of investigations of the Gorleben site in those years was considerably influenced by court orders to stop any construction work and higher court orders allowing further construction activities.

stepwise approach in decision making, a criteria based comparative selection process, the involvement of the public in each step as well as a comprehensive set of siting criteria. These results reflect (1) the international debate on nuclear waste, which was widely neglected in Germany before, (2) national developments towards more direct democracy in other sectors, (3) deficiencies of the siting attempt at Gorleben so far as well as the principals of transparency and (4) discussion with (mainly members of the interested) public. Although the recommendations of the AkEnd finally did not convince the responsible governmental organizations they set an important benchmark for defining siting criteria and designing a siting procedure. The lack of criteria and an insufficient decision-making framework contributed to the deadlocks in siting attempts. Criteria which are broadly agreed upon before explorations start, and a process of comparing sites would be a progressive solution.

At the end of its legislative term the Green Federal Minister for the Environment published the draft of a new law on radioactive waste management and site selection. This proposal was prepared without broader discussions in the political, societal or scientific arena and thus did not achieve a political majority. It was far from reaching the status of a binding national framework for decision making as defined by the AkEnd.

From 2000 to 2010 can be characterized as a period of stagnation in deliberative politics. Some attempts at dialogue and political participation were started, but they had no lasting influence on the waste management policy due to the distrust that existed among the key stakeholders and the unresolved role of the Gorleben site:

- In November 2008, the Social Democratic Minister for the Environment, Sigmar Gabriel, organized a stakeholder symposium on nuclear waste disposal that brought together a broad range of stakeholders as a first but very small and fragile step towards further dialogue.
- A Disposal Dialogue Forum (Forum Endlager-Dialog – FED) was established as an inter- and intra-disciplinary group of members of the planning team for the 2008 stakeholder symposium. This forum held regular meetings for about two years and ended with an unclear future after two members, both representatives of the Gorleben region, stopped participating in protest over the drastic changes in the nuclear policy enforced by the conservative government elected in late 2009.[27]

27 The changes in nuclear policy enacted in 2010 comprise mainly: the extension of the nuclear power plant operation times, the restart of the exploration of the Gorleben salt dome on the basis of the mining law instead of the atomic law. The continuation of shipments of high-level

- At the beginning of 2011 the "Gorleben Dialogue" was started by the BMU, and at this first stage, some online communication tools were implemented. The plans for the next stage, including the setup of different committees, finally became obsolete when the siting policy in Germany significantly changed after the Fukushima disaster in March 2011.

8 Information policy and new forms of participation including society

In analyzing the development of information policy three phases can be identified. "Bürgerdialog Kernenergie" and other early attempts to appease contradictions against the established radioactive waste management by talks, hearings and public communication, which were mainly dominated by the state, never became official policy. All attempts at radioactive waste management in the time from the mid-1970s to 1999 failed to generate a consensus or tolerance with regard to the central decisions. In particular, most political parties and the industries involved were not willing to invest in more than public relations (Mez 1997; Rüdig 2000; Mez 2009: 273-5). Parallel to this situation the social movement against the civil use of nuclear power (*Anti-AKW-Bewegung*) won support and stabilized itself with very limited resources.[28] Doubts concerning the role of science grew for different reasons. Salt lost its reputation as "the best host rock" vis-à-vis clay. The quality of the Gorleben site selection in the 1970s and of the explorations in the 1980s and 1990s came under debate and experiences with the Asse research mine caused public concerns about geological disposal. When the implementer and the directors of the responsible authorities did not react to these concerns in an open way, the delay was interpreted as incompetence and as a blockage in decision making. The struggles at the phase-out bargaining table in 1993 and the phase-out decision of 2002 showed that the pros and cons of nuclear energy intervened in a substantial way in constructive conflict resolution (Mez 2006).

The AkEnd committee working from 1999 to 2002 started a new phase as it was the first official attempt to terminate the established "decide-announce-defend" approach. This approach, which over decades has structured radioactive waste management not only in Germany, can be characterized as a top-down

waste from the La Hague reprocessing plant to the Gorleben interim storage facility was a further point of criticism expressed by environmental groups and citizens' initiatives.

28 See Rucht (1994), Rucht (2008) and Högselius (2009: 259-60). Social movement research differentiates between the peace movement and Anti-NPP-Movement (Roth and Rucht 2008). For this the links which are important in some other countries like France or the United Kingdom cannot be overstressed in the German case.

decision-making process, which does not consider discussions with civil society representatives as part of the procedure. As an impact of this unsuccessful one-way communication, the strategy changed and a restart with a new concept integrating the interested public in the beginning and later the affected people at the potential sites became attractive. Under conditions of relative transparency and openness, the experts met with decision makers and stakeholders to convince them that their general idea of a new comparative site selection process for a repository for heat-generating waste and to generally inform the interested public from the beginning were good ones (Kallenbach-Herbert and Minhans 2014; Radkau and Hahn 2013). Before the Fukushima accident there were some smaller attempts to mobilize interest in a restart (see Gabriel 2006) and some events to promote the new safety requirements for nuclear repositories (from 2008 to 2010) (Kallenbach-Herbert 2011: 50-1). The Ministry for the Environment´s Gorleben Dialogue and the offensive attempts of the Federal Minister for the Environment Norbert Röttgen led to discussions about the necessity of participation and stakeholder integration in a reformed process of RWM and – parallel to this – participation in technology governance in general was stressed as an innovative tool.[29] Except from the Asse site, where a committee of regional representatives takes part in the decision-making process on the closure of the unstable mine, substantial citizens' participation has not been realized at any other nuclear site (NPPs, interim storage facilities etc.).

The current ongoing legislation driven third phase was started by the Röttgen administration. Its parliamentary process on the national level reached a consensus on the issue of the "Repository Site Selection Act" after two years – a political, but not a societal consensus. This became possible as Baden-Württemberg, after the Fukushima accident, offered the possibility that areas in this state could become potential candidates for siting within a new site selection procedure.[30]

The German Ethics Commission for a Safe Energy Supply's discussions (March to May 2011) also stressed the expectations of public participation and deliberative forms of interest aggregation.[31] Shortly before the national election

29 A series of events in the Gorleben region failed as the regional public refused to participate, but in the end the Gorleben-Dialog campaign and the Preliminary Safety Case of the Gorleben site produced a number of relevant documents. For the 2014 interpretation of this see www.bmub.bund.de/P331/, last accessed 14 February 2014.

30 For this context see Jahn and Korolczuk (2012: 162-3).

31 The German Ethics Commission for a Safe Energy Supply (*Ethikkommission für eine sichere Energieversorgung*) was founded at the end of March 2011 and was headed by the former Environmental Minister Klaus Toepfer. It included leading players from politics, science, industry and religious groups. The committee recommended to the German government a limit

in September 2013, the draft of the StandAG was presented and negotiated within the official governmental procedures under an extraordinary deadline pressure. This process was flanked by a three-day national event in which the interested public was invited to comment on the published draft. The Commission for the Disposal of High-active Waste still has to work on the roughly delineated stepwise approach of this law.

So far, offers for a new dialogue and participation as laid out in the Stand AG and via the heterogeneous composition of the commission did not have the expected impact. A consensus, especially between the states and their politicians, the industry and the protest groups was not reached. Discussions about the right way to manage heat-generating waste in a professional way and protest against a number of fundamental regulations of the StandAG with its plan for participation are still going on.

9 Lessons learnt

The history of siting a highly active waste repository in Germany is characterized by a few major milestones – and sometimes turning-points – that considerably influenced the governance process. This section looks at these milestones and turning points and the triggering events by considering the intertwined political and societal developments and expectations. Our thesis based on our current research is that changes of political majorities on the one hand and external events on the other hand play an important role in changing concepts and modernizing nuclear waste governance.

One of the most frequent triggers for changes in German nuclear policy was the change of political majorities. In 1976 during the siting process for the nuclear waste management center, the exceptional position of Lower Saxony was developed because of the election of its new Prime Minister Albrecht. He decided to implement a state-dominated siting policy, which concentrated on a bilateral exchange between the state of Lower Saxony and the federal government. It was due to societal protest, political developments in the following years and the results of the Gorleben Hearing – the first open expert dispute in German nuclear history – that the original project was significantly changed. The rather broad political consensus (between political parties on the one hand and the federal and the state level on the other hand) that existed in the years of the Gorleben siting process dissolved considerably over the next years and since the 1980s nuclear issues have been a matter of conflicting political views in

on the use of nuclear energy and to phase it out within a decade (Jahn and Korolczuk 2012: 161-2).

Germany. Thus neither the attempt of the former Green Environment Minister Jürgin Trittin to revise the siting process by establishing the AkEnd, nor the measures to encourage public dialogue initiated by the Social Democrats under Environment Minister Sigmar Gabriel in 2008, nor the attempt of the Christian Democratic Minister Röttgen to start a participation process concentrated on the Gorleben site, found a political majority and failed in gaining broad societal tolerance.

Another important trigger for the nuclear policy and waste management amendments during the past decades was the influence of nuclear accidents. A severe accident in nuclear history, the Three Mile Island in Harrisburg, Pennsylvania, USA on March 28, 1979, happened exactly when the Gorleben Hearing was taking place. The broad societal and political debate that followed, supported by the national media, rising fears of the risks of nuclear technology and the doubts surrounding reprocessing were influencing factors which caused the plans for a reprocessing plant in Gorleben to be abandoned (Hatzfeldt et al. 1979). The Chernobyl accident in 1986 strengthened the wide spread nuclear scepticism in the German society and the political controversy. The Fukushima accident in March 2011 marked a further turning point. Impacted by this event political majorities were formed for new phase-out agreements and for restructuring nuclear waste management policy.

The German siting history proves that political majorities do not necessarily reflect the expectations of society in general and especially of stakeholders, advocatory interest groups and grass-root initiatives involved in nuclear waste management. The long lasting quarrel over the Gorleben site has formed a strong, well informed and well organized anti-Gorleben movement of citizens' groups and social movement organizations in that region. Also environmental organizations acting on a national or even international level, like Greenpeace or Friends of the Earth, represent the group of the Gorleben critics. The nuclear industry on the other hand has taken a pro-Gorleben position. During the second half of 2013, however, they seemed rather to avoid any clear statement on the site selection act. The communities with spent fuel interim storage facilities clearly expressed their interest in achieving progress in realizing the repository in order to get rid of the waste in their territory.

The possibilities of either of these groups or of a broader public to interfere in the law making process regarding the Repository Site Selection Act were rather limited. The three days hearing conducted before the law was passed was the government's attempt to make a move towards the stakeholders. But this attempt falls far behind the broad and early participation, openness and transparency that the act itself implies for the siting process and also behind public expectations. To this effect the revision of the siting process on the basis of the

Repository Site Selection Act as a political compromise is starting under fragile conditions. After long-lasting discussions among different environmental and civil society groups, the BUND (the German organization linked to “Friends of the Earth”) and the Deutsche Umweltstiftung sent a delegate to fill the two seats in the Commission for the Disposal of High-active Waste. The citizen initiative "aufpASSEn" from the region, in which the Asse mine is located, also applied for a seat, but has not been considered. All other environmental associations and initiatives rejected participation in the commission. It remains open how far credibility for the selection process can be gained within other stakeholder groups like the related industries or in communities with interim storage facilities.

The Repository Site Selection Act marks a new milestone in the long lasting and conflicting disposal process. But whether it will lead to a stable process of nuclear waste governance is still open. Until now it is only a published act but also includes the opportunity to balance the important interplay of societal and political expectations. The crucial question is whether the leading actors are willing to participate and integrate a high rate of differing and potentially contrasting positions from politics, industry and environmental associations and interested public. If such integration is possible, there will be a chance for a substantial consensus for an open and democratic procedure for the governance of nuclear waste disposal. But each step of this approach bears inherent challenges and conflicts and no one knows whether the affected people and the responsible authorities and politicians will be willing to accept and to moderate them in a constructive way.

References

AkEnd (2002). Site Selection Procedure for Repository Sites. Recommendations of the AkEnd (Commitee on a Site Selection Procedure for Repository Sites), Ms., Köln.

Appel, D. (2006). Historical Background of Decision Making for Repository Projects in Germany: The Gorleben Case – Example for Missing Participation of Stakeholders. In NEA, German Disposal of Radioactive Waste: Forming a New Approach in Germany (FSC Workshop Proceedings, Hitzacker and Hamburg, 5-8 October 2004). Paris, 55-62.

Ausgestrahlt (2014). Press Release "Politik sollte mit Umweltverbänden reden, nicht über sie", 2 February 2014. https://www.ausgestrahlt.de/presse/artikel/2de386e1df/atommuell-politik-sollte-mit-umwelt.html, last accessed 26 February 2014.

Bundesanstalt für Geowissenschaften und Rohstoffe (BGR) (2007). Endlagerung radioaktiver Abfälle in Deutschland. Untersuchung und Bewertung von Regionen mit potenziell geeigneten Wirtsgesteinsformationen. („Nuclear Waste Disposal in

Germany. Investigation and Evaluation of Regions with Potentially Suitable Host Rock Formations"), April 2007. Hannover/Berlin.

Bundesministerium für Umwelt und Reaktorsicherheit (BMU) (2010). Sicherheitsanforderungen an die Endlagerung wärmeentwickelnder radioaktiver Abfälle, Berlin.

BMU (2012). Joint Convention on the Safety of Spent Fuel Management and on the Safety of Radioactive Waste Management. Report of the Federal Republic of Germany for the Fourth Review Meeting in May 2012.

BMWi (2008). *Endlagerung hochradioaktiver Abfälle in Deutschland – Das Endlagerprojekt Gorleben*, Berlin (Brochure).

Bollingerfehr, W., Buhmann, D., Filbert, W., Keller, S., and Krone, J. (2013). Status of the Safety Concept and Safety Demonstration for an HLW Repository in Salt. Summary Report, GRS, Peine.

Cloosters, W. (2008). Atomaufsichtliche Aspekte der Zuverlässigkeit und Kommunikation. *Atw*, 53(11), 690-695.

Comel, G. (1982). Die nukleare Entsorgung in der Bundesrepublik Deutschland. Kurzer historischer Abriß über die technischen Entwicklungslinien und über die öffentliche Diskussion von 1955-1977 ("Nuclear Waste Management in Germany. Historical Synopsis of Technical Development Directions and the Public Discourse"), Ms., Hamburg.

DAPD (2011). *McAllister fordert Endlagergesetz* ("McAllister Calls for Disposal Act", in German), Retrieved from http://de.nachrichten.yahoo.com/mcallister-fordert-endlager-gesetz-041855525.html, last accessed 26 February 2014.

Deutsches Atomforum (2013). Kernenergie in Zahlen, Berlin (Brochure).

Fischer, F. (2006). Integrating Deliberation and Analysis in Radioactive Waste Policy. In Hocke, P., Grunwald A. (Eds.), *Wohin mit dem radioaktiven Abfall? Perspektiven für eine sozialwissenschaftliche Endlagerforschung*. edition sigma, Berlin, 143-154.

Fischer, C., and Boehnke, K. (2004). 'Obstruction Galore': A Case Study of Non-violent Resistance against Nuclear Waste Disposal in Germany. *Environmental Politics*, 13(2), 393-413.

Flauger, J., and Fockenbrock, D. (2011, July 30). Die Last mit der Altlast, *Handelsblatt*. http://www.handelsblatt.com/technologie/energie-umwelt/energie-technik/rueckbau-von-kraftwerken-die-last-mit-der-altlast/v_detail_tab_print/4445178.html-download 23.2.14, last accessed 28 February 2014.

Gabriel, S. (2006). *Verantwortung übernehmen. Den Endlagerkonsens realisieren*. Federal Ministry for the Environment, Nature Conservation and Radiological Protection, Ms., Berlin.

Gallego Carrera, D. (2013). Viele unterschiedliche Wege – ein gemeinsames Ziel. Öffentlichkeitsbeteiligung bei der Suche nach einem Endlager für hochradioaktive Abfälle. In Müller, M. C. M. (Ed.) Endlagersuche – gemeinsam mit den Bürgern! Information – Konsultation – Dialog – Beteiligung (Vol. 21, Issue 13, 99-120), Loccum, Rehburg-Loccum.

Hatzfeld, H., Hirsch, H., and Kollert R. (Eds.) (1979). *Der Gorleben-Report. Ungewissheit und Gefahren der nuklearen Entsorgung* ("The Gorleben Report.

Uncertainties and Risks of Nuclear Waste Management"). Fischer Taschenbuch Verlag, Frankfurt am Main.

Hocke, P. (2006). Expertenkommunikation im Konfliktfeld der nuklearen Entsorgung in öffentlichkeitssoziologischer Perspektive. Zum Wandel von Expertenhandeln in demokratischen Gesellschaften. In Hocke, P., Grunwald, A. (Eds.), *Wohin mit dem radioaktiven Abfall? Perspektiven für eine sozialwissenschaftliche Endlagerforschung*, edition sigma, Berlin, 155-180.

Hocke, P., and Gloede, F. (2006). Collective Action of Experts in a Stalemate Situation. Central Results of Evaluative Research on the Work of 'AkEnd'. In NEA, *Germany Disposal of Radioactive Waste: Forming a New Approach in Germany* (FSC Workshop Proceedings, Hitzacker and Hamburg, 5-8 October 2004). Paris, 91-96.

Hocke, P., and Renn, O. (2011). Concerned Public and the Paralysis of Decision-Making: Nuclear Waste Management Policy in Germany. In Strandberg, U., Andrén, M. (Eds), *Nuclear Waste Management in a Globalised World*. Routledge, Abingdon, 43-62.

Högselius, P. (2009). Spent Nuclear Fuel Policies in Historical Perspective: An International Comparison. *Energy Policy*, Vol. 37, 254-263.

Jahn, D., and Korolczuk, S. (2012). German Exceptionalism: The End of Nuclear Energy in Germany. *Environmental Politics*, 21(1), 159-164.

Kallenbach-Herbert, B. (2011). Germany. In Feiveson, H., Mian, Z., Ramana, M.V., and von Hippel, F. (Eds.), Managing Spent Fuel from Nuclear Power Reactors Experience and Lessons from Around the World, 43-51. IPMF, Princeton.

Kallenbach-Herbert, B., and Minhans, A. (2014 / forthcoming). Case Study on the History of Siting a Geological Repository for High Level Radioactive Waste in Germany – The Evolution of a Waste Management Concept over 40 years.

Kolb, F. (1997). Der Castor-Konflikt: Das Comeback der Anti-AKW-Bewegung. In *Forschungsjournal Neue Soziale Bewegungen*, 3, 16-29.

Kretschmer, W. (1988). Wackersdorf: Wiederaufarbeitung im Widerstreit. In Linse, U., Falter, R., and Rucht, D. (Eds.), *Von der Bittschrift zur Platzbesetzung*. Dietz Verlag, Berlin, 165-218.

Melucci, A. (1998). *Nomads of the Present*. Temple University Press, Philadelphia.

Meyer, B. (2012). Rückstellungen für Stilllegung / Rückbau und Entsorgung im Atombereich 2010. Thesen und Empfehlungen zu Reformoptionen. Greenpeace. Berlin/ Kiel.

Mez, L. (1997). Energiekonsens in Deutschland? Eine politikwissenschaftliche Analyse der Konsensgespräche – Voraussetzungen, Vorgeschichte, Verlauf und Nachgeplänkel. In Brauch, H.-G. (Ed.), *Energiepolitik*. Springer, Berlin, 433-448.

Mez, L. (2006). Zur Endlagerfrage und der nicht stattfindenden sozialwissenschaftlichen Endlagerforschung in Deutschland. In Hocke, P., and Grunwald, A. (Eds.), *Wohin mit dem radioaktiven Abfall?* edition sigma, Berlin, 38-52.

Mez, L. (2009). Expansion and Phasing-out Nuclear Power in Germany. In Mez, L., Schneider, M., and Thomas, S. (Eds.), *International Perspectives on Energy Policy and the Role of Nuclear Power*. Brentwood,Essex, 263-278.

Minhans, A., and Kallenbach-Herbert, B. (2012). Identifying Remaining Socio-Technical Challenges at the National Level: Switzerland, InSOTEC Report.

Müller-Dehn, C. (2008). Finanzielle Entsorgungsvorsorge aus Sicht der Betreiber. *Atw*, 53(6), 391-394.

Neumann, W. (2013). Bestandsaufnahme Atommüll 2013. Zur Sache, Bürgerinitiative Lüchow-Dannenberg, 2(8), 12-44.

Radkau, J., and Hahn L. (2013). *Aufstieg und Fall der deutschen Atomwirtschaft*, oekom verlag, München.

Roose, J. (2010). Der endlose Streit um die Atomenergie. Konfliktsoziologische Untersuchung einer dauerhaften Auseinandersetzung. In Feindt, P., Saretzki, T. (Eds.), *Umwelt- und Technikkonflikte*. VS Verlag, Wiesbaden, 79-103.

Roth, R., and Rucht, D. (Eds.) (2008). *Die sozialen Bewegungen in Deutschland seit 1945. Ein Handbuch*. Campus, Frankfurt am Main.

Rucht, D. (1980). *Von Wyhl nach Gorleben. Bürger gegen Atomprogramm und nukleare Entsorgung*, Beck, München.

Rucht, D. (1994). *Modernisierung und neue soziale Bewegungen*. Campus, Frankfurt.

Rucht, D. (2008). Die Anti-Atomkraftbewegung. In Roth, R., and Rucht, D. (Eds.), *Die sozialen Bewegungen in Deutschland seit 1945. Ein Handbuch*. Campus, Frankfurt am Main, 245-266.

Rüdig, W. (2000). Phasing Out Nuclear Energy in Germany. *German Politics*, 9(3), 43-80.

Ruetter & Partner (2005). Nukleare Entsorgung in der Schweiz. Untersuchung der sozioökonomischen Auswirkungen von Entsorgungsanlagen. Vol. 2: Fallstudien und Ergebnisse der Bevölkerungsbefragung, Rüschlikon/CH.

Smeddinck, U., and Roßegger, U. (2013). Partizipation bei der Entsorgung radioaktiver Reststoffe – unter besonderer Berücksichtigung des Standortauswahlgesetzes. *NuR*, 35, 548-556.

Solomon, B. D. (2009). High-Level Radioactive Waste Management in the USA. *Journal of Risk Research*, 12(7/8), 1009-1024.

Schönberger, U. (2013). Atommüll. Eine Bestandsaufnahme für die Bundesrepublik Deutschland. Sorgenbericht der Atommüllkonferenz. Arbeitsgemeinschaft Schacht KONRAD, Salzgitter.

Stahl, T., and Strub, E. (2012). Decommissioning of Nuclear Facilities. GRS-S-52, Köln.

Tiggemann, A. (2004). *Die 'Achillesferse' der Kernenergie in der Bundesrepublik Deutschland*. Europaforum Verlag, Lauf an d. Pegnitz.

Tiggemann, A. (2006). Der Weg nach Gorleben. Zur Geschichte der Endlagerung in der Bundesrepublik Deutschland (1955-1977). In Hocke, P., and Grunwald, A. (Eds), *Wohin mit dem radioaktiven Abfall? Perspektiven für eine sozialwissenschaftliche Endlagerforschung*. edition sigma, Berlin, 85-118.

Model or Muddle?
Governance and Management of Radioactive Waste in Sweden

Tomas Kåberger and Johan Swahn

1 Introduction[1]

The governance and management of radioactive waste in Sweden is often seen as a model for the world. Since the 1980s, the radioactive waste company SKB, which is owned by the Swedish nuclear operators and is legally responsible for radioactive management, has internationally encouraged the idea that Sweden "has solved the radioactive waste problem". The government has generally been pleased with this situation and has for many years presented the Swedish legislation as a governance model for other nations to follow. There have indeed been results, largely because Sweden has a financial system which pays for the final disposal of radioactive waste (RW) from nuclear power and for the decommissioning of nuclear reactors. The nuclear industry pays for future costs into a nuclear waste fund but the government is in full control of how resources are spent.

Today, the country has an operating repository for short-lived operational radioactive waste, a centralized temporary storage facility for spent nuclear fuel (SNF), and a ship-based transportation system for RW. Swedish nuclear industry has also managed to identify a site for a long-term repository for SNF.

As we enter the second half of the 2010s, the Swedish nuclear waste governance model no longer appears as efficient as it did before. The "model" is a muddle that has not adequately taken into account the future governance of radioactive waste in Sweden. There are plans to reform the legal system, but it is questionable whether this will be enough to solve the model's fundamental issues. In this chapter we describe how the Swedish system for the governance and management of nuclear wastes developed, consider the problems that have occurred, and ask whether it is it too late for a new model that could provide for an industry-financed, safe management of nuclear waste.

---

1 The content of this chapter is partly based on a chapter in Swahn (2011). An attempt to explain the Swedish nuclear policy development as a result of interacting economic interests is given in Kåberger (2002).

2 The context: A brief review of Swedish radioactive waste governance and management from 1945 to today.

As in other countries that were early nuclear adapters, military interests initially drove the Swedish nuclear programme. With time, it became apparent that it would be easier to develop a nuclear weapons capacity if the programme could be integrated in nuclear energy development. By 1985, there were 12 reactors operating at four nuclear power plant sites. The two reactors at Barsebäck, the fourth site close to Copenhagen and Malmö, were closed in 1999 and 2005, respectively.

While the number of reactors, capacity, and tonnes of radioactive waste are small in relation to those of the major nuclear countries in the world, with 10 GW nuclear power among less than 10 million people Sweden has more nuclear power and more nuclear waste per capita than any other country

2.1 1945-1970: A military-civil nuclear programme

The Swedish military-civil nuclear programme was based on heavy water technology. In 1947, the government created a company called AB Atomenergi that became central to the development of the bomb and nuclear energy programmes. A nuclear research facility was built in the late 1950s at Studsvik on the Baltic coast.[2] The reactor research programme resulted in the construction of a heavy water moderated reactor that delivered hot water for district heating and some electricity to the suburbs of Ågesta, situated South of Stockholm. The construction of the reactor, which was built underground in bedrock, was started in 1957; the reactor began operation in 1964.[3] The main purpose of the reactor was to produce plutonium for nuclear weapons in case the decision was made to start bomb production. The spent nuclear fuel (SNF) from the Ågesta reactor was thus considered to be an asset and not waste. After reprocessing, it could be used to produce plutonium for nuclear weapons or for nuclear fuel in breeder reactors. The high-level radioactive waste (HLW) from the reprocessing was considered

2 AB Atomenergi was responsible for building reactors while bomb construction and radiochemistry development was the responsibility of the Swedish Defence Research Establishment (FOA). Sweden gained knowledge about nuclear technology from the "Atoms For Peace" programme initiated by U.S President Eisenhower in 1953. For an historical account of the development of the Swedish military-civil nuclear programme see Johansson (1986), Jonter (2001; 2002) and Agrell (2002).

3 The Ågesta reactor was closed in 1974 (IAEA 2014).

to be small in quantity and useful for medical purposes.[4] Sweden was planning to build a large reprocessing plant underground in the bedrock in Sannäs on the West coast in the early 1960s. The plant could produce plutonium from SNF from the Ågesta nuclear reactor and from the planned, even larger heavy water reactor being built at Marviken on the Baltic coast south of Stockholm.[5]

By the 1960s, however, the military part of the Swedish civil-military programme was in trouble. Public opinion against Sweden's goal of acquiring nuclear weapons grew. For some years, the programme to construct nuclear weapons was carried out under the auspices of a programme to develop a *defence against* nuclear weapons. By the end of the 1960s, Sweden decided to reject nuclear weapons. Sweden joined the Nuclear Non-proliferation Treaty (NPT) in 1968, which was enforced in 1970. The heavy water reactor programme was abandoned after 12 commercial reactors were operational.

2.2 1970-1980: Environmental wake-up, political intrigue, end of reprocessing, and a referendum on nuclear power

In the 1960s there was a general awakening about environmental issues in Sweden. Rachel Carson's book "Silent Spring," written in 1962, was published in Swedish one year later and was as influential as the Club of Rome report, "Limits to Growth" (Carson 1962; Meadows 1972). The first United Nations Conference on the Human Environment was held in Stockholm in 1972, which showed that environmental issues had also become mainstream in political circles by the early 1970s. This environmental awareness led to more debates on the environmental risks of nuclear power in the 1970s. The problem with RW, which could no longer be claimed to be a minor nuisance, became a major part of the discussion.

In response to increasing public debate, a government commission was set up in 1972 to come up with a proposal to suggest how Swedish RW was to be managed. In 1976, this AKA commission delivered a report in three volumes

4 There was a general understanding in the 1950s and 1960s that the management and disposal of any waste from nuclear activities could easily be solved in the future by Swedish engineers. At this time there was also sea dumping of radioactive waste barrels, both in the Baltic Sea and along the Swedish west coast.

5 The Marviken reactor was built during the 1960s but was never loaded with nuclear fuel. There were technical problems with the reactor design, but the main reason for not going forward was that the decision had been made to abandon the Swedish heavy water reactor programme. It was too expensive to compete on a commercial market once plutonium for nuclear weapons was no longer needed. The Marviken plant was converted to run on oil until it was shut down in 2009 – the only nuclear reactor in the world to run on fossil fuel.

(AKA 1976). The main proposal was that SNF was to be reprocessed and the remaining waste should be encapsulated in canisters and deposited in tunnels a few hundred metres underground, in a bedrock without any, or with only very few and small, fissures.

With RW now identified as a liability rather than an asset, the problems of long-term RW management also became central in the political debate. Ethical arguments about the waste problems associated with using nuclear energy and discussions about alternative energy futures based on renewable energy became part of the political debate before the 1976 elections. These issues were picked up by the Centre Party and the party's gain in popularity allowed for a change in power to the first non-Social Democratic government in 40 years. The Centre Party shared power with the Liberal and Conservative parties, which were both pro-nuclear. Still, the government was able to pass new legislation through Parliament, namely through the Stipulation Act. The law decreed that no new nuclear reactors could be put into operation unless there was an *absolutely safe* way of managing the final disposal of the nuclear waste (SFS 1977). While the third reactor at the Ringhals NPP was awaiting a licence for operation, the Stipulation Act, which clearly connected RW solutions to the operation of new reactors, gave the waste problems an important political role for a few years. This act also forced the nuclear industry to act quickly; consequently it launched the KBS project to develop a repository concept for high-level reprocessing waste, or SNF. The project was developed in close collaboration with the Swedish Nuclear Fuel Supply Company (SKBF), which the nuclear utilities had created in 1972 to coordinate the Swedish nuclear fuel supply.

The 1977 KBS-1's report described how vitrified high-level nuclear waste from reprocessing, the industry policy at this time, could be disposed of (KBS 1977). It was approved in a controversial decision by a minority government after the Centre Party government fell, and enabled the operation of the Ringhals 3 reactor.

So, despite the ambitious criteria of the Stipulation Act, new reactors were allowed to become operational. The argument that an absolutely safe way to solve the RW issue had been found was very political. The proof that the KBS-1 method was safe depended on the existence of a crack-free bedrock. However, no such rock was found during the drilling programme for site selection. The argument supporting the decision was that there were no fissures in this part of the Sternö candidate site in southern Sweden, and therefore a crack-free rock existed. Yet, the reason why no fissures were found was that no test drillings were carried out in that particular part of Sternö. Hence, these have been called "political boreholes".

The political "logic" behind the decision to start Ringhals 3 was seen as very provocative by those who had supported the idea that no more RW should be produced until the waste problems were solved.[6]

The Stipulation Act also allowed for direct disposal of SNF without reprocessing. The 1978 KBS-2 report therefore provided a solution for the direct disposal of SNF, similar to the one for vitrified reprocessing waste (KBS 1978). The main difference was that the canister´s material had been changed from titanium to copper.

Because the main focus of the nuclear industry at the time of the Stipulation Act was to reprocess the SNF, reprocessing contracts had been signed with the French company Cogema and the British company BNFL. With the KBS-2´s method there was a possibility for the industry to also claim completely safe disposal for SNF that was not reprocessed. At the same time, it became clear that reprocessing was expensive compared to direct disposal. In addition, the U.S. government, under the Ford and Carter administrations, was working internationally to convince countries to reject reprocessing as a nuclear non-proliferation measure. Even though reprocessing is still legally possible in the Swedish nuclear legislation, by 1980 the industry and official policy was that Sweden would not reprocess SNF.

The political turmoil with regard to nuclear power did not end with the decision to declare the RW problem solved and the starting of new reactors. In the spring of 1979, following the Harrisburg nuclear accident, the nuclear debate became increasingly intense. As a result, the pro-nuclear Social Democratic Party changed its mind to allow for a referendum on nuclear power in 1980. The referendum was also politically manipulated because there was not a clear "yes" or "no" choice. However, the final result of the referendum was that 12 reactors would be allowed to operate. The Swedish Parliament decided that the reactors should be phased-out after 25 years of operation, i.e, by 2010 as the newest reactor (Forsmark 3) came on line in 1985.

2.3 1980-2006: Calm governance in a phase-out scenario

The radioactive waste company SKB continued to develop the KBS concept for final disposal of SNF. In 1983, the KBS-3 report became part of the licensing process for the last two Swedish nuclear reactors, Oskarshamn-3 and Forsmark-3 (SKBF 1983). The KBS-3 report had a more developed discussion of long-term

6 A good description of the political turmoil around the decision to start the Ringhals 3 can be found in Sundqvist (2002). A description from the environmental movement's viewpoint can be found in Åhäll et al. (1988).

safety and relied on the results from the geologic studies that have been done to date.[7]

The last two reactors were licenced in 1984 in accordance with the new Nuclear Activities Act, which replaced previous nuclear legislation including the Stipulation Act. After 1986, Sweden had 12 operating nuclear reactors at four NPPs. The new legislation placed all responsibility for developing and operating RW repositories on the nuclear industry. However, this legislation stipulated, as its only requirement for RW management and final disposal that the nuclear industry, through SKB, would provide a R&D programme every three years for the government to accept.

In 1995, the Äspö Hard Rock Laboratory was inaugurated near the Oskarshamn NPP and a number of experiments were started involving copper and clay, including a full-scale demonstration project with full-size copper canisters in the underground research site.

The siting process for a repository for SNF – the most long lived RW with a safety case necessary for hundreds of thousands or even millions of years – was not as easy. A voluntary process finally allowed the RW company SKB to start site investigations in 2000 at a site just south of the Forsmark NPP and at a site just adjacent to the Oskarshamn NPP.

As the twenty-first century arrived, the Swedish model seemed to be thriving. The economics, siting, repository design, and technology – all seemed fine. What could go wrong?

2.4 2000-2014: Moving towards a collapse of the Swedish model?

At the beginning of the 2000s it was clear that there were a large number of issues that were of great concern. These were raised by regulators and the Swedish Council for Nuclear Waste, but also by the academic community and the environmental movement. There was some concern about how the company would deal with threats to the repository during repeated glaciations – ice ages – that could threaten the artificial barrier system that now, without the illusion of crack-free bedrock, was seen as critical in isolating the waste from humans and the environment. The government was informed about these concerns, however since the legislation deemed the industry responsible for any necessary extra work, all the government could do in its decisions on the programme was to ask for some complementary work. It was the industry's risk if it did not do enough, because any work that was lacking or insufficient would not be reviewed until

7 In the rest of this chapter this will be called simply the KBS method or KBS concept.

the application for a licence. These were issues that the company had to deal with properly in its upcoming application for a licence.

The government that came into power in Sweden in 2006 included a very pro-nuclear party and changed the law in 2009 to allow the construction of new NPPs. This means that Sweden moved from a situation where nuclear power was seen as a technology to be phased out, with a finite amount of waste, to a situation where the future amount of RW is uncertain. As demonstrated in the early period, waste management regulation is not only a matter of managing an unavoidable limited problem. Again, the future production of waste may influence how RW should be managed.

In 2009, SKB finally chose an area south of the Forsmark NPP as the site for the repository for SNF. An encapsulation plant, for placing the SNF in copper canisters, was to be built at the Oskarshamn NPP. The siting choice was surprising to many, as the geology was very different at the two investigated sites. The experiments at the Äspö Hard Rock laboratory were seen as positive experiences and inspired confidence in how copper canisters and clay behaved in the bedrock near the Oskarshamn NPP. SKB, however, decided not to use the recommended criteria for the functioning of these barrier systems for final site selection. Instead the company returned, at this final stage, to using criteria involving fissures in the rock. The mean distance between major fissures was higher in Forsmark than in Oskarshamn. Thus, Forsmark was chosen mainly because it had a "crack-free" rock. In March 2011, SKB handed in its application to build a SNF repository in Forsmark to the Environmental Court and the Swedish Radiation Safety Authority (SSM). The application will be reviewed in parallel by both bodies. As of September 2014, the process was still ongoing and it was not yet decided which additional requirements SKB must fulfil in order to complete the application and review its merits. There were many requests for additional changes on a large number of issues, from SSM, from the Swedish Council for Nuclear Waste, and also from members of the environmental movement. SKB has given SSM additional tasks, but the company's interaction with the environmental court has not been so successful. The large volume of demands for amendments and the time it is taking SKB to answer, as well as its refusal to acknowledge many demands, has prompted the environmental court to make a ruling between January and October 2015 on whether a "process hindrance" exists. If such a ruling is made, SKB may be forced to make further amendments. Ultimately, a "process hindrance" could mean that the licence application will be rejected.

Perhaps the major issue that has led to this situation is the scientific controversy that emerged in 2007 regarding copper corrosion in a repository environment. Researchers from the Royal Institute of Technology in Stockholm have published a number of scientific articles questioning SKB's view that copper in an environment without oxygen in gas form – as is the situation in a repository – corrodes very slowly. The researchers have shown that water corrodes copper even if oxygen is not available (Szakálos 2007; Hultquist 2009; Hultquist 2011; Hultquist 2013). If this is the case, the SKB safety case for the KBS method will have to be questioned. SKB is fighting the issue, but as of June 2014 the controversy had not been resolved and SSM is continually demanding more information from SKB. For the environmental court, it is certainly unreasonable to have to try and judge a case mired in scientific controversy. The court is so far mainly waiting for the results of interaction between SKB and SSM.

3 Overview of the nuclear power programme and radioactive waste volumes

The Swedish nuclear programme includes 13 commercial reactors listed in the IAEA Power Reactor Information System (IAEA/PRIS 2014). In 2014, 10 of these 13 reactors were operational. They are located at three NPPs, Oskarshamn with three reactors, Ringhals with four and Forsmark with three. The two reactors at the Barsebäck NPP, close to Copenhagen and Malmö, were closed in 1999 and 2005. All other reactors are being refurbished to ensure increased safety and life-time extension.

The average production cost for electricity from the three reactors at the Oskarshamn nuclear power plant, OKG, over the last five years is significantly higher than the electricity market price for the coming 10 years. In January 2014, OKG announced that it was preparing to file an application to put the oldest reactor on the site, Oskarshamn 1, out of operation. The company's rationale for this was that it was expecting demands for safety upgrades as a result of evaluations of the Fukushima accident that would be too expensive to carry out. Oskarshamn 2, the second oldest reactor, is under major modernisation that is expected to take about one year. In June 2014, OKG told the power market that there was no longer a start date scheduled.

The nuclear legislation was changed in 2009 to allow the construction of new nuclear reactors if old ones were shut down. Since the summer of 2012, the state-owned electricity company Vattenfall announced that it was considering building one or two reactors at the Ringhals NPP on the West coast. There is an "application" in the early stages to force the regulator SSM to start working on the regulations for new-build. A decision to build a new reactor is expected soon. There is presently an overcapacity in Sweden, as wind power generation has been expanding rapidly in the last 10 years. The life-time extensions of the present reactors are said by the industry to ensure operations for 50 or even 60 years. Finally, it would not be possible to transfer the investment to a new nuclear reactor in Sweden without clear and broad political support for nuclear new-build, one that would probably require subsidies and financial guarantees. This is not the situation in Sweden today, and may be difficult to achieve in the future.

3. 1 Radioactive waste from NPPs[8]

The RW from the Swedish NPPs, including the initial small Ågesta reactor that is considered part of Sweden's commercial nuclear programme, is primarily stored in the intermediate storage facility Clab. There is a total of 5,740 tonnes of SNF in the facility; 5,695 tonnes come from the nuclear power reactors, 20.2 tonnes from the Ågesta nuclear reactor, 2.4 tonnes is spent fuel from tests which were done at the Hot Cell Laboratory at Studsvik, and 22.5 tonnes is German spent MOX fuel traded for Swedish fuel exported to France in transferred reprocessing contracts from the 1970s.[9] In addition, there were 556 tonnes of SNF in the cooling pools at the 10 operating reactors on December 31, 2013. In the cores of the operating reactors there is approximately an additional 1,000 tonnes of fuel that will become SNF. The total amount of SNF to dispose of will be about 12,000 tonnes if the existing Swedish reactors are in operation for the 50 to 60 years that the industry is presently planning for.

The repository for short-lived operational waste, SFR, contains 34,953 m^3 of short-lived low- and intermediate-level radioactive waste from the operations of the NPPs. Some of this waste is also from medical and industrial sources. It is expected that about a total of 53,000 m^3 of this type of short-lived waste will be produced from reactor operations.[10] The operation of Clab and the planned encapsulation facility for SNF is expected to add another 400 m^3. In addition,

8 These and the following figures are taken from the Swedish Ministry of the Environment (2014).

9 The reprocessing contracts that had been signed with France were transferred to Germany; 28 tonnes of German spent MOX fuel was transported to Sweden and still awaits final disposal. The British contracts were never broken and 140 tonnes of Swedish SNF was reprocessed at Sellafield. The high-level reprocessed waste was not returned to Sweden and in the spring of 2014 it was made known that the plutonium would also be transferred to British ownershipThe Hot Cell Laboratory at the Studsvik facility has been in operation since the late 1950s and has, for many years, been operating as a commercial test facility for nuclear fuel. The facility is now owned by Studsvik Nuclear AB. The fuel rods are moved to Clab after the tests and are stored in special steel containers with the same form factor as spent fuel assemblies. Studsvik Nuclear AB has a permit to also store foreign SNF that has been tested in Clab and will be disposed of along with the Swedish SNF.

10 The data from the rest of this section comes from SKB (2013).

73,000 m^3 of short-lived decommissioning waste from the nuclear reactors is expected. This waste is to be stored in a new facility, SFR 2, which SKB wants to build as an expansion of the SFR facility. The decommissioning of Clab and the planned encapsulation facility for SNF is expected to add another 3,400 m^3 of waste to SFR2 as well.

It is also expected that the decommissioning of the NPPs will produce 3,700 m^3 of long-lived intermediate level waste (SFL). A new facility for this waste has to be built, but the method and the siting for this construction have still not been decided on. The NPPs at Ringhals, Forsmark, and Oskarshamn, as well as the Studsvik nuclear facility, also have shallow land burial facilities for very low-level waste. In total, Sweden is expected to have approximately 12,000 tonnes of SNF, 154,000 m^3 of short-lived, and 15,500 m^3 of long-lived intermediate nuclear waste to dispose of at the end of the nuclear era, unless there are new plant constructions.

3.2 Historic radioactive waste

Most of the historic waste from the early Swedish military-civil programme was reconditioned, put into new containers in the 1990s, and then stored underground at the Studsvik facility.[11] The SNF from the Ågesta reactor is in Clab, but there is still 40 kg of spent fuel from the R-1 research reactor in Sweden. Most of the fuel from the reactor, 4.8 tonnes, was exported in 1997 to the United Kingdom for reprocessing, but a small amount that was considered too damaged to move remains in Sweden.[12] This fuel is metallic uranium and cannot be disposed of using the KBS method. It is possible that this SNF will be put in a future

11 There are 7,750 containers in this facility. Judging mainly from documentation, the containers at the Studsvik site could contain too much long-lived RW, which should be deposited in SFR. A programme was carried out to try and ascertain in more detail what was in each container. It found that the documentation of the containers was not accurate. There were liquids in 20-30% of the containers though no liquids should have been in them. There was mercury in some containers and some containers contained fissile material that was not accounted for in the safeguard registers. In the original reconditioning process 2,844 canisters were considered only to contain short-lived waste and were placed in 75 containers that were then deposited in SFR. These containers with the canisters now have to be retrieved because the documentation apparently is not good enough to classify the canisters as only containing short-lived wastes. A new facility for reconditioning and classifying the "historic" waste in the canisters is planned for the Studsvik site. Sweden kept 3.3 kg of separated plutonium from different sources in the military-civil programme stored in a vault at the Studsvik research facility for many years. The plutonium was exported to the United States in March 2012 as part of the U.S. Global Threat Reduction Initiative.

12 Ownership of the plutonium from the reprocessed R-1 fuel is to be transferred to the United Kingdom, along with with the plutonium fromSwedish SF reprocessed in Sellafield.

repository for intermediate-level, long-lived radioactive waste. The SNF from the other research reactors was returned to the fuel source in the United States.

The amount of historic low-level RW from the early military-civil programme, together with the operational short-lived waste from the present Studsvik Nuclear AB activities, is expected to reach 11,500 m3. Decommissioning waste from all the facilities will add an additional 13,000 m3 of short-lived waste.

Long-lived intermediate-level RW from the early military-civil programme, together with operational waste of the same type from the present Studsvik Nuclear AB activities, is expected to reach 11,800 m^3.

4 Overview of operating and planned radioactive waste facilities

At present, the nuclear industry is planning three new final repositories in Sweden, in addition to the existing SFR and Clab facilities.

4.1 Centralised intermediate storage facility for SNF (Clab)

In 1985 the centralized intermediate storage site for spent fuel (Clab) started operations at the site of the Oskarshamn NPP on the South-East Baltic coast. A special ship had been commissioned to move both short-lived operational RW to the SFR facility in Forsmark and SNF to the Clab facility at the Oskarshamn NPP. The storage is in a pool about 30 metres underground in granite bedrock. This facility was expanded and a second pool was put into service in 2008. Currently, the SNF inventory in Clab is about 5,800 tonnes. The present storage capacity is 8,000 tonnes. If a licence to build a repository for SNF is not granted, it is possible to store some SNF in Clab. It would also be possible to expand the facility with a third storage pool or build a facility for dry storage. According to the nuclear industry, if necessary, the SNF can be stored safely in the pools for at least 100 years.

4.2 Repositories for short-lived operating radioactive waste, SFR and SFR 2

By 1988, the nuclear industry had built a repository for short-lived operational RW from the nuclear power reactors. It was the first of its kind in the world. The repository, called SFR, was situated 50 metres under the sea-bed outside the Forsmark NPP on the Baltic coast north of Stockholm. This repository is for operational waste from the NPPs, but each year 10-20 m^3 of waste from hospitals, industry and research is also deposited. The storage capacity is approximately 63,000 m^2, of which 35,000 m^3 is used.

In the early 2000s, it was discovered that the RW that had been deposited in SFR had excessive long-lived radioactive elements, most significantly carbon-14 and plutonium. The regulators therefore forced SKB to conduct a new safety analysis. Because the safety case for SFR included a release of radioactivity from the repository into the Baltic Sea, it was not easy to get an acceptable result from this analysis. SKB and the regulator have been under public criticism for the safety case ever since.[13]

Another problem for SFR is that the repository is deteriorating faster than initially foreseen. Waste containers are corroding faster than anticipated and the concrete, including steel reinforcements, is breaking apart. There is an inflow of water into parts of the operating repository that is hindering operations. In 2005, it was discovered that there were stray electric currents from the direct current transmission line going from Forsmark to the Finnish side of the Baltic. These stray currents passed through the SFR repository and it is not yet clear if they have been causing corrosion problems or if there are other causes for the corrosion.[14] These SFR issues are a problem for SKB, which is currently preparing an application for a licence to expand SFR with a new addition, SFR 2. The new expansion will be a little deeper than the original repository in order to make it safer, at 120 m below sea level. The original SFR will also be reviewed along with the new facility; it remains to be seen whether the old and new repositories, which are both part of the safety case with a deliberate release of radioactivity into a recipient, will be accepted in accordance with the modern Swedish nuclear and environmental legislation. The total waste volume needed for both repositories is about 200,000 m^3.

13 The repository was designed according to the environmental standards of the late 1970s, as dilution into a recipient was still considered to be an acceptable part of a safety case for a new facility as long as the releases were below certain limits.

14 In June 2005, a leak was discovered from corroding drums that had been deposited in SFR. The radioactive material was caught in the drainage system. The drainage water has since that time been collected and delivered to the Forsmark NPP for further treatment. The repository roof has also been covered to decrease the inflow of water.

4.3 Planned repository and encapsulation facility for SNF

Since the 1970s the Swedish nuclear industry has been planning to build a repository for SNF using the KBS method. The 1983 KBS-3 report looked in greater detail at the theoretical understanding of corrosion processes and the required thickness of the copper canister. Although no recommendation for the thickness of the copper in the canister wall was given, a decision was later made that a thickness of 5 cm would be sufficient. This assumption is still in effect today. The KBS-3 report was the last report in this series and the proposed disposal method is sometimes called the KBS-3 method.

An analysis of the KBS method was carried out in two projects, SKB 91, around 1990, and SR-97, in the second half of the 1990s (SKB 1992; SKB 1999). The SKB 91 report was important in that it shifted the focus of the safety case from the importance of a crack-free bedrock to the importance of the artificial barriers of copper and clay. This allowed SKB to move over into a voluntary siting process. It was assumed that there was a sufficient amount of suitable bedrock for a repository and this assumption played an important role in the subsequent siting process.

The SNF will be put into a cast iron insert and then encapsulated within a 5 cm thick copper canister. The canisters are then to be lowered into bored holes in mined tunnels at a depth of about 500 metres in granite bedrock. A bentonite clay buffer will be placed around the copper. The deposition tunnels are then filled with more clay.

The safety case of the KBS method relies on the integrity of the copper canister and the bentonite clay buffer. The copper canister should be corrosion resistant in the repository environment which, when completed, should contain water but no molecular oxygen. Biological and chemical processes are supposed to consume the oxygen after the deposition holes are sealed. The bentonite clay buffer absorbs water from the surrounding bedrock and swells. The water-saturated clay should prevent the movement of ground water and corrosive substances in the water in the vicinity of the copper surface. Thus, for the safety case to be valid, the clay has to swell and protect the copper. In addition, the safety case has to ensure that the repository will not be affected by mechanical and chemical changes that could take place during repeated ice ages.

On March 16 2011, SKB filed an application for a licence to build and operate a SNF repository south of the Forsmark NPP. The application also included an encapsulation plant to be built as an addition to Clab at the Oskarshamn NPP.

4.4 Planned repository for long-lived intermediate-level radioactive waste (SFL)

A new, final repository for long-lived intermediate-level radioactive waste is also planned, but these plans are vague. The regulator and the government have criticised the industry for its lack of planning, but it is generally believed that the industry wants to build SFL as an extension with a few more tunnels once the licence is granted for a repository for SNF. At a depth of 500 metres, SFL would not need much encapsulation or buffering to make it safe.

5 The Swedish concept for radioactive waste governance and the legal framework

The Swedish state takes the ultimate responsibility for the management of RW, but differently from many other countries, has tried to shift this responsibility to industry, including historic, pre-commercial RW, management and final disposal, as well as for the financing of all activities and regulation. The Swedish concept for RW governance places the entire responsibility for RW governance on the owners and operators of the NPPs. The nuclear industry, in turn, has transferred this responsibility to their co-owned radioactive waste company SKB. The waste company has been tasked with developing and operating all nuclear waste facilities. The company also has to develop new waste management solutions and final repositories, as well as manage the decommissioning of nuclear facilities.

The main legislation regulating the industry's work on nuclear waste is the Nuclear Activities Act (SFS 1984a) and the Nuclear Activities Ordinance (SFS 1984b). These regulate the way in which applications for licences should be reviewed. In addition to the Nuclear Activities Act and Ordinance, there are a Radiation Protection Act and an Ordinance that set up requirements for radiation protection and, in the case of repositories, the criteria for long-term release of radiation (SFS 1988a, SFS 1988b).

The nuclear power operators are also obligated to carry out a R&D programme every three years which is to be reviewed by the regulator(s) and the Swedish Council for Nuclear Waste, the government's scientific advisory board. These programmes are sent out for public comment. Each programme has to be approved by the government, which can also require additional work on certain issues. The government decides if the programme is sufficient to allow further operation of the nuclear power reactors. These so-called "Fud" programmes essentially represent the only way that the industry's efforts in RW management can be influenced by the state.[15] The government can make requests that the industry has to answer, but since the only way to really make the industry do what the government wants is to stop the nuclear reactors, this mechanism seems very blunt. No government so far has refused to give a go-ahead to the programme. In practice, the system has allowed the industry to do as it wants. Independent researchers, critical of SKB, cannot use funding from the nuclear waste fund because it can only be used by the industry. There are no economic incentives offered to trace any problems with the plans. Any problems with the RW programmes must be identified and dealt with by the licence reviewers. This is true for both selection of method and siting processes.[16] It is all done by the industry, which then has to try and get the necessary permit. The present Fud programme was started in 2013 and the government will make a decision regarding the programme in late 2014.

The Swedish concept for governance of RW is dependent on a strong regulator. In the middle of the 2000s, the government and parliament decided to reform the regulatory system by joining the two regulators, SKI and SSI. The new regulator, the "Swedish Radiation Safety Authority" (SSM), was created in 2008. SKI and SSI had very different cultures. During the 1980s and 1990s the main regulator SKI, which analysed the safety of the barrier systems and reviews of the RW department, was "captured" by the industry; they did little to challenge the industry's work. The secondary regulator SSI was more critical on many issues but did not have direct access to the government in the Fud process.[17] SKI could filter criticism from SSI in the reports reaching the government. Political interest in RW issues during the last two decades of the last century was low. Successive governments, therefore, were happy to agree with the regulator that everything was acceptable. As a result, nuclear industries around the world hailed the Swedish model for radioactive waste governance.

The new regulator SSM put immediately in 2008 new effort into analysing the financial system. A separate department for financial control was created and

15 SKB issued R&D reports every three years, from 1986 onwards, describing a programme for management and final disposal of all RW as well as for the decommissioning of all reactors and other nuclear facilities. The reports got the name "Fud report" – the D stands for demonstration – and the latest report is the Fud-13. The reports are available in English at the SKB web site http://www.skb.se.

16 One of the authors described how the economic incentives made all involved parties interested in hiding, rather than discovering, problems in the early phase of the programme. This resulted in significant cost escalations later, when the industry was no longer able to provide the funds required (Kåberger 1992).

17 SSI had a responsibility for long-term safety with regards to the release of radioactive elements. It took this issue very seriously and has had a more open culture. SSI also followed the development of the environmental legislation closely and recognized the importance of investigating siting and choice of method.

it was soon discovered that the cost estimates that SKB had been giving over the years were not correct. The financial collapse of 2008 caused large problems for the nuclear waste fund. It became clear that there was a big risk that the fund would be insufficient to cover costs.

A problem with the Swedish system of governance has been that crucial problems have been swept under the carpet. Not until an application for a licence for a spent fuel repository has come under review has the regulator had the power to force the industry to explain issues that may be problematic. This opportunity for questioning is much too late for the system to be effective.

The governance system might have been able to continue to muddle through had the regulatory system not been reorganized in 2008. But the new regulator, SSM, has taken many critical issues raised historically by the previous Swedish Radiation Protection Institute (SSI) under its wing and has tried to build a culture of integrity regarding interactions with industry. The licence applications are now reviewed with a greater focus on problems that might previously have remained unaddressed. This includes implications for applications that are being reviewed by SSM at a level that SKB may not have anticipated when preparing the applications.

5.1 The Environmental Code and licencing procedures

Sweden adopted a separate Environmental Code in the late 1990s, which was a step in the wrong direction for the nuclear industry (SFS 1998). Nuclear activities have to undergo review also under this act, and applications for licences are reviewed both by the nuclear regulator and the Environmental Court. The Nuclear Activities Act and the Radiation Protection Act have been changed and have the same or similar effect regarding the general principles that are in the Environmental Act, including the use of the precautionary principle, the use of the best available technology, and the need to show that the best method and site have been chosen.

The importance of the Environmental Code with regard to other laws should not be underestimated. There is on-going work to integrate the Nuclear Activities Act and the Radiation Protection Act into one chapter in the Environmental Code. Presently, licences for nuclear waste repositories are reviewed by both the regulator and the court according to the different acts, albeit with careful coordination. In the future it will be the main responsibility of the Environmental Court to manage the review.

The current review of the licence application to construct a SNF repository in Forsmark shows how the present legislation with a parallel review by SSM and the Environmental Court has made things difficult for SKB. The company has for decades avoided working on alternative methods for siting, as well as any issues that could be problematic for the industry's methods. The regulator and other actors in the review process have not had an opportunity to address all of these issues.

Comments have been made by a number of parties, including the nearby communities, the Swedish Council for Nuclear Waste, and the environmental movement. The regulator also reports to the court. There are a number of issues where more facts and analysis are requested.[18] If and when the application is found to be complete, it will be officially announced and then reviewed on its merits. Finally, the regulator and the court will give their recommendations regarding the application to the government, which has the final say.

It is not certain, however, if the government will get the chance to decide on the licence application. The nuclear waste company SKB is very reluctant to do more work to address the issues that have been raised. The licence application was submitted in March 2011 and SKB announced in September 2014 that the company would supply more information in January 2015. Still, the Environmental Court would then have to decide on the additional work which is requested by the different parties for the review. If such a work would show that the safety case would not hold together, the review process may have to be halted.

5.2 Transparency and public participation

In 1980, the referendum on nuclear power and the political decision to put a final date on the use of the Swedish nuclear reactors led to a sudden loss of political interest in nuclear power, but, the referendum opened up a broader discussion on nuclear waste issues. With a final date set for the production of nuclear waste, it became possible to know what waste amounts would be produced in the Swedish nuclear programme. There was also an understanding of what facilities would be needed. This allowed a financial system to be developed to pay for the disposal of waste and the decommissioning of the nuclear reactors.

The fact that a private company created by the nuclear industry is in charge of nuclear waste issues could be seen as an additional problem for the transparency of radioactive waste governance. There is a freedom of information

18 The issues include: a) problems with the barrier systems of copper and clay, especially in the dry rock at Forsmark; b) issues with the siting in Forsmark; c) safeguards and the need for long-term surveillance; and d) a lack of investigation of alternatives, including using deep boreholes for the final deposition of SNF. See also Swahn (1992) and Swahn (1995).

system in Sweden when it comes to authorities and the government, but a private company, like SKB, has no obligation to show material other than what it wants to become public. This makes it difficult, even for the regulator, to get information about problems that may have already been documented by the company.

Transparency and public participation are, however, important in order to increase the safety of nuclear facilities (IAEA 2006). Historically, transparency and public participation in Sweden has developed well. Sweden has also implemented the Aarhus Convention in its environmental legislation (UN 1998). The Environmental Act and the Nuclear Activities Act both include the need for formal consultations with a broad range of stakeholders before the submission of a licence application. The aim of the consultation process is to bring onboard issues that have to be dealt with in the environmental impact statement, which is a part of the application. From 2003 onward, SKB has carried out a consultation process according to Swedish nuclear and environmental legislation in preparation for a licence application for a repository for SNF. The legal aim of this process was to get input that could improve the environmental impact report in the company's application. SKB held a large number of meetings until spring 2010. Sweden has already implemented Art.10 on Transparency and Public Participation of the 2011 Euratom Radioactive Waste Directive, and thus the regulator, who is responsible for developing the Swedish national programme on RW, will have to consult with stakeholders.

In the field of Swedish RW management there is even support for environmental organisations to take part in the consultation processes. Since 2005, some environmental organisations have received support from the nuclear waste fund. Three organisations now receive a total of €0.39m per year (§§ 32-33 in SFS 2008).

6 The financing system

In accordance with the polluter pays principle, the Swedish nuclear industry is responsible for the costs of management and for the final disposal of radioactive waste. Before the referendum on nuclear power in 1980, the costs of waste management were the responsibility of the industry and were supposedly handled by internal accounts with sufficient money. When Sweden was revising its nuclear legislation at the beginning of the 1980s, a piece of financial legislation was introduced. By the mid-1980s, a financial system for the management and disposal of radioactive wastes and the decommissioning of the reactors was also established. A nuclear waste fund, controlled by the government, was in place and the nuclear industry was paying a fee per kWh of nuclear electricity produced into the fund. A special legislation forced the industry to also pay a smaller fee for the management and final disposal of the historic wastes from the military-civil past.

It has been observed that these economic conditions make it unlikely that environmental risks will be discovered. The implementer will have to pay for measures to reduce the risks that are discovered, but not the damages of risks that are realized sometime in the future (Kåberger 1993).

In 1993, the Nuclear Power Inspectorate published a report from projects where stakeholders, including environmental citizen groups and researchers, had discussed scenarios for the future waste management process. In this report, it was suggested that the risk of a lack of funds in the coming decades might also result in environmental risks more significant than those of the current programmes (SKI 1993).

The principles behind the financial system are the same today, but the Nuclear Waste Financial Act and Ordinance was also revised in 2006 (SFS 2006, SFS 2008). The Nuclear Waste Financial Act and Ordinance requires the reactor operators to provide estimates of the cost for RW management and final disposal, as well as for decommissioning the nuclear reactors. The regulator SSM then reviews the estimates and recommends a fee per kWh to the government for the nuclear electricity that is produced. Second, the level of additional financial securities is recommended. The government then decides on the fee, which is paid into a nuclear waste fund, and on the financial securities required. The fund's present investment policies allow foremost placement in Swedish government bonds. The financial resources in the nuclear waste fund are controlled by the government. Historically, the fee and securities were revised and changed every year, but it is now a three-year process coordinated with the review of the industry's R&D programme and governmental decision on the programme. The regulator is now reviewing the fees and securities for the next three-year period, from 2015-2017. The decision about the new fees and securities will be made by the government in late autumn 2014.

Historically, the fee was close to 0.1€ cent. The nuclear waste fee was raised by the government for the period 2012 to 2014 to only €0.22 per kWh, despite the regulator's recommended fee of €0.33 (Swedish Ministry of the Environment 2011). For the year 2015, the regulator has recommended a fee of 0.4 €-cent (SSM 2014).

Since the new regulator was created in 2008, the resources allotted for the regulation on financial issues were increased several times. This has allowed the regulator to examine the financial information from SKB more closely. Apparently, previous estimates for costs were too low, especially as at the end of the last century long-term interest rates became lower. The contribution from interest on accumulated funds now appears to be less than before. Taken together, the resulting analysis is that there may be €3 to €5bn lacking in the fund. In 2011, the government asked the regulator to investigate a way to reduce the risk the state will take if it has to pay for the industry's costs in the future. This was done in cooperation with the Swedish National Debt Office and the Swedish Nuclear Waste Fund Authority. The initial assessment of the fee paid by the nuclear power companies might have to be raised from the present 0.22€ cent to between 0.66€ cent and 1.1€ cent per kWh.

This strategy, however, would not necessarily work. The production costs of the NPPs in Sweden are nowadays high compared to electricity prices. Successful development of renewable electricity from wind and biomass has "out-competed" expensive electricity generation and lowered electricity prices, not only in Sweden but also in Denmark and Germany. In addition, a large new-build of wind power in Sweden is producing a surplus of electricity on the Nordic market. As a result, the average production costs of the last five years for some nuclear reactors are above the electricity prices in the futures market for the coming 10 years. Increasing the fee may thus contribute to the NPPs being closed immediately, and mean no income for the waste management fund.

In the final report of the review, published in spring 2013, the regulator tried to "solve" this problem through creative, new principles for the legislation. First, a longer lifetime for the reactors and for fee-paying was assumed. Second, the authority proposed that the fund should be allowed to invest in shares and other more risky financial instruments, as this could yield a higher rate of return. Third, the real rate of return should be expected to increase in the future, even though this is evidently not the case. The regulator´s solution to the industry's problems has been criticized in the consultation review of the proposals. Even the National Debt Office that took part in this process did not have a final say in the recommendations made by the report and has questioned the ideas presented by SSM to dissuade a fee increase. The office instead suggested a more serious consideration of the issues involved. The government now has to decide how to go forward before introducing new legislation.

The companies who own the nuclear reactors in Sweden are drained by low electricity prices and bad assets. The competitiveness of the reactors themselves may disappear if fees for waste management are raised enough to cover expected costs. In this situation, Sweden is not unique. However, because of the openness in the public administration it may be the first country where the collapse of the waste finance system becomes clearly visible.

7 Conclusions: Time for a new model?

It appears that much of the success story of the Swedish model is disintegrating and its financial system is in disarray. A final scrutiny of the filed application shows that the quality of the research on the KBS method, as well as the actual safety of the KBS method, appear to be less impressive than its international reputation has postured for the last 30 years. A licence for a repository for SNF in Forsmark according to the submitted application cannot be taken for granted. The safety case for the existing repository for short-lived waste in Forsmark can be questioned. It may be time for a new, or at least revised, model for RW management in Sweden. Such a model could include:

- Explicit opportunities for the regulator and the government to force the industry to carry out, or use the wage management fund to finance, research required to support the licence application reviews.
- Allow the use of reserves from the nuclear waste fund for R&D independent of the nuclear industry, in order to allow better critical evaluation of the industry's work.
- Force the industry to organise research on RW management so that it is transparent and under the auspices of freedom of information, similar to that of the Swedish authorities.

Finally, the financial system should be revised in order to reduce the risk of future generations having to pay for the costs that today's nuclear operators are supposed to pay.

References

Agrell, W. (2002). Svenska förintelsevapen. Historiska Media, Lund.

Åhäll, K.-I., Linström, M., Holmstrand, O., Helander, B., and Goldstick, M. (1988). Nuclear Waste in Sweden – The Problem is Not Solved! The Peoples' Movement Against Nuclear Power and Weapons. http://www.folkkampanjen.se/nwfront .html#Contents.

AKA Commission (1976). Spent Fuel and Radioactive Waste, Government reports SOU 1976:30, SOU 1976:31 and SOU 1976:32.

Carson R. (1962). Silent Spring. Houghton Mifflin.

Hultquist, G., Szakálos, P., Graham, M. J., Belonoshko, A. B., Sproule, G. I., Graåsjö, L., Dorogokupets, P., Danilov, B., AAstrup, T., Wikmark, G., Chuah, G.-K., Eriksson, J.-C., and Rosengren, A. (2009). Water Corrodes Copper. *Catalysis Letters*, 132, 311-316.

Hultquist, G., Graham, M.J., Szakalos, P.G. , Sproule, I., Rosengren, A., and Gråsjö, L. (2011). Hydrogen gas production during corrosion of copper by water. *Corrosion Science*, 53(1), 310-319.

Hultquist, G., Graham, M. J., Kodra, O., Moisa, S., Liu, R., Bexell, U., and Smialek, J. L. (2013). Corrosion of copper in distilled water without molecular oxygen and the detection of produced hydrogen. Swedish Radiation Safety Authority, Stockholm.

International Atomic Energy Agency (IAEA) (2006). Stakeholder Involvement in Nuclear Issues. Report by the International Nuclear Safety Group. INSAG-20. International Atomic Energy Agency.

International Atomic Energy Agency (IAEA)/ Power Reactor Information System (PRIS) (2014). Sweden. http://www.iaea.org/pris/CountryStatistics/CountryDetails.aspx?current=SE, last accessed 6 October 2014

Johansson, T. B. (1986). Sweden's Abortive Nuclear Weapons Project. *Bulletin of the Atomic Scientists*, 42(3), 31-34.

Jonter, T. (2001). Sweden and the Bomb: The Swedish Plans to Acquire Nuclear Weapons, 1945–1972, Swedish Nuclear Power Inspectorate, SKI Report 01:33.

Jonter, T. (2002). Nuclear Weapons Research in Sweden, The Co-operation between Civilian and Military Research, 1947-1972, Swedish Nuclear Power Inspectorate, SKI Report 02:18.

Kärnbränslesäkerhet (KBS) (1977). Handling of Spent Nuclear Fuel and Final Storage of Vitrified High Level Reprocessing Waste (KBS-1). Kärnbränslesäkerhet. Stockholm. Available at http://www.skb.se.

KBS (1978). Handling and Final Storage of Unreprocessed Spent Nuclear Fuel (KBS-2). Kärnbränslesäkerhet. Available at http://www.skb.se.

Kåberger, T. (1993). On the Economic Conditions of Suitability Assessments. Presentation at Endlager Hearing, Braunschweig September 21-23. Niedersächsisches Umweltministerium.

Kåberger, T. (2002). Swedish Nuclear Power and Economic Rationalities. *Energy and Environment*, 13(2), 185-200.

Meadows, D. H., Goldsmith, E. I., and Meadow, P. (1972). *The limits to growth* (Vol. 381). London: Earth Island Limited.

Svensk Författningssamling (SFS) (1977). Lag om särskilt tillstånd att tillföra kärnreaktor kärnbränsle, SFS 1977: 140.

SFS (1984a). Lag om kärnteknisk verksamhet. SFS 1983:4 (Act on Nuclear Activities).

SFS (1984b). Förordning om kärnteknisk verksamhet. SFS 1983:14 (Ordinance on Nuclear Activities).

SFS (1988a). Strålskyddslag. SFS 1988:220 (Radiation Protection Act).

SFS (1988b). Strålskyddsförordning. SFS 1988:293 (Radiation Protection Ordinance).

SFS (1998). Miljöbalk. SFS 1998:808.

SFS (2006). Lag om finansiella åtgärder för hanteringen av restprodukter från kärnteknisk verksamhet. SFS 2006:647 (Financial Act).

SFS (2008). Förordning om finansiella åtgärder för hanteringen av restprodukter från kärnteknisk verksamhet. SFS 2008:715 (Financial Ordinance).

Svenks Kärnbränslehantering (SKB) (1992). SKB 91 – Slutlig förvaring av använt kärnbränsle. Berggrundens betydelse för säkerheten. Svensk Kärnbränslehantering AB.

SKB (1999). SKB TR 99-06 Main Report Summary – Deep repository for spent nuclear fuel SR 97 – Post-closure safety. Svensk Kärnbränslehantering AB.

SKB (2013). SKB Plan 2013, Kostnader från och med år 2015 för kärnkraftens radioaktiva restprodukter, Underlag för avgifter och säkerheter åren 2015–2017. Svensk Kärnbränslehantering AB. Available at http://www.skb.se.

SKBF/KBS (1983). Final storage of spent nuclear fuel: KBS-3. Swedish Nuclear Fuel Supply Co. Available at http://www.skb.se.

Strålsäkerhetsmyndigheten (SSM) (2014). Yttrande avseende kärnavfallsavgifter samt finansierings- och kompletteringsbelopp för 2015-2017 enligt lagen (2006:647) om finansiella åtgärder för hanteringen av restprodukter från kärnteknisk verksamhet, Swedish Radiation Safety Authority, October 13, 2014.

Sundqvist, G. (2002). *The Bedrock of Opinion: Science, Technology and Society in the Siting of High-Level Nuclear Waste*. Dordrecht: Kluwer Academic Publishers.

Swahn, J. (1992). *The Long Term Nuclear Explosives Predicament*. PhD Thesis, Institute of Physical Resource Theory, Chalmers University of Technology. http://tinyurl.com/nvomlwp

Swahn J. (1995). Retrievability and Safeguards Concerns Regarding Plutonium in Geological Repositories. In Merz, E. (Ed.), *Disposal of Weapons Plutonium: Approaches and Perspectives*. Kluwer. 9-22.

Swahn, J. (2011). Sweden/Finland. In Feiveson H., Mian, Z., Ramana, M.V., and von Hippel, F. (IPFM). Managing Spent Fuel from Nuclear Power Reactors Experience and Lessons from Around the World. International Panel on Fissile Materials. Princeton.78-91.

Swedish Ministry of the Environment (2011). Kärnavfallsavgifter samt finansierings- och kompletteringsbelopp för 2012-2014 enligt lagen (2006:647) om finansiella åtgärder för hanteringen av restprodukter från kärnteknisk verksamhet. Decision, December 22, 2011.

Swedish Ministry of the Environment (2014). Sweden's Fifth National Report Under the Joint Convention on the Safety of Spent Fuel Management and the Safety of Radioactive Waste Management, Ministry of the Environment, Ds 2014-32.

Szakálos, P., Hultquist, G., and Wikmark, G. (2007). Corrosion of copper by water. *Electrochemical and Solid-State Letters*. 10, C63-C67.

Statens kamkraftinspektion (SKI) (1993). The Dialogue Project, Report from the Actors Group, Swedish Nuclear Power Inspectorate SKI Report 93:41.

United Nations (UN) (1998). Convention on Access to Information, Public Participation in Decision-Making and Access to Justice in Environmental Matters. The United Nations Economic Commission for Europe (UNECE).

A Final Solution for a Big Challenge
The Governance of Nuclear Waste Disposal in Finland[1]

Burkhard Auffermann, Pertti Suomela, Jari Kaivo-oja, Jarmo Vehmas and Jyrki Luukkanen

1 Introduction

Finland is one of the first countries to try to solve one of the most important and pressing problems facing mankind: the secure final disposal of high-level radioactive waste. Fortum Oyj and Teollisuuden Voima Oyj (TVO), the two companies that own the existing nuclear power plants in Finland, formed in 1996 a joint company, Posiva Oy, to deal with their nuclear waste. The site and the project to create the final disposal solution is called Onkalo.[2] The construction of the site at Olkiluoto in South-Western Finland began in 2004; the operating company Posiva is aiming to start the disposal of high level waste in 2020.

This article presents the Finnish policy on the disposal of high-level radioactive waste and analyses specific Finnish conditions, problems and solutions in the present institutional and political contexts. Four contentious issues and points of debate in Finland regarding nuclear power are: 1) the severe delay in the scheduled completion of the Olkiluoto-3 project, 2) the validity of Fennovoim's decision to build a new nuclear power plant, 3) the amount of influence the Russian energy company, Rosatom, has in Fennovoima, and 4) the practically limited liability of Finland's nuclear power companies. The final disposal solution, Onkalo, is proceeding with very little public debate. The directive 2011/70/EURATOM is not an issue in Finland because the Finnish legislation was already well aligned with the directive and needed only minor amendments. There was already a reasonably solid plan for nuclear waste management and the nuclear power companies' operations are compliant with the legislation.

---

1 All views presented in this article are personal opinions of the authors and do not reflect the official view of the University of Turku or the Finland Futures Research Centre.

2 The word "onkalo" means in Finnish hollow, cave or cavity. There is another word in Finnish with nearly the same meaning: "luola". The notation of onkalo means a hole in the earth that is more hidden, "safer" than a normal cave. Choosing this word for the final disposal site for highly radioactive waste was a very smart idea, as it suggests safety.

2 The institutional and legal framework

The main actors in the field of nuclear power are Fortum and TVO, the owners of the nuclear power plants in Loviisa and Olkiluoto, respectively. These two companies formed in 1996 a joint company, Posiva Oy, to carry out the requirements of the amended Nuclear Energy Act from 1994 and to deal with the nuclear waste their nuclear power plants produce.

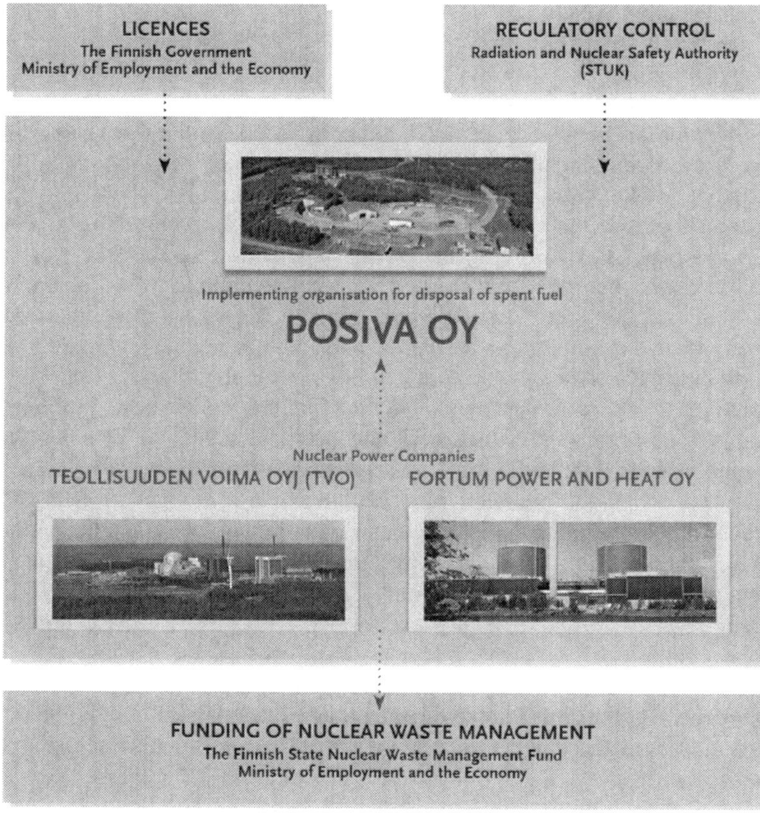

Figure 1: Organisation of nuclear waste management in Finland.

Source: Posiva (2011).

Political guidance in matters of nuclear energy in Finland is carried out by the Ministry of Employment and the Economy (until 2008, the Ministry of Trade and Industry), which is responsible for preparing and monitoring national legislation and negotiating international agreements.[3] The Radiation and Nuclear Safety Authority (Säteilyturvakeskus (STUK)) is in charge of the management of nuclear and radiation safety (Nuclear Energy Act 2008).

The European Council issued the Directive 2011/70/EURATOM on 19 July 2011 establishing a community framework for the responsible and safe management of spent fuel and radioactive waste. To implement this directive, the Finnish government prepared a proposal to amend the Nuclear Energy Act. As the legislation was already well aligned with the directive it needed only minor and non-controversial amendments. A proposal was submitted on 10 October 2012 to the industry for comments. The government received feedback from Fortum, Fennovoima, Posiva, some smaller companies, STUK and the relevant ministries. TVO did not respond. Most of the responses agreed with the amendments, suggesting only minor "rewording". Consequently the government issued on 28 June 2013 the act to amend the Nuclear Energy Act. There were three subsections added to the Nuclear Energy Act, with the following unofficial translation and summary: a) safety requirements and measures shall be planned according to the risks, b) the amount of nuclear waste shall be minimized regarding both the activity and the volume, and c) the Ministry of Employment and Economy will organize regularly an internal and an international review of the national framework of nuclear safety including nuclear waste management. The Radiation and Nuclear Safety Authority will organize an internal review of its own operations

The Finnish Nuclear Energy Act, the Nuclear Energy Decree and the Nuclear Liability Act are together a reasonably clear legal framework for nuclear waste management in Finland (Nuclear Energy Act 2008; Nuclear Energy Decree 2008; Ydinvastuulaki 2008). The producers of nuclear waste are responsible for all measures needed for its safe management and costs entailed. State authorities monitor the waste management and issue regulations when needed. Since the amendment of the Nuclear Energy Act in 1994 (put into effect in 1996), all radioactive waste produced in Finland must be handled and disposed of in the country itself. The Nuclear Liability Act states that nuclear liability lies with the operator of the nuclear installation which includes the final repositories of the spent nuclear fuel as clarified in the government bill HE 2/2005 (Finnish government 2005).

According to the Nuclear Energy Act, the producers of nuclear waste are responsible for the management and all costs of nuclear waste management. The

3 The sphere of authority of the Ministry of Employment and the Economy includes industrial policy; energy policy and integration of the national preparation and implementation of climate policy; innovation and technology policy, internationalization of enterprises and technical safety. See http://www.tem.fi/en/ministry.

companies have to present annual cost estimates for the future management of their nuclear waste to the Ministry of Employment and the Economy. The estimates should be based on the latest knowledge of the power companies, and have to include the costs estimated for the decommissioning of the nuclear power plants. In 1988, a fund was set up to collect, save and reliably invest the money needed to finance nuclear waste management in the future. The fund is now being administered by the Ministry of Employment and the Economy; all decisions are made by a board.[4] Based on cost estimates and expert advice, the ministry decides annually on the contributions which should be made by the energy industry to the fund. In 2013, the contributions from the operators of the nuclear plants were €5.7m and from the operators of waste management €1.7m (Ministry of Finance 2013). In January 2014, the fund included €2.1bn – not included in or taken from the annual state budget (Ministry of Finance 2013). In this way the Finnish government provides a financial guarantee for the handling of nuclear waste "under all circumstances" (Nuclear Energy Act 2008). Since the ministry bases its decisions on the quantity of the payments largely on the data provided by "the parties liable for nuclear waste management" (referring to the energy industry) (Nuclear Energy Act 2008), one can say that the industry decides to a great extent for itself how much the society should set aside for nuclear waste management. When a nuclear power plant has been decommissioned and the spent nuclear fuel has been disposed as required, the ownership of the disposed spent nuclear fuel and the responsibility for its management is transferred to the state of Finland. Posiva has done a lot of research on radionuclide release with regards to the repository systembut solid conclusions are difficult to draw as far as the probability of an incident in the final repository and the damage it could cause.[5] Posiva's research reports have been interpreted in very positive (Korhonen 2013) and slightly sceptical (Greenpeace 2011) tones.

The Nuclear Waste Fund lends 75% of the money back to the nuclear power companies and 25% to the government. So in practice, the nuclear power companies collect money that they then borrow and pay interest equal to the 12 month Euro prime level.

The Minister of Environment, Ville Niinistö, has criticized a number of issues concerning nuclear liability (Niinistö 2012). First, even though the liability is unlimited according to the Nuclear Liability Act, that law has not yet

---

4 The fund is managed by its board and a director appointed by the government. The board makes all of the most important decisions concerning the operative functions of the fund. The members of the board represent the Hanken School of Economics, the Ministry of Finance, the Ministry of Employment and Economy and the State Treasury. See http://www.tem.fi/?l=en&s=1550.

5 http://www.posiva.fi/en/databank/posiva_reports.

been ratified by the government and when it is ratified, the nuclear energy companies need to place a deposit of only €700m and the "unlimited" nuclear liability is practically limited to the assets of the nuclear energy companies. The assets of the Finnish nuclear energy companies are a few percent of the estimated cost of the Fukushima disaster – up to US$623bn (Morton 2012). Possible solutions to this dilemma include extending the liability to the shareholders of the nuclear energy companies or pooling the liabilities across the industry or countries. It seems unlikely that an insurance company can provide a policy that would cover the full cost of a large nuclear disaster, and even if this was available, it would be so expensive that it would render the nuclear energy business unprofitable. The country of Finland provides the majority of this insurance implicitly and free of charge to the nuclear energy companies. Additionally, the Finnish Nuclear Liability Act excludes nuclear liability in the case of an "unusual natural catastrophe", leaving the interpretation of such a disaster quite open. Fortunately, Finland is relatively safe from most serious natural catastrophes like earthquakes and tsunamis.

A potentially grey area in Finland´s nuclear legislation relates to the responsibilities and arrangements pertaining to the final disposal solution. The Nuclear Waste Fund is defined only in the Nuclear Energy Act. The Nuclear Liability Act does not explicitly say that its jurisdiction covers the decommissioning and final disposal phase. This issue should be clarified in the Nuclear Liability Act.

3 Nuclear energy production

The first nuclear power reactor in Finland, Loviisa-1, started its commercial operation in 1977 on the island of Hästholmen near the city of Loviisa. The state-owned Imatran Voima Oy (IVO) chose the Soviet reactor manufacturer V/O Technopromexport for political reasons. Another reactor, Loviisa-2, situated at the same site started operating in 1981. The reactors were upgraded in 1998 from 440 MWe (net electric output) to 488 MWe each. The power plant is owned by Fortum Oyj (former IVO).

Another nuclear power plant, owned by TVO, was built in Olkiluoto. The first reactor, Olkiluoto-1, started operating in 1979 and the second, Olkiluoto-2, in 1982. Olkiluoto is a small island in the Baltic Sea just off the Finnish coast and belongs to Eurajoki. The reactors were both upgraded in several phases between 1984-2006 from 660 MWe to 880 MWe.

There have also been two unsuccessful applications for additional nuclear power. The first attempt was in March 1986 when Perusvoima Ltd, a joint company of

Imatran Voima (IVO; currently Fortum) and Teollisuuden Voima (TVO) submitted an application to the government. The application, however, was frozen after Chernobyl and the amendment to the Nuclear Energy Act. The second attempt was a joint application by IVO and TVO in 1990. The project was supported by the government but rejected by parliament in September 1993.

In 2002, the Finnish government issued the decision-in-principle to TVO authorizing them to construct a new nuclear reactor, a third unit in Olkiluoto (Olkiluoto-3). This reactor is supplied by a consortium formed by AREVA NP and Siemens AG. The consortium is lead by Areva. It is the first European Pressurized Reactor (EPR) and is to have a net electric output of 1600 MWe, the highest in Europe. The construction of Olkiluoto-3 started in 2005 and was to have been ready for energy production in 2009. However, the project has suffered many delays and disputes between TVO and Areva. TVO intends to start production of electricity in 2016 but Areva has stopped trying to set a date for the completion of the project.[6] In 2010, a comprehensive economic analysis of Olkiluoto-3 was reported in a refereed journal (see Karppinen et al. 2010). It was based on updated regional input-output analyses and revealed considerable employment problems resulting from the projects in the Satakunta region. Many uncertainties and risks concerning future income were also reported.

A new nuclear energy company, Fennovoima Oy, was set up in 2007. In July 2010 the Finnish parliament granted Fennovoima Oy a license (called a "decision-in-principle") to build a new nuclear power plant (Ministry of Employment and the Economy (MEE) 2010a). As per the request of the Ministry of Employment and the Economy, Simo and Pyhäjoki stated in 2009 that they were willing to host Fennovoima's plant. The national nuclear regulator (STUK) found two proposed greenfield sites suitable for a nuclear power plant. In October 2011, Fennovoima announced in a press release that the nuclear power plant would be located on the Hanhikivi peninsula in the municipality of Pyhäjoki on the coast of Bothnian Bay (Fennovoima 2011). Thus, the nuclear power plant was given the name, Hanhikivi-1. The German energy company E.ON was one of the biggest shareholders of Fennovoima with 34% of the shares, but in October 2012 they decided to divest from all their business interests in Finland, including Fennovoima. As a result, Fennovoima had to find a new major investor. They accepted a bid from the Russian energy company

6 This is not the place to discuss the reasons for the problems related to the construction of the Olkiluoto-3. The building of the ERP reactor started by AREVA (France) and Siemens (Germany) is in delay due to quality and safety problems. Additionally the cost exploded from the original turnkey contract of €3.2bn to more than €8bn, as estimated by AREVA's chief executive in December 2012. Political opposition against the reactor is not the reason for the delay.

Rosatom to take a 34% share of Fennovoima. Additionally, Fennovoima and Rosatom signed a contract stating that Rosatom would construct the nuclear power plant with their technology (Fennovoima 2013). Fennovoima submitted an application in March 2014 to revise the decision-in-principle so that it could cover the changes in the ownership of the company and in the builder of the nuclear power plant.

Since the signing of this contract between Fennovoima and Rosatom, the Finnish government and parliament have been arguing whether the granted decision-in-principle with a revision is still valid or if Fennovoima should apply for a new decision-in-principle because of the changes. If the government and parliament decide that the granted decision-in-principle is valid, Fennovoima can carry on with the nuclear power plant project as planned. If the government and parliament decide that the decision-in-principle is no longer valid and a new one is needed, then there could be long delays. When setting up the government, the parties decided that new decision-in-principles for nuclear power plants would not be granted by this government. If that decision holds, the processing of the new decision-in-principle will be transferred to the next government and the progress would then be uncertain. Another potential issue with Rosatom to be taken into account is the law which governs foreign ownership in companies that have very important national interest (Laki ulkomaalaisten yritysostojen seurannasta 2012). If it is interpreted that a company owning and operating a nuclear power plant is of very important national interest, the foreign ownership in that company could be limited significantly, causing a big problem for Fennovoima.

Figure 2: Schedule of the Finnish nuclear power plants

1977: Operation of Loviisa-1 started with 440 MWe
1979: Operation of Olkiluoto-1 started with 660 MWe
1981: Operation of Loviisa-2 started with 440 MWe
1982: Operation of Olkiluoto-2 started with 660 MWe
1984-2006: Olkiluoto-1 and -2 were upgraded in several phases to 880 MWe each
1998: Loviisa-1 and -2 were upgraded to 488 MWe each
2002: The Finnish government gave the decision-in-principle to TVO to construct a new 1600 MWe nuclear reactor (Olkiluoto-3).
2010: The Finnish government gave the decision-in-principle to Fennovoima to construct a new nuclear reactor (Hanhikivi-1) and to TVO to construct a new nuclear reactor (Olkiluoto-4).
2016: Start of energy production in Olkiluoto-3 (estimate by TVO)

The nuclear reactors produce all together about 25% of the electricity used in Finland (on the nuclear history of Finland, see Vehmas 2009 and, Kojo and Litmanen 2009). In 2012, the 4 units generated 22 TWh of electricity, 32.6% of the total domestic Finnish power production in that year.

4 The evolution of nuclear waste disposal in Finland

European countries have chosen different strategies to deal with nuclear waste management: some countries like France and the United Kingdom dispose of their waste after reprocessing while others like Sweden and Finland have opted for direct disposal solutions. The Finnish Nuclear Energy Act states that nuclear waste generated in Finland shall be handled, stored, and permanently disposed of in the country.[7] This principle concerns only Finnish waste; the handling, storage, or permanent disposal of spent nuclear fuel produced anywhere else is excluded. This means that the import of nuclear waste from other countries is forbidden as was agreed in the membership negotiations when Finland joined the EU in 1995. This concerns other EU Member States, too. The responsible ministry has issued a long-term timetable for disposal.

In Finland, high level nuclear waste is currently being produced by the four reactors of the Loviisa and Olkiluoto nuclear power plants and small amounts of nuclear waste is produced by the Otaniemi research reactor (FiR 1) operated by the Technical Research Centre of Finland (VTT) in Espoo, near Helsinki. No major decommissioning projects are on the way or planned for the near future.
Between 1981 and 1996, the spent nuclear fuel originating from the reactors in Loviisa was brought to the Soviet Union (later Russia) for reprocessing and storage. This "waste management" was based on an agreement between Finland and the Soviet Union signed in 1969 on the use of nuclear energy in peacetime, based on the principle, that the Soviet Union takes back the spent nuclear fuel arising from the Soviet made reactors. IVO's transports to Russia ended in 1996, when the export ban for nuclear waste (part of the Nuclear Energy Act amendment in 1994) was enforced. Since then, spent nuclear fuel has been stored temporarily in pools in the area of the nuclear power plants in Loviisa. The repository at Loviisa was built between 1993 and 1997 in Rapakivi granite and is used for low and intermediate level waste resulting from the two nuclear reactor units. Excavation has been to a depth of 120 meters with a gross capacity of 113,000 m³. An expansion is planned which will be large enough to store the

7 Only exception is the waste produced at the small research reactor.

waste that will be generated from the decommissioning of the plant. The repository is operated by Fortum, the owner of the nuclear power plant.

The situation of the high level waste produced in Olkiluoto is more complex. The agreement with the Soviet Union only covered the waste that resulted from the Soviet made reactor units. Since the Olkiluoto-1 and Olkiluoto-2 reactors came from Sweden, the owner company TVO had to decide for itself how to manage the nuclear waste produced. The TVO decided on interim storage in pools at the Olkiluoto power station.[8] The repository at Olkiluoto for low- and intermediate-level waste was built between 1988 and 1992 with two excavations of 60 to 100 meters depth. The net capacity covers about 9,500 m^3 of waste to be embedded in blocs of concrete with a total capacity of about 60,000 m^3. This repository can be enlarged to store more waste from any new additional units.

As explained in more detail in the table below, Fortum and TVO have been screening and studying the potential sites for the final disposal of spent nuclear fuel for over three decades. The company Posiva Oy was founded in 1995 to manage the final disposal of spent nuclear fuel generated from the reactors of TVO and Fortum. Posiva filed decision-in-principle applications to enlarge the Onkalo final repository to accommodate spent fuel from the proposed new reactors (Olkiluoto-4 and Loviisa-3). The Ministry of Employment and the Economy processed all five decision-in-principle applications during 2009-2010 and the government made its decisions in May 2010. All applications fulfilled all safety and environmental requirements. As specified in the Nuclear Energy Act, decisions on all decision-in-principles were based on the projects' overall good-for-society, projected national energy needs in 2020 and the then agreed to limit of two new nuclear power plants. The TVO's Olkiluoto-4 and Fennovoima's Hanhikivi-1 new building projects received positive decision-in-principles in 2010, as did Posiva for its repository enlargement project to store spent fuel from Olkiluoto-4. Fortum applied for a new reactor Loviisa-3 but did not get it, and Posiva's application to expand Onkalo to accommodate spent fuel from Loviisa-3 was denied.

In December 2012 the final disposal of high-level radioactive waste reached a breakthrough, when Posiva submitted an application to the Finnish government (the Ministry of Employment and the Economy) to get permission to build a repository and facilities at Olkiluoto, to deal with encapsulation and disposal of spent nuclear fuel. The disposal repository Onkalo will be designed and constructed in such a way that its safety will not need to be monitored after the repository is closed. The decision-in-principle requires that the final disposal canisters may be removed from the repository after its closure (STUK 2009).

8 See details on alternative strategies discussed over the years in Kojo (2009: 163f).

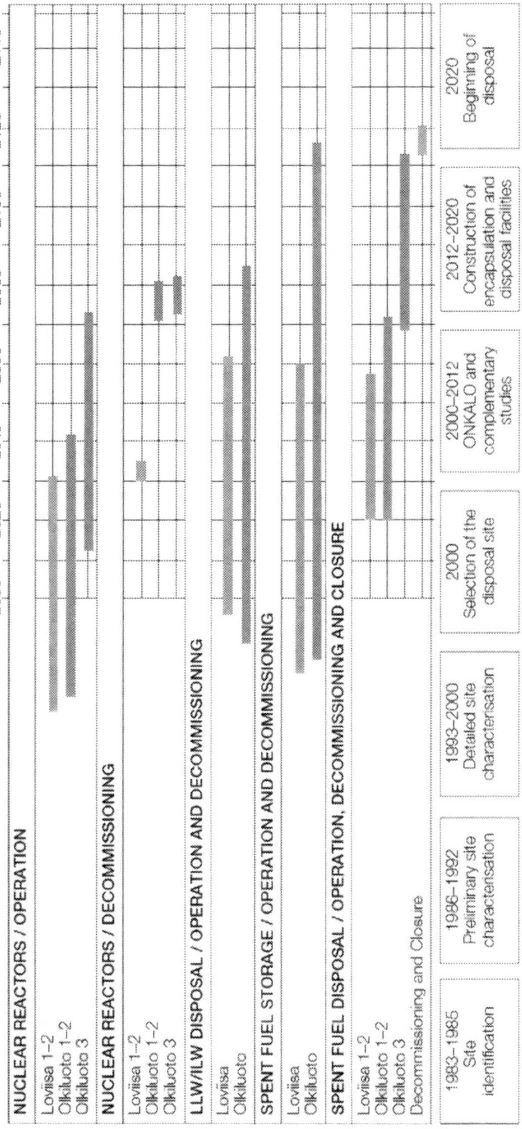

Figure 3 Timetable for the disposal of spent nuclear fuel from the nuclear power plants at Loviisa and Olkiluoto.
Source: MEE (2011).

Table 1: Schedule for Finnish spent nuclear fuel disposal as planned and realized

1978:	Interim storage of spent nuclear fuel starts in Loviisa.
1983:	Government sets schedule for nuclear waste management.
1983-1985:	Start of screening of potential sites for final disposal of spent nuclear fuel.
1986-1992:	Preliminary site investigations.
1987:	Interim storage of spent nuclear fuel starts in Olkiluoto.
1987:	Field research starts in 5 municipalities for the selection of a final repository
1988:	Construction of LLW and ILW repository starts in Olkiluoto.
1992:	Final disposal of LLW and ILW starts in Olkiluoto.
1992:	Detailed site characterization for final disposal of spent nuclear fuel starts in Eurajoki, Kuhmo, and Äanekoski.
1993-2000:	Detailed site investigations.
1993:	Construction of LLW and ILW repository starts in Loviisa.
1994:	Amendment to the Nuclear Energy Act: Ban of import and export of nuclear waste.
1995:	Posiva established to manage final disposal of spent nuclear fuel.
1997:	Loviisa selected as fourth potential site for a final repository.
1998:	Final disposal of LLW and ILW started in Loviisa.
2000:	Olkiluoto selected as site for final disposal of spent nuclear fuel/high level waste.
2000:	Government gives decision-in-principle to Posiva for disposal of spent nuclear fuel from Loviisa-1, Loviisa-2, Olkiluoto-1 and Olkiluoto-2 in Onkalo.
2002:	Government gives decision-in-principle to Posiva for extending Onkalo to cover disposal of spent nuclear fuel from Olkiluoto-3.
2003:	Eurajoki municipality issues a building permit for underground characterization facilit, Onkalo.
2004-2009:	Ground-level service and monitoring buildings under construction; excavation of access tunnel reaches depth of 420 m.
2009-2011:	Excavation of Onkalo reaches 520 m. Research on the mechanical characteristics of the bedrock, and adaptation of repository layout continue.
2010:	Government gives decision-in-principle to Posiva for extending Onkalo to cover disposal of spent nuclear fuel from Olkiluoto-4.
2012:	Posiva submits license application to the government for final disposal facility construction.
2015:	Construction of final disposal facility to start (PLANNED).
2020:	Final disposal to start (PLANNED).
2050-2060/2100:	Closing of the disposal site (PLANNED.)

When Fennovoima was granted the decision-in-principle for the nuclear power plant, Hanhikivi-1, they had to start planning a disposal solution. Although the final disposal project at Onkalo was already under construction and was planned only to be used by its owners, Fortum and TVO, the Ministry of Employment and the Economy appointed a working group in March 2012 to steer the nuclear power companies' joint investigation into alternatives for a final disposal of nuclear fuel. Their report was published on 10 January 2013. It compares construction alternatives and recommends utilizing the competence and experience accumulated in the field during Posiva Oy's project (MEE 2013a). There have been negotiations between Fennovoima and Posiva to decide how to organize the disposal of the high level waste but no agreement has been reached.

Quantaties of HLW are reported in metric tonnes of heavy metal (tHM). Quantaties of LLW and ILW are reported in cubic meters. The current quantities of different types of radioactive waste are listed in the table below.

Table 2: Radioactive waste in Finland at the end of 2013

Waste type	Facility	Quantity (Activity)
Spent fuel from NPPs	Loviisa NPP Olkiluoto NPP **Total**	560 tHM 1424 tHM **1984 tHM**
LILW from NPPs (excluding activated metal waste)	Loviisa NPP Olkiluoto NPP **Total**	3438 m^3 (17 TBq) 6118 m^3 (71 TBq) **9556 m^3 (88 TBq)**
Small user waste	Central storage	56 m^3 (50 TBq)

Source: Country Profile 2014 of Finland at the Radioactive Waste Management Programmes in OECD/NEA Member Countries.[9]

The yearly production of HLW from Finland's four reactors is about 70 tHM. The total amount of HLW produced during an operation time of 50 to 60 years is estimated at about 4,000 tHM (MEE 2011). The estimated quantity of HLW from Olkiluoto-3 and Olkiluoto-4 after 60 years of operation is 2,500 tHM per unit (MEE 2011). The decision-in-principle for Onkalo from 2010 takes into account the waste from the Olkiluoto-4 reactor, and allows the storage of 9,000 tonnes of spent nuclear fuel (MEE 2010b). In 2013, the ministry's working group reported that Onkalo's capacity will be increased to about 12,000 tonnes

9 See http://www.oecd-nea.org/rwm/profiles/index.html.

(MEE 2013b). This figure appears to include the estimated 2,500 tHM of HLW from Fennovoima's planned nuclear power plant Hanhikivi-1 generated during 60 years of operation (MEE 2013b). Thus, the 12,000 tHM is only the latest capacity plan for final disposal in Finland. The report states that it is not yet known if Onkalo can be expanded to that capacity because the quality of the bedrock can only be conclusively verified once the building of the repository is in progress. Posiva has not yet provided any estimate of the waste storage capacity of Onkalo in its current shape and size.

Table 3: Total quantity of HLW produced until decommissioning

Plant	Total quantity of HLW until decommissioning (tHM)
Loviisa-1 and -2, Olkiluoto-1 and -2	4000
Olkiluoto-3 and -4	5000
Hanhikivi-1	2500
Total	**11500**

Source: MEE (2011).

5 Nuclear waste disposal concept

The final disposal concept for the depository in Olkiluoto is based on a concept developed by Posiva Oy together with the Swedish company SKB (Svensk Kärnbränslehantering AB), called KBS-3. After cooling down the nuclear fuel assemblies for 20 to 30 years, they are transported to an encapsulation plant for spent nuclear fuels that is located above the repository. The waste is then encapsulated in boron steel canisters that are encapsulated in copper capsules. The canisters and capsules are then welded closed. Subsequently the canisters are transported to the final disposal tunnels at a depth of about 450-500 meters in the rock, where they are placed in a layer of bentonite clay in a hole that is eight meters deep and two meters wide. Finally the hole is closed. Bentonite absorbs water very efficiently. It isolates the canisters from water and protects them from any minor movements of the bedrock. After 100,000 years of storage, the radioactivity level of the waste is at the same level as that of uranium ore that is mined to make fuel (Posiva 2011).

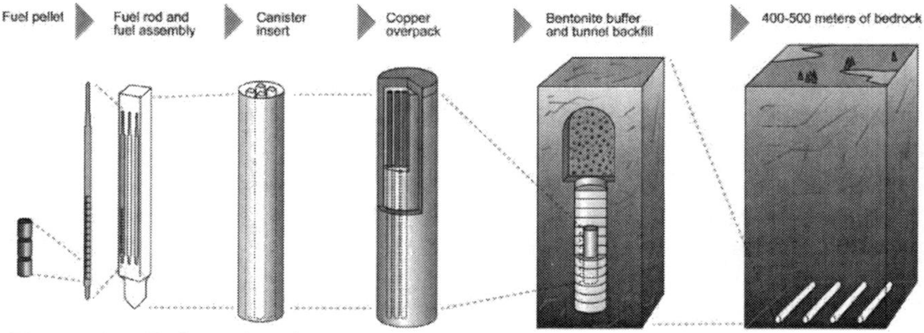

Figure 4: Release barrier system.
Source: Posiva (2014).

The spent nuclear fuel must be retrievable from the repository at any stage of the process, even after the final closure of the facility. The discussion about this principle did not last long and was very technical in Finland, unlike in France or the UK (Lehtonen 2010: 153-154). Issues about what to do with the possibly damaged waste packages and what would be the real costs of retrieving the waste packages were left unaddressed (Lehtonen 2010: 155).

The safety of the copper capsules has been recently questioned. A research group at the Royal Institute of Technology in Stockholm suggests that the copper capsules are not as safe and long-lasting as Posiva Oy and SKB argue. Instead of 100,000 years, the group claims, the capsules will only last about 1,000 years. Reacting to these questions, STUK asked POSIVA for further explanation, but POSIVA dismissed the results and referred to their own research (Granath 2012). At the time of the writing of this article in the summer of 2014, the conflict in Sweden over the quality of the copper capsules is rising again and might put the whole Swedish final disposal project in question (see Kåberger and Swahn in this volume). Such a development would also have major consequences for the Finnish Onkalo project.

6 Siting and participation of the public

According to Kojo (2009: 168ff), analysis of bedrock in search of suitable sites for the final spent nuclear fuel disposal have been on-going since the latter half of the 1970s. In the first part of the 1980s, geological conditions throughout Finland were analysed to verify their suitability as final disposal sites. A more detailed analysis was made during the 1990s at six locations. Geological suitability was presented as the main criteria, but also issues related to transport logistics and the political situation in the municipalities played a role. In 1999, Posiva Oy proposed the municipality of Eurajoki (Olkiluoto) as its first choice for a final disposal location, when it submitted an application for a decision-in-principle. All four areas that had been investigated as possible sites (Eurajoki (Olkiluoto), Loviisa, Kuhmo, and Äänekoski) were presented as suitable. They "show sufficiently large and sufficiently integrated rock capacities, where the conditions are chemically and mechanically sufficiently suitable and stable to provide a sufficient barrier to prevent the release of radioactive substances, and which are suitable for the construction of final disposal facilities" (Posiva 1999: 28, as quoted in Kojo 2009: 173).

The suitability of geological conditions was not the only important factor, however. Posiva argued that the four locations were all suitable but did not want to rank them. Most probably, a ranking would have led to a different choice than the one in favour of Eurajoki, a coastal location where there is a more saline quality to the groundwater than is the case in inland locations. Eurajoki and Loviisa, the two "nuclear municipalities", are located on islands in the Baltic Sea. Äänekoski and Kuhmo are located far away from the coast. Posiva used political arguments in favour of Olkiluoto: "In evaluating the suitability of the site, attention must be paid not only to the geological conditions related to long-term safety, but also to the aspects of the implementation of the final disposal. An essential factor regarding the implementation is also to gain local acceptance for the operation" (Posiva 1999: 28, as quoted in Kojo 2009: 173).

When preparing the environmental impact analysis programme for the final repository site, Posiva organized over 50 meetings during the autumn of 1997 with citizens at the prospective locations in Eurajoki, Kuhmo, Loviisa and Äänekoski (MEE 1999). Via the feedback forms which were distributed with the bulletins, Posiva received the opinions of about 700 people. In their responses to the environmental impact assessment reports and on other occasions, environmental NGOs, like The Finnish Association for Nature Conservation (Suomen Luonnonsuojeluliitto) and Greenpeace, questioned the safety and reliability of Posiva's research and plans, referring to research from other experts (e.g. Suomen luonnonsuojeluliitto 2009).

In 1999, Posiva published research stating that 59% of the residents of Eurajoki would approve of final disposal of spent nuclear fuel in their municipality (Posiva 1999). However, Litmanen, Kojo and Kari (2010) researched the acceptance of the spent nuclear fuel repository in Eurajoki and concluded: "The survey indicated that less than half (42%) of the residents of the municipality of Eurajoki are willing to accept the expansion of the repository for the needs of the 'older' nuclear operators, TVO and Fortum. The disposal needs of possible newcomers are less tolerated." It seems that local acceptance has changed significantly or it varies between research results. Another study from the same group of researchers indicates that active involvement and communication is also needed after a site has been selected to achieve local acceptance (Kojo et al.: 2010).

7 Experts, lobbying and consensus

Ruostetsaari (2010: 238) a long-term analyst of Finnish energy politics estimates that power structures have remained stable since the late 1980s. According to Ruostetsaari, the nucleus of the energy elite is made up of the Ministry of Employment and the Economy and the governmental agencies it controls. A special characteristic of Finnish nuclear energy politics is that nuclear energy companies have the responsibility for and largely also the power to determine the approach to the final disposal of spent nuclear fuel. That is seen as one of the reasons for the fast development of a final disposal solution in Finland (Ruostetsaari 2010: 239).

Another special characteristic of Finnish energy politics is that the energy elite is exclusive. The number of people within the group is limited. With the informal relationships built up among them during their long careers they are able to wield power and shape decision making (Ruostetsaari 2010: 242). Environmental NGOs are consulted more than before but their actual influence has not grown that quickly. Consulting NGOs can be seen as a way of winning legitimacy for their decisions.

Ruostetsaari (2010: 258) also finds that citizens trust energy experts but not politicians. Citizens do want alternative views from those they receive from the energy elite and for this they turn to environmental NGOs even though the actual influence of NGOs has remained very limited. In the early 2000s, an energy elite alliance formed by the Ministry of Trade and Industry, Posiva, STUK and VTT dominated discussions concerning the final disposal of spent nuclear fuel in Finland. It is noteworthy that no other final disposal alternatives than long-term placement in bedrock were researched. Alternative proposals were quickly dismissed by the elite experts (Litmanen 2009; Lehtonen 2010).

The very close cooperation between the experts and leaders in the ministries and energy industry could be regarded as a nuclear-industrial complex. They have supported progress in the nuclear industry at the expense of alternatives, which have been given less consideration than they might have been. Teräväinen, Lehtonen and Martiskainen (2011: 3441) coined the term "technology-and-industry-know-best" to describe Finland's orientation on nuclear questions.

Since the late 1960s, collective agreements about national income policies were made in a close tripartite co-operation among the trade unions, employers' organizations, and the government. Together they decided yearly about a wide range of economic and political issues, such as salaries, taxation, pensions, unemployment benefits, and housing costs (Pekkarinen and Vartiainen 1993; EIROnline 2013). During this time, a strong drive to reach broad consensus on central national questions among key players in society was born (Teräväinen et al. 2011: 3436). The same principle has been applied implicitly to the matter of nuclear waste management, because in addition to the energy elite described above, also other interest groups such as employer's organisations and trade unions have joined the consensus within this topic while most of the other players have remained silent. The opposition has not strongly repudiated the government's plans and NGOs have only protested mildly.

8 Conclusions

Governance of nuclear waste disposal in Finland is characterised by a low intensity of public discussion. The argument of politicians, officials and industry is basically: "we already have the appropriate solution" with final disposal in Posiva's Onkalo in Olkiluoto, Eurajoki. They see this as the best answer and argue research and development are progressing with no problems. The reality, however, is that despite the strong institutional consensus and the industry's plans, Onkalo is still only at a development and testing phase. It has not yet been approved. The key open questions are the safety of the nuclear waste capsules, the safety of the bedrock and the capacity of the repository. First, some researchers claim that the capsules could start leaking after 1000 years instead of the planned 100,000 years. This, of course, would be a serious problem and as a result, that risk is now being checked. Second, some researchers claim the quality of the bedrock is less safe than estimated. However, more and stronger research results backing up the concerns are needed before the siting can be reconsidered. The selection of Olkiluoto in Eurajoki as the site for final disposal was not quite transparent because the alternatives were not rated openly. The third potential issue is the storage capacity. The requirements for storage space keep growing. The latest decision-in-principle permits the capacity of 9,000 tHM in Onkalo. However, the projected quantity of spent fuel from the reactors that have been granted a decision-in-principle is already 11,500 tHM. It is not really known to which capacity Onkalo can be expanded, because the quality of the bedrock can be conclusively verified only once the building of the repository is in progress. Thus, it may be that Onkalo's expansion potential may be limited.

We will only learn as the construction progresses if and when another repository will be needed in addition to Onkalo.

Public discussion regarding nuclear matters in Finland revolves primarily around plans for new nuclear power plants and not about nuclear waste. There are two major projects for new nuclear reactors: TVO's Olkiluoto-3 and Fennovoima's Hanhikivi-1. Both have quite severe problems. The Olkiluoto-3 project is delayed and over budget. It is difficult to estimate their progress and status because the supplier Areva has stopped updating its construction schedule. Both TVO and Areva are making claims against each other for huge compensation. Fennovoima's lost their biggest investor, E.ON; they had to replace it with the Russian Rosatom. There is an on going political debate about the validity of the granted decision-in-principle for the new nuclear power plant and uncertainty as to what will happen if it is determined invalid. Another political debate is about Rosatom and the amount of power they have in the project, being a large shareholder as well as the supplier of the plant.

The institutional and legal frameworks for nuclear waste disposal in Finland are mostly clear and well-defined, but there are some areas that deserve more discussion. Some of the most important open questions are the quantification of payments to the Nuclear Waste Fund, the practically limited liability of the nuclear energy companies, and the coverage of the Nuclear Liability Act for the final disposal phase.

Some researchers say that decision-making power in nuclear matters in Finland is held by a limited and exclusive elite consisting of energy experts from energy companies and key officials in the Ministry of Employment and the Economy, supported by the national institutes, Posiva, STUK and VTT. Their strong co-operation can be described as a nuclear-industrial complex where there is a "technology-and-industry-know-best" attitude which is often adopted by other stakeholders. If criticism about nuclear plans is presented, it is quickly dismissed by the nuclear energy elite. Another peculiarity of Finnish energy politics is a strong internal drive for consensus which is reflected in relatively one-sided public discussion and decision-making "without alternatives". These factors also tend to dampen public discussion on nuclear waste management.

According to many research findings, the Finnish plan for nuclear waste management using direct disposal in bedrock seems to be a feasible solution. Currently known risks are seen as relatively small and thus, there are few concerns among the elites. Possible future problems and obstacles due to the limited quality of the bedrock in Olkiluoto and to the intended copper capsule solution are pushed aside and neglected in public discourse.

References

Decamps, F., and Dujacquier, L. (1997). Overview of European practices and facilities for waste management and disposal. *Nuclear engineering and design*, 176(1), 1-7.

European Industrial Relations Observatory On-line (EIROnline) (2013). Finland: Industrial relations profile. Pay and working time developments . http://www.eurofound.europa.eu/eiro/country/finland_5.htm, last accessed 9 May 2014.

Fennovoima (2011). Fennovoima nuclear power plant to be constructed in Pyhäjoki. http://www.fennovoima.fi/en/news/news/fennovoima-nuclear-power-plant-to-be-constructed-in-pyhajoki, last accessed 12 May 2014.

Fennovoima (2013). Fennovoima and Rusatom Overseas signed plant supply contract. http://www.fennovoima.fi/en/news/news/fennovoima-and-rusatom-overseas-signed-plant-supply-contract, last accessed 12 May 2014.

Finnish government (2005). Hallituksen esitys Eduskunnalle eräiden ydinvastuualalla tehtyjen kansainvälisten pöytäkirjojen hyväksymisestä sekä laeiksi näiden pöytäkirjojen lainsäädännön alaan kuuluvien määräysten voimaansaattamisesta ja ydinvastuulain muuttamisesta. http://www.finlex.fi/fi/esitykset/he/2005/20050002, last accessed 12 May 2014.

Granath, J. (2012). Ydinjätteen loppusijoitus ajautumassa vaikeuksiin, Ajankohtainen kakkonen, *YLE* . http://yle.fi/uutiset/ydinjatteen_loppusijoitus_ajautumassa_vaikeuksiin/6421859, not available in English, last accessed 12 May 2014.

Greenpeace (2011). Reverse Logic – Safety of Spent Nuclear Disposal. http://www.greenpeace.org/finland/fi/kampanjat/ydinvoima/Ongelmat/ydinjate/Safety-of-spent-nuclear-disposal, last accessed 12 May 2014.

Karppinen, A., Oikarinen, E., and Kaivo-oja, J. (2010). Olkiluoto 3 –ydinvoimalaitosyksikön rakennusprojektin alueelliset tuotanto- ja työllisyysvaikutukset. *Kansantaloudellinen aikakauskirja*, 106 (2), 7-22.

Kojo, M. (Ed.) (2004). *Ydinvoima, valta ja vastarinta*. Keuruu: LIKE.

Kojo, M., and Litmanen, T. (Eds.) (2009). *The renewal of nuclear power in Finland*. Basingstoke: Palgrave Macmillan.

Kojo, M., Kari, M., and Litmanen, T. (2010). The socio-economic and communication challenges of spent nuclear fuel management in Finland: the post site selection phase of the repository project in Eurajoki. *Progress in Nuclear Energy*, 52(2), 168-176.

Korhonen, J. M. (2013). "Graph" of the Week: What happens if nuclear waste repository leaks? http://jmkorhonen.net/2013/08/15/graph-of-the-week-what-happens-if-nuclear-waste-repository-leaks/, last accessed 9 May 2014.

Laki ulkomaalaisten yritysostojen seurannasta [Law on follow-up of company purchases by foreigners] (2012). http://www.finlex.fi/fi/laki/alkup/2012/20120172, last accessed 12 May 2014.

Lehtonen, M. (2010). Opening Up or Closing Down Radioactive Waste Management Policy? Debates on Reversibility and Retrievability in Finland, France, and the United Kingdom. *Risk, Hazards & Crisis in Public Policy*, 1(4), Article 6.

Litmanen, T. (2009). The temporary nature of societal risk evaluation: understanding the Finnish nuclear decisions. In Kojo, M. and Litmanen, T. (Eds.), *The renewal of nuclear power in Finland*, 192-217. Basingstoke: Palgrave Macmillan.

Litmanen, T., Kojo, M., and Kari, M. (2010). The rationality of acceptance in a nuclear community: analysing residents' opinions on the expansion of the spent nuclear fuel repository in the municipality of Eurajoki, Finland. *International Journal of Nuclear Governance, Economy and Ecology*, 3(1), 42-58.

Ministry of Employment and the Economy (MEE) (1999). EIA procedure 1998-1999. https://www.tem.fi/en/energy/nuclear_energy/a_final_disposal_plant_project_concerning_nucl ear_fuel/environmental_impact_assessment/eia_procedure_1998-1999, last accessed 12 May 2014.

MEE (2010a). *Handling of the application for a Decision-in-Principle 2010.* https://www.tem.fi/en/energy/nuclear_energy/handling_of_the_new_nuclear_power_projects_decision-in-principle/a_decision-in-principle_on_fennovoima_nuclear_power_project/decision-in-principle_2010, last accessed 12 May 2014.

MEE (2010b). Decision in principle procedure on Posiva's extension. Ministry of Employment and the Economy. https://www.tem.fi/en/energy/nuclear_energy/a_final_disposal_plant_project_concerning_nuclear_fuel/decision_in_principle_procedure_on_posiva_s_extension, last accessed 12.05.2014.

MEE (2011). Nuclear Energy in Finland. http://www.tem.fi/ajankohtaista/julkaisut/julkaisujen_haku/nuclear_energy_in_finland.98249.xhtml, last accessed 9 May 2014.

MEE (2013a). Joint use of know-how on nuclear waste management to be utilised. http://www.tem.fi/en/current_issues/press_releases/press_release_archive/year_2013?113931_m=109189, last accessed 09.05.2014.

MEE (2013b). Ydinjätehuoltoyhteistyön selvitys – Työryhmän loppuraportti. http://www.tem.fi/files/35334/TEMrap_1_2013.pdf, last accessed 9 May 2014.

Morton, O. (2012). The dream that failed. *The Economist*, March 10, 2012. London. http://www.economist.com/node/21549098, last accessed 12 May 2014.

Niinistö, V. (2012). Ydinenergiateollisuuden aika ottaa oma vastuunsa kannettavakseen. http://www.vihreat.fi/blogit/ville-niinisto/ydinenergiateollisuuden-aika-ottaa-oma-vastuunsa-kannettavakseen, last accessed 12 May 2014.

Nuclear Energy Act (2008). http://plus.edilex.fi/stuklex/en/lainsaadanto/19870990), last accessed 12 May 2014.

Nuclear Energy Decree (2008). http://plus.edilex.fi/stuklex/en/lainsaadanto/ 19880161, last accessed 12 May 2014.

OECD/ Nuclear Energy Agency (2013). Nuclear Energy Data 2013. NEA No: 7162, OECD. Paris.

Pekkarinen, J., and Vartiainen, J. (1993). Talouspolitiikan pitkä linja [The long line of economic policy]. Porvoo: WSOY.

Posiva (1999). Posiva jättää loppusijoituslaitoksen periaatepäätöshakemuksen toukokuussa – Enemmistö Loviisan ja Eurajoen asukkaista hyväksyy ydinjätteen loppusijoituslaitoksen. http://www.posiva.fi/media/tiedotteet/posiva_jattaa_loppusijoituslaitoksen_periaatepaatoshakemuksen_toukokuussa_enemmisto_loviisan_ja_eurajoen_asukkais-ta_hyvaksyy_ydinjatteen_loppusijoituslaitoksen.123.news#.U3Cd8CgkodM, last accessed 12 May 2014.

Posiva (2011). Final disposal of spent nuclear fuel in Olkiluoto. http://www.posiva.fi/en/databank/publications/brochures/final_disposal_of_spent_nuclear_fuel_in_olkiluoto.2272.xhtml#.U32v0Cgav44, last accessed 22 May 2014.

Posiva (2014). Encapsulation Plant. http://www.posiva.fi/en/final_disposal/final_disposal_facility/encapsulation_plant#.U32ydSgav46, last accessed 22 May 2014.

Raittila, P., Hokkanen, P., Kojo, M., and Litmanen, T. (Eds.) (2002). *Ydinjäteihme suomalaisittain*. Tampere: Tampere University Press. http://tampub.uta.fi/bitstream/handle/10024/65721/951-44-5485-5.pdf?sequence=1, last accessed 11 April 2014.

Rasilainen, K. (Ed.) (2002). Nuclear Waste Management in Finland. Final Report of the Public Sector's Research Programme JYT2001 (1997-2001), Ministry of Trade and Industry. Studies and Reports 15/2002, Energy Department, Helsinki.

Ruostetsaari, I. (2010). *Energiavalta: Eliitti ja kansalaiset muuttuvilla energiamarkkinoilla*. Tampere: Tampere University Press.

Säteilyturvakeskus – STUK (2009). Encapsulation and the final disposal facility. http://www.stuk.fi/ydinturvallisuus/ydinjatteet/loppusijoitus_suomessa/kapselointi_ja_ls_laitos/en_GB/laitos/, last accessed 9 May 2014.

Suomen luonnonsuojeluliitto (2009). Ympäristövaikutusten arviointiselostukseen ydinpolttoaineen loppusijoituslaitoksen laajennuksesta. http://www.sll.fi/ajankohtaista/liitto/2009/ymparistovaikutusten-arviointiselostukseen-ydinpolttoaineen-loppusijoituslaitoksen-laajennuksesta-posiva-oy, last accessed 12 May 2014.

Teräväinen, T., Lehtonen, M., and Martiskainen, M. (2011). Climate change, energy security, and risk – debating nuclear new build in Finland, France and the UK. *Energy Policy*, 39(6), 3434-3442.

Ministry of Finance (2013). Talousarvioesitys 2014, Valtion ydinjätehuoltorahasto. http://budjetti.vm.fi/indox/sisalto.jsp?year=2014&lang=fi&maindoc=/2014/tae/hallituksenEsitys/hallituksenEsitys.xml&id=/2014/tae/hallituksenEsitys/Yleisperustelut/8/1/25/25.html, last accessed 12 May 2014.

Vehmas, J. (2009). The role of nuclear power in Finnish energy and climate policies. in: Mez, L., Schneider, M., and Thomas, S. (Eds.), *International Perspectives of Energy Policy and the Role of Nuclear Power*, 37-54. Multi-Science Publishing.

Ydinvastuulaki [Nuclear Liability Act] (2008). http://www.finlex.fi/fi/laki/ajantasa/1972/19720484, last accessed 12 May 2014.

The Trouble with Democracy
The Challenges of Nuclear Waste Governance in the Czech Republic

Martin Bursík

1 Introduction

Since the mid 1980s, the Czech Republic has been trying in vain to find a site to deposit four to nine thousand tonnes of spent nuclear fuel from the Dukovany and Temelín nuclear power plants. The total amount of nuclear waste to be deposited became clear when the electricity company ČEZ, which is owned predominantly by the Czech State, decided to stop its tender for a supplier of equipment for new nuclear reactors at Temelín or Dukovany on April 10, 2014. Besides the official reasons given by the State, namely the instability of the European energy sector, the real reason that the tender was stopped was the economics behind new nuclear reactors, which cannot compete with the market price of electricity at €33.55/MWh (EEX 2015 price by June 3, 2014). ČEZ decided to stop its tender when the company received an official Government refusal to provide new NPP a fixed-price guarantee. It then became clear that a continuation of the nuclear project would send ČEZ into the red.[1] An important and strategic side effect of this decision was that the ROSATOM consortium, which participated along with Westinghouse in the Czech nuclear project, has no more influence in the future Czech energy system. This is a rather significant moment, given the recent military intervention by the Russian armed forces in Ukraine and the political, strategic and security aspects which dictate that the Czech Republic dispense of its energy dependence on Russia.

The Radioactive Waste and Spent Nuclear Fuel Handling Policy of the Czech Republic was passed by the Government in 2002, in spite of the disapproval of the Strategic Environmental Impact Assessment (SEA) from the Ministry of Environment (MoE) (Anděl 2001). However, the siting for the nuclear waste repository has come up against strong resistance from environmental NGOs, which, together with municipal self-governments, organised 28 valid local referenda between 2003 and 2013. In all of these referenda, the

1 See http://www.pxe.cz/Produkty/Detail.aspx?isin=FCZBLY151231#KL, last accessed 3 June 2014.

inhabitants of the concerned municipalities refused the exploration for and/or the construction of a nuclear repository in their territories.

In 2004, the Government was forced by the public pressure to suspend the exploration works for five years. In 2009, the Ministry of Industry and Trade (MIT) came up with the idea to build the repository on the premises of former military grounds, which are not under the jurisdiction of municipal authorities; in this way the State could avoid a conflict with local self-governments and their communities. However, the MoIT abandoned the idea soon after analysing the geological conditions at the military sites. A "Working group for dialogue on the deep repository involving representatives of municipalities" was set up in 2010. It proposed an amendment to the law pursuant to which a municipality's rejection of a repository siting could only be overruled by a Senate resolution.[2] However, the proposed amendment was not considered in the legislative process whatsoever and the Working Group was disbanded by the Ministry of Industry and Trade (Sequens 2013).

In 2011, State representatives changed their strategy and decided to "buy" municipalities' consent with so-called compensation payments, instead of supporting a dialogue and an amendment to the law that would empower the municipalities. As a result, an amendment to the Atomic Act has been in effect since 2012; pursuant to it, municipalities are given monetary compensation while explorations for siting for the nuclear waste repository are being done. The fixed "blanket" sum is CZK 600,000 per municipality and an additional CZK 0.30 per m^2 of exploration territory, but a cap is set at no more than CZK 4 million per municipality (i.e., set blanket payment of approx. €22,000 plus €0.01/m^2; no more than €145,000 per municipality). The Radioactive Waste Repository Authority (RAWRA) expected that the monetary compensation to municipalities would change their position on the repository, and therefore it promised that explorations would not be done without the municipalities' consent. The last of a series of local referenda took place in January 2013; in spite of the compensation, the inhabitants again disagreed with the proposed exploration. In response, the MoIT ruled that the condition of municipal consent was no longer valid, and that exploration areas would be delineated in all currently short-listed sites, despite the municipalities' disapproval.

In March 2013, a group of 131 mayors of the affected municipalities initiated a proclamation to the Czech Government – "Be fair about the deep repository."[3] The first of the eight exploration areas was delineated on the Kraví hora site by MoEs Decision of 30 May 2013 where the most recent local referenda took place and where the locals refused the repository exploration and

2 The Senate is the upper chamber of the parliament.

3 See http://www.temelin.cz/images/PDF/vyzva_uloziste_ferove.pdf, last accessed 3 June 2014.

siting. However, the delineated exploration area did not exist for long; the Minister of Environment approved the appeal made by NGOs that same year, and stopped the exploration. The main argument posed by the NGOs was that the Ministry had not at all dealt with the question of whether the public interest in the repository exploration and future construction would outweigh the interests of local communities.

The management for planning the deep nuclear waste repository has long been unsuccessful and was even criticised by the State Office for Nuclear Safety (SONS), which stated in its 2012 annual report: "The SONS does not regard the current situation around the deep repository as satisfactory. Although the Radioactive Waste Repository Authority RAWRA publicly declares the safety of the deep repository as its priority, the progress of the programme of locating the site for the deep repository does not correspond to that. The entire programme is greatly ineffective in its parts on which the SONS sees itself as authorised to comment. If it continues as it has proceeded so far, the SONS will most probably not have sufficient relevant information and analyses required for issuing a permit to locate the deep repository in 2025" (Drábová 2012).

2 The present policy

2.1 Radioactive waste disposal

Two intermediate storage facilities for nuclear waste (within the Dukovany Temelín nuclear power plants) and two repositories for institutional waste produced in research and development and in industry (Richard near Litoměřice and Bratrství near Jáchymov in the Ore Mountains) are currently in operation.

Dukovany intermediate storage facility

High-level radioactive nuclear waste was previously transported from the Dukovany nuclear power plant to an intermediate storage facility inside the Jaslovské Bohunice nuclear power plant in Slovakia. From there, the then-Soviet Union was supposed to take the waste, as it was laid out in an international agreement. Following the disintegration of the Soviet Union, its successor, the Russian Federation, withdrew from this commitment. After the division of Czechoslovakia, the spent fuel ended up abroad and had to be transported little by little to the country's own intermediate storage facility, built inside the Dukovany nuclear power plant. The dry storage facility has a capacity of 600 tonnes and comprises 60 dual-purpose (transport and storage) Castor 440/84 containers which were supplied by the German company GNS Nukem. The intermediate storage facility has been in permanent operation since 1995; its storage capacity was used up in 2005 and another storage facility was built for an additional 1,330 tonnes of nuclear waste. This capacity covers the planned operation and waste generation of the Dukovany power plant. The storage facility consists of 112 ferro concrete reservoirs. The total capacity of the storage area is 55,000 m^3 (approx. 180,000 barrels).

Temelín intermediate storage facility

The permit to operate the intermediate storage facility inside the Temelín nuclear power plant was issued in December 2011. The spent nuclear fuel is stored in CASTOR 1000/19 dry transport and storage container units. The Temelín intermediate storage facility has a designed capacity of 1370 tonnes of heavy metals (HM). The intermediate storage facility, commissioned by ČEZ, was built by an unknown company, CEEI, which is without a transparent ownership structure and without any experience in building intermediate storage facilities. It is interesting that the original project design, made by the Nuclear Research Institute in Řež, anticipated costs amounting to one-third of the price offered by CEEI. Given that CEEI is a company with totally non-transparent ownership, this indicates that the company might be in the hands of politicians and former ČEZ CEO Martin Roman. The total contractual cost of the construction was CZK 1.5bn, whereas the cost of an equivalent facility in the Isar nuclear power plant in neighbouring Bavaria (completed in 2007) was CZK 810m according to E.ON, the power plant owner.

Bratrství repository

The Bratrství repository stores waste produced in the institutional domain (research and development, industry, healthcare, etc.). Most of the institutional radioactive waste produced in the Czech Republic is handled by the Nuclear Research Institute in Řež, a.s. (ÚJV), which is also the biggest producer of this type of waste in the country; it produces about 60% of the total quantity of radioactive waste. The repository is located inside the former Bratrství uranium mine. It was put into operation in 1974. The total capacity of the storage area is 1200 m^3. The repository capacity was at than 80% in 2008, and a closure of the repository is currently being planned.

Richard repository

This repository consists of the former Richard I, II, and III mine compound, which has more than 40 km of tunnels and cross cuts. It was built in a "sub-horizontal" slab of a clayey limestone approximately 5 m thick. Both the roof and floor of the mine compound are made of impermeable marl stones over 50 m thick. There are fine-grained sandstones underneath the marl stones. The underground repository is situated above the groundwater table. Institutional waste has been deposited here since 1964. More than 24,000 container units are in storage here at the moment. The total capacity of the utilised spaces exceeds 17,000 m^3; the actual waste reposition capacity is half of that. The rest of the space comprises tunnels for operators and waste handling.

The operating costs of the intermediate storage facility and repository managed by the RAWRA are paid from the so-called nuclear account, to which each producer of radioactive waste in the Czech Republic contributes.[4]

2.2 The policy issue of 2014: reviving uranium mining

President Zeman proposed in January 2014 that the possibility of reviving uranium mining in Jáchymov should be considered. Mining took place there between 1945 and 1964, when the workforce of political prisoners was abused. A total of 7,200 tonnes of uranium was extracted, and the mines were closed after the extractable reserves were exhausted. Mining in Jáchymov is not a part of the published uranium reserves of the Czech Republic.

In April 2014, the Prime Minister Bohuslav Sobotka supported the proposal to open new extraction sites at the B and H V deposits near Polná in the Jihlava District, with an expected 3,100 tonnes of uranium. This quantity would ensure a mere five years of operation of the Temelín and Dukovany nuclear power plants, at the expense of billions of crowns of State investment. The reason for this high cost is that the exploration well at Brzkov is backfilled and flooded. The two million tonnes of ore would have to be transported to the existing GEAM plant at Dolní Rožínka for processing. However, the settlement pools there lack additional capacity for the storage of sludge and other waste from the plant. The DIAMO state enterprise does not have the resources for investment in extraction or renovation of the settlement pools; the current sale of uranium concentrate roughly covers only the costs of mining and processing.

Additional potentials for uranium mining exist in the Liberec Region, but the geological conditions make the uranium extractable only by chemical leaching. This would endanger the underground reserves of drinking water and the environment. In the course of 30 years of chemical uranium extraction at the Stráž pod Ralskem deposit, located on the site of one of the most important sources of drinking water in the Czech Republic, the underground aquifier

4 See http://surao.cz/cze/Uloziste-radioaktivnich-odpadu/Soucasna-pripovrchova-uloziste, last accessed 3 June 2014.

absorbed almost 4.1 million tonnes of sulphuric acid, 320,000 tonnes of nitric acid, 111,000 tonnes of ammonia, and 26,000 tonnes of hydrofluoric acid. At present, the area contains approximately 400 million m^3 of contaminated groundwater. The overall consequences of mining may threaten the underground drinking water reserves, and the residual chemical solutions are threatening an area of 160 km^2. The Government has committed to paying CZK 31.3bn to eliminate the harmful consequences of the chemical leaching at Stráž pod Ralskem in the next 30 years.[5]

3 Waste storage concept

The planning of the deep repository started in the mid-1980s with a 700-meter borehole on the site of the Temelín nuclear power plant. The first reference design for the deep geological repository on a hypothetical site in the Czech Republic was made in 1999 (Mandík et al. 1999). Since 1999, the Radioactive Waste Repository Authority (RAWRA) has implemented a number of projects that have resulted in the narrowing down of choices for potential sites to six locations. All of the potential sites are located in areas rich in granite rock. The Government approved the Radioactive Waste and Spent Nuclear Fuel Handling Policy in 2002. The Nuclear Research Institute in Řež updated the reference design for the deep repository for radioactive waste in 2012 (Pospíšková et al. 2012).

Table 1: Reference design and construction time table

2010-2015	Exploration phase and geological surveying
2015-2025	Rock exploration and research and development phase
2025-2050	Work on the final location, including the construction of an underground laboratory and research work in the laboratory
2050-2064	Construction of the underground and surface repository compounds
2065	Start of operation

5 See http://calla.cz/index.php?path=hl_stranka/tiskovky/2014&php=tz140115.php, last accessed 3 June 2014.

The planned deep repository for high-level radioactive waste and spent nuclear fuel should be made up of three sections of underground spaces for: 1) storing and handling containers of spent fuel and high-level radioactive waste, 2) access wells and tunnels, and 3) an above-ground surface compound.

The storage chambers are to be built approximately 500 m below ground (depending on the type of bedrock) in a stable geological formation, and the waste is to be placed in special containers. Depending on the quantity of waste to be stored, the underground tunnels should extend over several km^2; the layout of the tunnels will be based on the repository´s design, which is still not clear (in terms of single or multiple storeys, and horizontal or vertical storage of containers). The storage spaces should be connected to the surface compound by vertical access wells and a helical tunnel. The surface repository compound should extend over several hectares.

4 Costs and financing

The available information about the expected investment and operating costs of the repository is very limited. In 1999, the total costs for the deep repository planning and construction were calculated to be CZK 47bn (Mandík1999).

Thirteen years later, in 2012, the investment costs, without the costs of exploration, were estimated at CZK 30bn, whereas the costs of construction and operation according to the updated reference design in the base option are CZK 106bn (€3.85bn). Further details are not available.

The investment costs and the operation of the repository should be paid from a nuclear account maintained by the Czech National Bank, to which every producer of radioactive waste should contribute in keeping with Decree No. 416/2002 Coll. The sole operator of nuclear power plants in the Czech Republic, ČEZ Group, deposits CZK 50/MWh for electricity generated in nuclear power plants to this account. The contributions of the other radioactive waste producers are negligible. The nuclear account amounted to roughly 13bn in 2009, and approximately CZK 1.4bn accrues each year. Assuming that the service life of the Dukovany NPP will extend until 2027, the power plant will have contributed another close to CZK 10bn to the nuclear account before the end of its life cycle. Assuming that the theoretical operation of the Temelín NPP will be until 2055, a total of CZK 46bn will have been deposited into the nuclear account. Unless the operation of the power plants is terminated sooner, the nuclear account will amount to about CZK 60bn by 2055. This means that approximately CZK 60bn will be available for the construction and operation of the deep repository for nuclear waste, instead of the projected CZK 106bn.

Besides contributing to the nuclear account, each holder of a permit to operate a nuclear facility has to create a financial reserve for the disposal of the facility, in keeping with the Atomic Act. According to the Background Study of the Radioactive Waste and Spent Nuclear Fuel Handling Policy of the Czech Republic the financial resources for decommissioning the Temelín and Dukovany NPPs should be CZK 32.3bn (2008 prices, Vokál et al. 2013).

5 Legal framework

The legal foundation for permitting and construction of the repository is the Building Act (No. 183/2006 Coll.), specifically its component regarding land-use planning. The Spatial Development Policy (SDP) details the tasks of land-use planning. When developing the draft, the SDP pursuant to Section 33, Para. 1 of the Building Act, stated that the Ministry of Regional Development has no legal obligation to cooperate with municipalities, but only regional authorities – even though the impact of the SDP on municipalities is obvious. Regional authorities may state their positions on the drafted SDP, but municipalities do not have that option and may only comment on the draft like any other members of the public pursuant to the legal provision. The Ministry of Regional Development is required to reflect the position of the regional authorities, to hear public comments, issue the results of the impact assessment on the sustainable development of the territory, and then adjust the drafted SDP accordingly. Unfortunately, it only discusses the adjusted SDP draft with representatives of ministries and other top-level administrative and regional authorities, not with representatives of municipalities.

Municipalities are therefore required to indicate areas for projects proposed in the Spatial Development Policy (SDP) in their land-use plans, regardless of their agreement or disagreement with the proposal.

The section on waste management in the Spatial Development Policy, approved by Government Resolution No. 929 of 2009, includes the plan to "delineate areas for a deep repository for high-level radioactive waste and spent nuclear fuel." The SDP proposes the spatial protection of such areas, for the purpose of continuing geological surveying of the rock environment and examination of the territorial and environmental conditions of the potential repository. The tasks for ministries and other administrative authorities include selecting the two most appropriate sites for repository construction by 2015, a process which should involve the participation of the affected municipalities. However, the form of this involvement is not specified.

Geological exploration is regulated by Act No. 62/1988 Coll. Municipalities cannot influence the geological projects in their territories during the design phase, even though the subsequent work will affect their territories and/or ownership rights. They have no other choice but to rely on the position of the regional authority, which may take into account protected interests and impose a compulsory expert assessment, but is not obliged to do so. The ownership rights of municipalities and citizens are theoretically protected by Section 14 of the Act, which rules that organisations doing geological studies are required to sign a written agreement to report their findings to the land owner before entering another owner's land. However, if an agreement is not made, the regional authority should again decide about restricting the land owner's or tenant's ownership rights. It will then be obliged to accept the geological findings. This leads one to think that the position of the municipalities in the search for a location for deep geological nuclear waste repositories is totally inadequate (Bouda et al. 2010).

The Euratom Member States have committed themselves to putting legal and administrative regulations in effect in order to comply with the Directive 2011/70/Euratom by 23 August 2015 (since they have to report the contents of the national programme to the European Commission, as stated in Article 12 of the Directive). In addition, the Directive requires that Member States develop a national spent nuclear fuel and radioactive waste handling programme, covering all of the different stages involved from production to disposal. The Nuclear Research Institute in Řež developed a Background Study on the Radioactive Waste and Spent Nuclear Fuel Handling Policy of the Czech Republic in February 2013 (Vokál et al. 2013). The updated national programme should be presented to the Government by June 2014 at the latest. The policy should include a strategic EIA (SEIA). No information about this programme is available as of May 2014.

6 The institutional framework

The planning and construction of the deep repository for high-level radioactive waste and spent nuclear fuel is coordinated by the Radioactive Waste Repository Administration (RAWRA), established on 1 June 1997 as a state organisation based on the Atomic Act (Section 26 of Act No. 18/1997 Coll. on Peaceful Utilisation of Nuclear Energy and Ionising Radiation). Since 2000, the RAWRA has been an organisational component of the State, in keeping with Section 51 of Act No. 219/2000 Coll.

The RAWRA's tasks and activities are to:

- ensure the planning, construction, commissioning, operation, and decommissioning of radioactive waste repositories and monitoring of their influences on the surroundings;
- ensure radioactive waste handling;
- ensure reprocessing of spent or irradiated nuclear fuel into a form suitable for storage or reuse;
- keep records of radioactive waste accepted and its producers;
- manage the charges from radioactive waste producers at the nuclear account;
- develop proposals for setting the charges to the nuclear account;
- ensure and coordinate research and development in the area of radioactive waste handling;
- inspect permit holders' reserves for decommissioning of their facilities;
- provide services in the area of radioactive waste handling;
- handle radioactive waste imported to the Czech Republic from abroad that cannot be returned; and,
- ensure temporary management of radioactive waste that will be passed into state ownership.

7 Siting (search) procedures

The selection of a suitable site has been moving along unsuccessfully since the mid 1980s. The following overview summarises the process involved in siting the deep geological repository:

- 1990–1993: Based on several geological criteria, the Czech Geological Institute (CGI) proposed 27 sites broadly recommended for further surveying.
- 1990–1998: The Nuclear Research Institute in Řež selected the 13 most promising areas and collected and analysed archived geological information. Based on this analysis, the selection was narrowed to 5 areas, in which 8 sites were short listed.
- 2002–2003: The RAWRA expanded previous siting work with a regional survey based on recommendations of the International Atomic Energy Agency; it recommended 11 sites based on the new criteria.
- 2003–2005: The RAWRA concluded the initial regional phase of surveying by choosing 6 sites that it recommended for a second phase of surveying using airborne geophysical surveying and satellite pictures. Areas of poten-

tially homogeneous rock environment were defined and polygons for future survey areas on each of the sites were proposed. In addition, feasibility studies for the location of the surface compound of the future deep repository, and its connection to the underground, technical, and transport infrastructures were elaborated.

- 2005–2009: Because of local referenda which demonstrated the rejection of the survey by inhabitants of affected municipalities and the negative positions taken by the municipalities, further work on the six exploration sites was put on hold until the end of 2009 by a decision made by the Government of the Czech Republic. A geological survey was commissioned in 2008, aiming to analyse archived geological information on five military areas in the Czech Republic. The purpose of the survey was to potentially extend the list of suitable sites in areas of less public conflict.
- 2010–2015: The siting phase of the geological exploration includes detailed geophysical and geochemical mapping. A detailed geological map of the site will be elaborated, with the goal of proposing a place for a deep bore (800-1000 m deep) and ultimately delineating an area, 500-600 m deep in a relatively homogeneous rock environment covering approximately 3-5 km^2, for the actual future storage. At the same time, a technical design for the surface compound, the underground compound, and their connecting infrastructure will be made (including a system for transporting future waste underground and storage). The suitability of the site will be confirmed by a feasible technical design and a preliminary safety assessment. The outcome of the siting phase should be the designs for a future primary deep repository site and a back-up site.
- 2015–2025: The suitability of the selected primary site should be confirmed in the exploration stage, based on detailed rock boring surveys and other surveys in progress. The back-up site will only be examined in the case that the primary site fails to conform to the original assumptions.
- 2025–2050: Work on the final site will continue into the detailed surveying stage, including mining and the construction of an underground laboratory. The properties and suitability of the rock environment will be surveyed directly on the site in the "rock massif". The primary objective will be to acquire a sufficiently trustworthy set of data to prove the safety of the site, in order to then be able to apply for a permit to build the deep repository.
- 2050–2065: Construction of the underground repository and the surface compound, and commissioning (Slovák 2013).

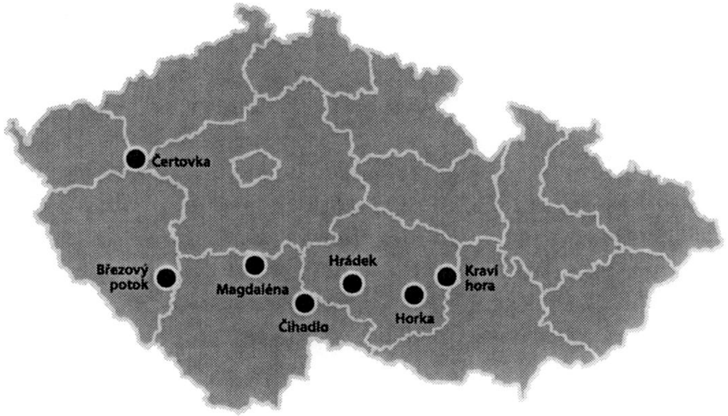

Figure 1: Sites considered for deep repository construction.
Source: RAWRA (2013).

The above overview documents a considerable variability over time in the proposals for execution of the deep nuclear waste repository. The number of sites has changed from 27 to 8, and then again from 6 to 7. In June 2004, the Government decided to put the surveying on hold for at least 5 years, based on a proposal from the Minister of Industry. The reasons for the decision were public opposition, clear results from local referenda, petitions, opinion polls, demonstrations, and the disapproval of local and regional self-governments.

Moreover, it turned out that in 2012 the RAWRA was planning to build a storage facility for spent fuel in the deep repository compound, which should include a series of tunnels in which containers are to be stored. The planned capacity of the underground intermediate storage facility is similar to the surface facility at Temelín, and should double that if a planned back-up is constructed. The implementation of such a plan would result in not only considerably increased transportation of radioactive waste, but also the many-year deposition of containers not half a kilometre underground, as they should be, but right beneath the surface.

8 Information policy and forms of participation including civil society

The unilateral siting decision by the State has led to strong opposition from the population, as well as local self-governments. A total of 28 valid local referenda took place between 2003 and 2013, in which the inhabitants overwhelmingly refused the location of the nuclear waste repository and even rejected the geological surveying (see Table 2).

In January 2007, the Green Party, which had just become part of the Government coalition, pushed a commitment into the Government's programme which stated that: "Further steps in searching for a deep geological repository for spent nuclear fuel will be taken in a transparent manner; consent of affected municipalities will be an inviolable criterion." [6]

Although the Government that had made this commitment to the municipalities terminated its office in March 2009, the Ministry of Industry and Trade made a (repeated) public promise, via the RAWRA, to the mayors and the inhabitants of the municipalities affected by the planned repository that until the end of 2012, it would not commence any geological exploration without their consent. However, a fundamental turn came after that. In December 2012, the Minister of Industry, Martin Kuba, transferred the right to acquire a surveying permit on the Kraví hora site near Žďár nad Sázavou to the state enterprise DIAMO, GEAM Dolní Rožínka unit, which extracts and processes uranium in the area. By doing that, Kuba formally divested the RAWRA of its public commitment that municipality consent was an inviolable condition for the commencement of any survey.

The State also decided to disregard the results of the local referendum held in Bukov on 11 and 12 January 2013 and ignore opinion polls in Bor and Sejřek, in which most of the voters refused the exploration. In February 2013, the Brno territorial section of the Ministry of Environment started a process (which extended until May 30) for identifying the first exploration site for special interference with the Earth's crust, in order to search for the ultimate deep repository for spent nuclear fuel. This site extends over more than 17,000 hectares and lies in the territories of the municipalities of Sejřek, Drahonín, Moravecké Pavlovice, Věžná, Olší, Střítež, Milasín, and Bukov.

Since the RAWRA failed to obtain the consent of the affected municipalities, it decided to apply for the delineation of exploration areas on all the short listed sites, regardless of the results of the referenda and the municipal representatives' opinions. Another six places in the Czech Republic – around Lubenec and Blatno near Louny, Pačejov near Plzeň, Jistebnice near Tábor, Lodhéřov near Jindřichův Hradec, Rohozná near Jihlava, and Budišov in the Bohemian-Moravian Highlands – face an ordeal similar to that of Kraví hora. Another site being considered is Boletice in South Bohemia, and DIAMO has even claimed that it has a brand new site near Dukovany in store.

6 See http://www.vlada.cz/scripts/detail.php?id=20780, last accessed 3 June 2014.

Table 2: Results of local referenda on deep repository location in the affected municipality's territory.

No.	Municipality	year	Participation in referendum	Agree with repository location	**Disagree with repository location**
1	Oslavička	2003	80.0%	1.5%	**98.5%**
2	Nadějkov	2003	67.5%	2.8%	**95.7%**
3	Přeštěníce	2003	76.4%	0.6%	**99.4%**
4	Božetice	2003	73.0%	2.8%	**95.7%**
5	Hodkov	2004	72.5%	4.0%	**96.0%**
6	Rudíkov	2004	71.9%	4.0%	**95.7%**
7	Budišov	2004	51.1%	12.1%	**81.8%**
8	Nárameč	2004	71.7%	5.2%	**93.4%**
9	Lodhéřov	2004	84.0%	0.6%	**99.4%**
10	Deštná	2004	63.4%	1.8%	**93.1%**
11	Zhoř	2004	69.1%	0.0%	**99.3%**
12	Jistebnice	2004	51.2%	1.8%	**93.7%**
13	Pačejov	2004	80.8%	1.7%	**95.5%**
14	Maňovice	2004	89.5%	0.0%	**97.1%**
15	Olšany	2004	95.1%	0.0%	**99.4%**
16	Rohy	2004	65.2%	9.6%	**89.0%**
17	Hojkov	2007	92.4%	0.0%	**98.3%**
18	Opatov	2007	67.5%	2.8%	**95.4%**
19	Dušejov	2007	65.3%	3.7%	**95.8%**
20	Jedlov	2007	68.7%	1.9%	**98.1%**
21	Milíčov	2007	86.4%	1.1%	**95.5%**
22	Dvorce	2007	73.3%	4.5%	**95.5%**
23	Hubenov	2007	82.6%	1.1%	**98.9%**
24	Cejle	2008	72.6%	15.1%	**79.4%**
25	Rohozná	2009	75.0%	2.6%	**96.6%**
26	Okrouhlá Radouň	2009	59.5%	5.7%	**93.2%**
27	Lubenec	2012	53.4%	17.2%	**79.7%**
28	Bukov	2013	87.9%	42.7%	**51.9%**

Source: Calla (2009) and Bursík (2013).

The RAWRA has scheduled exploration "borings" until 2015. The results should lead to a narrowing of the selection to four sites on which deep excavation (up to a kilometer) will be made and other survey work will be done. The pressure on the local municipalities is accompanied by monetary compensations for those whose cadastral territories the exploration areas will be delineated on. Within five years, the RAWRA is planning to locate the final sites, and will locate the ultimate site by 2025. Eleven years from now, it should therefore be clear where the Czech Republic will store its nuclear waste.

The failure of the State's policy in planning the deep geological nuclear waste repository thus far is worsened by the fact that the State's representatives have regularly succumbed to the temptation to take advantage of, or abuse, the weak position of municipalities, instead of starting a dialogue with the affected inhabitants and local self-governments. A hope for a change came from the establishment of the Working Group for a dialogue on the repository, involving representatives of all the sites and the national authorities in charge. After two years, the Group proposed amendments to the law that would increase the currently insufficient powers of the municipalities. If adopted, municipalities would have the option to refuse further surveys when the two final sites are selected (in 2018), but only conditionally and with the possibility of being outvoted by the Senate.

It is now obvious that the bill arising from the Working Group became an "unwanted child" for the MoIT, which then began to ignore the Working Group. The legal amendment is also no longer wanted by the RAWRA. In a new national Radioactive Waste and Spent Nuclear Fuel Handling Policy currently in preparation, there is a proposal for a procedure that contradicts the Working Group's opinion. It would only permit municipalities to veto the surveying at the very end of the process, when the ultimate site is being selected in 2025. That is far enough into the future to allow the legal amendment to be completely forgotten.

It is doubtful whether such an unsound attitude will achieve any results. Disregard for the opinions of people and communities threatened by the repository cannot succeed in the long term. It has been the case abroad, and those in charge in the Czech Republic will find out as well.

9 Conclusions

The process of searching for the ultimate repository for nuclear waste has progressed in a non-transparent way without any clear criteria. The State's inconsistent approach has led to constant changes in the list of potential sites. The response of the inhabitants of the affected municipalities is unanimous; in all 28 local referenda that have taken place so far, the inhabitants have refused the exploration connected to the repository siting process. The State has therefore changed its strategy; it has decreased the involvement of municipalities in decision-making and enforceability has become the criterion for selecting the repository location, instead of safety and environmental impact. The inconsistent approach of the Radioactive Waste Repository Authority has been criticised, not only by citizens' associations and affected self-governments, but also by the State Office for Nuclear Safety, which has declared that everything seems to suggest that it will not have sufficient information to issue an approval on the repository siting in 2025. Inconsistency and a democratic deficiency are the greatest weaknesses of the Czech Republic's repository planning process.

References

Anděl, P. (2001). *Stanovisko EIA ke Koncepci nakládání s radioaktivními odpady a vyhořelým jaderným palivem v ČR.*

Bouda, P., and Nezhyba, J. (2010). Právní analýza k otázkám zapojení a účasti obcí v procesu vyhledávání hlubinného úložiště vysoceradioaktivního odpadu v ČR a návrh na změnu příslušných zákonů. Advokátní kancelář Šikola a partneři, s.r.o.

Drábová, D. (2012). Výroční zpráva SÚJB 2012. http://www.sujb.cz/dokumenty-a-publikace/vyrocni-zpravy/vyrocni-zpravy-sujb/, last accessed 3 June 2014.

Mandík, F., Vavřina, V., Lietava, P., Blažek, J., Tarasová, J., Čihák, F., (1999). Referenční projekt hlubinného úložiště. EGP Invest s.r.o.

Pospíšková, I., Vokál, A., Fiedler, F., Prachař, I., Kontour, P., (2012). Aktualizace referenčního projektu hlubinného úložiště radioaktivních odpadů v hypotetické lokalitě, průvodní zpráva. SÚRAO.

Sequens, E. (2013a): Vyhledávání hlubinného úložiště v České republice, CALLA – Sdružení pro záchranu prostředí.

Sequens, E (2013b). Jaderný odpad? Děkujeme, nechceme! Bulletin sdružení Calla, Hnutí Duha.

Slovák, J. (2013). Rámec postupu výběru kandidátních lokalit HÚ v ČR. SÚRAO.

Vokál, A., Trtílek, R., Tarasová, J., Popelová, E., Podlaha, J, Krmela, J., Vojtěchová, H., Uhlíř, J. (2013). Podkladová studie pro koncepci nakládání s vyhořelým jaderným palivem a radioaktivním odpadem v ČR. Finální verze.

"Yucca Mountain is Dead"
The Challenge of Nuclear Waste Governance in the United States

Richard A. Forrest

1 Introduction

The United States faces unique and complex challenges in relation to the management of its large volume of high-level radioactive waste (HLW). This is due in part to the especially long history and large-scale nature of the country's nuclear research and power generation programs. Radioactive waste has been generated for more than half a century in connection with both civilian and military activities (and has been managed under a bifurcated system of separate civilian and military management procedures). Not only is a large volume of HLW produced on an ongoing basis, but there also remains a complex legacy of different types of accumulated wastes, which are being stored at a wide variety of aging facilities throughout the country. The U.S. situation is further complicated by heightened concerns – in the wake of the September 11, 2001, terror attacks – regarding the safety and security of HLW as a potentially weaponizable material which could be utilized by terrorists to create a "dirty bomb".

The U.S. management regime for HLW is embedded within the broader governance system of the United States, which is characterized by its multi-level nature and the existence of a variety of access and leverage points where diverse actors can potentially confound processes set in motion to address a public policy concern. The nuclear waste issue has demonstrated this vividly. A decades-long process had proceeded methodically, if slowly, toward the realization of a unitary national deep geological depository of the type recommended by experts – the Yucca Mountain Nuclear Waste Repository – but progress toward this goal was brought to an abrupt halt in recent years as a result of the intervention of a few key political actors. Thus, the development of an enduring solution to the HLW issue – one consistent with a well-established "scientific consensus" – has for the time being been stalled, with the result that the United States currently lacks a permanent disposal site for its high-level radioactive waste.

2 The present policy for disposal of high-level radioactive waste

High-level radioactive waste in the United States has accumulated as a result of both military and civilian activities. HLW from military sources has been produced in connection with the development, manufacture and dismantling of nuclear weapons and related facilities and materiel since World War II, as well as the decommissioning of nuclear-powered submarines and the operation of defense-related power generation plants and research sites. Waste from civilian sources has accumulated ever since commercial energy production at nuclear power generation facilities began in 1958. The total amount of HLW in the United States now stands at more than 70,000 metric tons, the largest amount of any country, with some 3,000 additional tons of HLW generated annually.

The waste from military sources is largely concentrated at the locations where uranium and plutonium were processed for use in weapons: the Hanford Nuclear Reservation in the state of Washington and the Savannah River Plant in South Carolina. At these sites, liquid waste amounting to approximately 70 million gallons is stored in underground tanks. There are 177 separate disposal tanks at the Hanford site alone (IAEA 2013a). Problems associated with these sites include leaks of radioactive material; multiple leaks were reported in 2013 at the Hanford Reservation.

The United States has the world's oldest and largest civilian nuclear industry, which supplies roughly 20% of U.S. electricity demand. HLW from commercial nuclear power plants consists almost exclusively of the used ("spent") fuel rods from the country's 104 licensed civilian nuclear reactors located at 65 power stations nationwide. There are also two reactors currently in the process of being decommissioned, as well as ten reactors at nine locations that have already been decommissioned (NRC 2013).

Spent fuel is first stored in spent fuel pools (also known as "wet pools"), which are ca. 12-m deep open-top, cement and stainless steel receptacles located inside nuclear power generating facilities (although often not within the reactor containment building itself). Once depleted of fissionable material inside reactors, spent fuel rods are submersed in spent fuel pools, having a depth of some 9 m of water above their highest point; this water acts as a coolant and a barrier to the spread of radiation. Many such disposal facilities are approaching or have exceeded their planned capacity.

After a period of several years – or when needed to alleviate overcapacity within wet pool disposal locations – waste is transferred to dry casks (6 m high, 3.3 m diameter cylindrical concrete and metal containers having 0.6 m-thick side walls, which are filled with an inert gas). Evidently, no radiation releases have been associated with such dry cask disposal, although there are concerns that the salt air environment at some facilities along ocean shorelines may lead to corrosion of the casks. Dry casks are intended for short-term disposal of roughly 10-20 years. Dry casks are stored at power plant sites, including those plants that are in the process of being decommissioned.

The amount of accumulated spent fuel from commercial reactors was reportedly almost 63,000 metric tons in 2009, with roughly 78% stored in wet pools and 22% in dry cask disposal. In addition, there are 63 "independent spent fuel disposal installations" in 33 states (NRC 2013).

Because many civilian nuclear power generation plants are located near population centers (such as the Indian Point plant, only 60 km north of New York City), concerns have been expressed regarding the currently utilized temporary disposal approach. Risks include the possibility that such waste material may be vulnerable to natural disasters or terrorist attacks or sabotage, including the potential draining of disposal pools or disabling of water circulation equipment.

3 Waste disposal concept and site selection

The dry cask and wet pool disposal methods are officially considered safe on the order of decades; disposal at reactor sites is therefore considered improper for longer periods. Geologic disposal is considered preferable to above-ground disposal, given that HLW may be hazardous for tens of thousands or possibly millions of years.

To realize a deep geological disposal solution, in 1982 the U.S. Congress passed and President Reagan signed the Nuclear Waste Policy Act (NWPA). The NWPA outlined procedures for determining disposal methods for nuclear waste and provided an initial timetable for establishing a permanent underground repository, envisioned at first to be ready by the mid-1990s. The NWPA entrusted the Department of Energy (DOE) with responsibility for studying candidate sites for geological disposal. Initial scoping considered approximately 120 sites; five sites were to be assessed in detail, with three sites to be formally recommended by 1985. After the designation of the first repository, a second repository was to be selected.

In 1987, the NWPA was amended, adding the obligation to open a repository by 1998, while restricting research to a sole candidate site for a "terminal" or a long-term HLW repository, Yucca Mountain in Nye County, Nevada. Department of Energy geoscientist Abraham Van Luik has stated that the decision to restrict the selection process to the single Yucca Mountain site resulted from cost considerations faced by Congress: "They started looking at the huge bills associated with site-specific studies – excavation is not cheap – and they said: let's just do one site and see if it's suitable. If it is not, then we'll go back and see what else we can do" (Quoted in Manaugh 2009).

The Yucca Mountain site underwent intensive research, to the extent that it has been called (as in the title of a Senate committee report) "the most studied real estate on the planet" (U.S. Senate 2006). Some 10 km long, Yucca Mountain is a ridge of igneous rock, primarily welded tuff, created by an extinct super-volcano. It is located some 150 km northwest of Las Vegas (which has grown several times in size since the original consideration of the Yucca Mountain site, to its current population of approximately 600,000), a city which is of paramount economic importance for Nevada, serving as it does as a key tourist destination and center for employment. The repository location lies within an uninhabited area of desert on federally-owned land under the control of the Bureau of Land Management (an agency under the Department of the Interior) and is adjacent to an area known as the Nevada National Security Site (formerly, the Nevada Test Site) which was previously used for above- and below-ground testing of nuclear weapons.

Various issues have been raised concerning the suitability of Yucca Mountain, including the potential for rare but potentially significant earthquakes or volcanism, as well as intrusions of water that could travel along fissures within the tuff, undermining the integrity of the disposal facilities sooner than the one-million-year timeframe required for the project.

The DOE formally recommended the Yucca Mountain site in 2002 to serve as the Yucca Mountain Nuclear Waste Repository (YMNWR). Initial site preparation work commenced thereafter, with the expectation that the repository would open in 2017. In 2008, the George W. Bush Administration formally applied to the Nuclear Regulatory Commission (NRC) for approval to place YMNWR into operation. At long last, it looked like the country would have its hoped-for solution to the nuclear waste disposal challenge.

The repository was originally planned to have a capacity of 70,000 metric tons. With the failure to designate a second repository site, the capacity was raised to 125,000 tons. The area underneath the mountain that would be prepared for the disposal of waste lies approximately 300 m deep and 300 m above the water table. Some 11,000 containers, known as "waste packages," were envisioned to be fabricated of highly corrosion-resistant alloy and sheltered by drip shields made of titanium; these waste packages would be moved into the site on railroad-like tracks, entering 91 "emplacement drifts" – 600 to 800 m long dead-end tunnels (Manaugh 2009). Although detailed specifications for the waste packages were not developed, the overall plan envisioned that the waste packages would not be surrounded by fill material. The site was intended for final disposal of the waste, with no plans designed for potential future recovery of the waste (although reversibility and retrievability have become a goal in the design of HLW disposal sites under consideration in other countries).

HLW was to be transported mainly by rail, but also by truck, raising the possibility of accidents along lengthy transportation routes spanning several thousand kilometers from distant facilities as far away as the Atlantic coast, along routes which would pass through densely populated areas.

Despite some $15 billion spent on research, construction and preparation for the site, politicians in the host state of Nevada, as well as presidential candidates in the Democratic Party (especially during the 2008 election), voiced their opposition to the use of Yucca Mountain. Barack Obama, who won the election in November 2008, was supported by Harry Reid, a Democratic Party Senator from Nevada who had become Senate Majority Leader in January 2007. Reid was a long-time opponent of the Yucca Mountain project. The first official budget proposal of the Obama Administration, submitted in early 2009, stated in a terse section regarding radioactive waste: "The Yucca Mountain program will be scaled back to those costs necessary to answer inquiries from the Nuclear Regulatory Commission, while the Administration devises a new strategy toward nuclear waste disposal" (OMB 2009: 65).

Evidently as a result of agreement between Senate Majority Leader Reid and President Obama, funding for the YMNWR was thus effectively halted through the tactic of not including any further funding for the project within federal budget plans. No detailed explanation or justification for the decision to halt funding for the repository was provided in the budget proposal, or any other readily available document. However, the document which formally suspended the project, the "U.S. Department of Energy's Motion to Withdraw" of March 3, 2010, alluded to advancements in science and the lack of "broad public support" as reasons to "focus on alternative methods" of dealing with nuclear waste (DOE 2010).

Nevertheless, various stakeholders had developed plans that depended on the use of the Yucca Mountain repository. For example, it had been envisioned that the Yucca Mountain repository would accept waste from military sources in addition to civilian ones. The Hanford site, utilized since the 1943 Manhattan Project for development of the atomic bombs dropped on Japan, accounts for some two-thirds of national HLW in terms of volume. In response to the loss of a potential final repository for this waste, the state of Washington sued the federal government to force compliance with the revised NWPA, which had specifically mandated the opening of the Yucca Mountain national repository. They were joined in the lawsuit by the state of South Carolina, home to the Savannah River Plant, and others.

A first lawsuit was dismissed, but a subsequent suit resulted in a decision on August 13, 2013, in favor of the plaintiffs. In its ruling, the U.S. Court of Appeals stated that the NRC had been "simply flouting the law" (U.S. Court of Appeals 2013: 5) by not processing the DOE application for licensing of the repository, and that the process of making an official determination either for or against licensing the project must move forward. The small amount of funding available for this purpose, some $11 million, is likely to be insufficient for the NRC to conduct significant further work. Nevertheless, the court's decision was significant in that it stated clearly that the executive branch may not simply elect to not implement a law that it finds inexpedient, in this case the NWPA.

Retrievability and Reversibility

Given the many uncertainties surrounding long-term disposal and management of HLW, planners recognize the importance of ensuring that future decision-makers have the option to reconsider HLW disposal approaches and to retrieve waste that is emplaced in disposal sites. Over the mid-term and long-term, there is likely to be significant evolution in terms of scientific, technological and economic conditions and knowledge, which could justify the adoption of alternative and even currently unforeseen approaches and measures for handling HLW, thus reversing past decisions that have been made. This is especially germane in light of the fact that high-level radioactive wastes can also be considered to be a potential future fuel source for nuclear reactors.

With regard to options for the retrievability of HLW, some countries have thus chosen to build retrievability and reversibility into their basic site planning processes. The Yucca Mountain site was not designed specifically with the intention of facilitating future retrieval of the spent fuel. However, the nature of the design does, in effect, leave open the ability to retrieve the HLW up to the final stage, as the sealing of the emplacement drifts with rock backfill would occur only as the final step of the site preparation process. Because the full details of plans for the repository have not been developed, a step-wise approach is inherent in the current approach to disposal of HLW.

Relevant laws and regulations do envision that any potential interim disposal facilities (where waste may be stored prior to emplacement within a "terminal" repository such as Yucca Mountain) must allow for waste retrievability. Despite the potential importance of retrievability, however, the very term itself remains undefined in a precise manner and practical, operational guidelines have yet to be delineated. For this reason, a review process incorporating public comments on these matters is being undertaken to clarify the definition of retrievability and to identify related technical considerations and needs.

4 Legal Framework

The relevant U.S. legal framework includes both federal and state level laws. At the federal level, the key law is the Nuclear Waste Policy Act of 1982, mentioned above. A Nuclear Waste Fund (NWF), established in 1982 as part of the NWPA, receives payments from entities generating nuclear waste, mainly commercial electric power utilities. Fees paid to the NWF are based on the total amount of electricity generated; in effect, this fee is paid by electricity ratepayers. The total amount that has been paid into the NWF is over $30bn: more than $9bn has been used for program costs (O'Keefe 2013). The NWF is to be used, in part, for development and maintenance of national geologic disposal facilities.

Other policy instruments, regarding issues such as safety criteria, monitoring and compensation measures, are found in a wide range of laws and regulations that apply to the handling and disposal of nuclear and other hazardous forms of waste or that regulate pollution-causing activities. According to information provided by the U.S. government to the International Atomic Energy Agency (IAEA 2013a), these include various laws concerning energy policy and related matters (such as the Atomic Energy Act of 1954, and the Energy Policy Act of 2005) and federal environmental laws regarding pollution and environmental quality (such as the National Environmental Policy Act, and the Clean Water Act, among others). There are a number of laws and regulations specifically related to radioactive waste (including the "Licensing requirements for Land Disposal of Radioactive Waste" of 1960 and the "General Guidelines for the Recommendation of Sites for Nuclear Waste Repositories" of 1999).

Compensation for nuclear waste-related incidents is governed in accordance with the Price-Anderson Nuclear Industries Indemnities Act, which regulates liability for civilian nuclear facilities. Originally enacted in 1957 for a period of ten years, the Price-Anderson Act has been repeatedly extended, and currently remains in effect through 2025. This law created a "no-fault" insurance fund (amounting to $12.6bn as of 2011) that would supplement mandated industry-purchased private insurance coverage of $375m per facility (as of 2011), which is supplied through the American Nuclear Insurers (ANI) insurance pool formed by 60 U.S. private insurance companies (NRC 2011). The government fund, financed by mandatory payments from power companies, would also cover incidents in connection with the transportation of certain HLW waste to an eventual repository. According to the NRC, "[c]laims can include any incident (including those that come about because of theft or sabotage) in the course of transporting nuclear fuel to a reactor site; in the disposal of nuclear fuel or waste at a site; in the operation of a reactor, including the discharge of radioactive effluent; and in the transportation of irradiated nuclear fuel and nuclear waste from the reactor" (NRC 2011). However, the NRC also clarifies that "Price-Anderson does not require coverage for spent fuel or nuclear waste stored at interim disposal facilities, transportation of nuclear fuel or waste that is not either to or from a nuclear reactor, or acts of theft or sabotage occurring after planned transportation has ended."

Laws and policies at the state and local level are also relevant. In 1987, the Nevada state legislature created a new county, the short-lived Bullfrog County, as a means to recover a greater amount in tax revenues from economic activities related to the Yucca Mountain repository. This zero-population country was ruled to be in violation of the state constitution (Associated Press 1988).

A key provision of the NWPA granted state governments the ability to veto a federal choice of a nuclear waste depository site within their state. The State of Nevada did exercise its veto power in response to the Yucca Mountain determination in early 2002, but, as provided for in the NWPA, the state's veto was overridden by a congressional vote later that year (Patel and Ewing 2013).

5 Institutional framework

Regulatory functions relevant to HLW are performed by numerous agencies: at the federal level by the Nuclear Regulatory Commission, the Environmental Protection Agency, the Department of Energy and the Department of Transportation, and at the state-level by governmental agencies responsible for radiation protection and hazardous waste management (IAEA 2010c).

Various actors are involved in licensing and approval of siting and implementation. Of central importance is the U.S. Department of Energy (DOE), which performs most activities related to the disposition of spent fuel and other materials. The Office of Civilian Radioactive Waste Management (OCRWM), a unit within the DOE, had been responsible for management of the Yucca Mountain Project, but this office was eliminated after funding for the project was discontinued in 2010 (AllGov.com 2013).

The Nuclear Regulatory Commission (NRC), an independent government agency, develops regulations relevant to the planning and operation of repositories and the handling of nuclear materials. The Environmental Protection Agency (EPA) establishes environmental standards, including radiation standards in connection with the management of spent fuel and radioactive wastes. Another relevant body is the Nuclear Waste Technical Review Board (NWTRB), an independent agency that provides technical assistance and advice to the DOE.

Other relevant actors include the owners and operators of commercial reactors and other waste-generating locations, as well as corporations that are contracted to manage and transport waste. The interests of the nuclear power industry are represented in part by organizations such as the Nuclear Energy Institute (NEI).

As has been seen, a key role is played by the courts, in that they may consider challenges to proposed rules and plans made by governmental agencies, allowing for issues to be revisited even after definitive-seeming decisions have been made by the legislature.

In 2010, President Obama established the "Blue Ribbon Commission on America's Nuclear Future" (BRC). The BRC presented a final report on January 26, 2012, which called for immediate work to establish a geological repository, which should be in operation by 2048 (Patel and Ewing 2013) and to also develop a "pilot interim disposal facility." Key recommendations of the BRC report also included that a new organization be formed to provide oversight for related transportation, disposal, and disposal matters; that planning be adaptive, staged, consent-based and transparent; and that decisions be "science-based." (BRC 2012: iii-viii). In light of the failure to secure political support for the Yucca Mountain project, the BRC highlighted the importance of a "consent-based approach" to siting that secures host community approval.

6 Information policy and form of participation including civil society

On its website, the NRC emphasizes its open approach to governance. Regarding participation procedures, the NWPA required consultations with affected states and tribes. The original indigenous inhabitants of the Yucca Mountain site are the Western Shoshone and Southern Paiute groups; although there are records of community consultation events that were held with tribal representatives, there may have been inadequate prior informed participation of affected groups in project planning.

Regarding future involvement of stakeholders in nuclear waste governance, the Blue Ribbon Commission recommended that "all affected levels of government (local, state, tribal, etc.) must have, at a minimum, a meaningful consultative role in all other important decisions. Additionally, states and tribes should retain – or where appropriate, be delegated – direct authority over aspects of regulation, permitting, and operations where oversight below the federal level can be exercised effectively and in a way that is helpful in protecting the interests and gaining the confidence of affected communities and citizens" (BRC 2012: vi).

The issue of nuclear waste management is governed by the same policies that have been useful for achieving public access to information and participation in decision-making on other key environmental issues; these include the Freedom of Information Act (FOIA), the Administrative Procedures Act (APA), and the National Environmental Policy Act (NEPA).

Civil society organizations appear to have not been very actively involved in the Yucca Mountain issue or other issues of nuclear waste management. A few specialized groups, including the Union of Concerned Scientists (UCS) and the Natural Resources Defense Council (NRDC), however, had expressed criticism of the selection of the repository site, as well as regarding procedures and activities related to the Yucca Mountain Project. The fact that environmental groups are in general more supportive of the Democratic Party may account for some reticence in challenging the decision made by Majority Leader Reid and President Obama to end federal funding for the repository.

7 Conclusion and lessons learned

Governance in the United States is characterized by multiple levels of potential intervention and veto opportunities. While policies are set by the Congress at the federal level, with specific implementation steps developed by relevant administrative agencies, there remain opportunities for political actors to challenge or reverse decisions once they have been made.

The ongoing inability of the United States to move forward on the policy of developing an operational national deep geologic repository for HLW appears to be a case in which powerful political actors have decided to derail what they see as a politically inexpedient course of action, thereby ignoring a clear "scientific consensus" justifying actions that conflict with their perceived political goals.

The 2008 DOE recommendation to proceed with the Yucca Mountain project emphasized that in adopting the NWPA the Congress had "recognized the overwhelming consensus in the scientific community that the best option for such a facility would be a deep underground repository" (DOE 2002: 1). It further noted that there was "a worldwide scientific consensus that deep geological disposal…is the best option" (DOE 2002: 2) and claimed that such an approach was in fact "the only scientifically credible, long-term solution" (DOE 2002: 26) as well as "the only answer that has any degree of realism" (DOE 2002: 29). Responding to this scientific consensus, the political consensus in the United States appeared to have coalesced around the need for a deep geological repository, specifically at Yucca Mountain. Nonetheless, building a HLW repository requires political determinations, not just scientific ones. The overall process, which requires sustained commitments over many decades, can be at the mercy of changing perceptions and attitudes, as well as the vicissitudes of an ever-evolving political calculus.

The U.S. experience over the past three decades does not bode well for the longer-term governance of HLW, for which management should proceed in a prudent and predictable manner, based on sound science – and which indeed must do so for many, many millennia. Given the current situation, it seems unlikely that the United States will be able to designate, implement and operate a long-term geological waste disposal repository within the coming decades.

Much like the tacit consensus that prevails in the United States, for the time being, to neither significantly expand nor reduce the scale of nuclear power generation, there also appears to be a consensus to not actively address and solve the long-term waste disposal issue. The management of HLW in the United States is likely therefore to continue to exist in a state of limbo, without being settled decisively. In the absence of any obvious and urgent crisis, there remains little discussion of the issue by civil society organizations, politicians or the media.

The decades of accumulated actions and decisions – the billions of dollars invested, millions of man-hours worked, thousands of pages of studies produced, and hundreds of test holes bored into Yucca Mountain's layers of rock – adds up, for the time being, only to a lack of decision, or rather a decision to not decide.

Asked what he thought about the August 2013 U.S. Court of Appeals ruling, Senator Reid said: "As a result of a political compromise, we put some really bad judges on the (Washington) D.C. circuit court and they produced a 2-1 decision requiring the Nuclear Regulatory Commission to license Yucca Mountain. Their opinion means nothing. Yucca Mountain is dead. It's pad-locked. There's nothing going on there" (Quoted in Velotta 2013).

While Senator Reid may feel that he has had the final word on the matter, the country's large and ever-growing volume of high-level radioactive waste remains scattered among more than one hundred vulnerable sites, often near population centers. This waste will remain a hot issue that will need to be dealt with eventually, somehow. Perhaps it is too soon to know for certain if Yucca Mountain is indeed truly dead.

References

AllGov.com (2013). Office of Civilian Radioactive Waste Management. http://www.allgov.com/departments/department-of-energy/office-of-civilian-radioactive-waste-management?agencyid=7408, last accessed 25 August 2013.

Associated Press (1988, February 13). Nevada County is Held Illegal. *The New York Times*. 1988. http://www.nytimes.com/1988/02/13/us/nevada-county-is-held-illegal.html, last accessed 6 June 2013.

Blue Ribbon Commission on America's Nuclear Future (BRC) (2012). Disposal Subcommittee Report to the Full Commission – Updated Report. http://cybercemetery.unt.edu/archive/brc/20120620220845/http://brc.gov/sites/default/files/documents/disposal_report_updated_final.pdf, last accessed 24 August 2013.

International Atomic Energy Agency (IAEA) (2010a). United States of America 2008 – Laws/Regulations. http://newmdb.iaea.org/admin/Reports/Published/Laws.aspx?Pub=1&IsoCode =US&PeriodID=8&ByCountry=US&ByYear=, last accessed 5 October 2013.

IAEA (2010b). United States of America 2008 – Policy Survey. http://newmdb.iaea.org/admin/Reports/Published/Policies.aspx?Pub=1&IsoCode=US&PeriodID=8&ByCountry=US&ByYear=, last accessed 25 August 2013.

IAEA (2010c). United States of America 2008 – Regulators. http://newmdb.iaea.org/admin/Reports/Published/Regulators.aspx?Pub=1&IsoCode=US&PeriodID=8&ByCountry=US&ByYear=, last accessed 23.08.2013.

IAEA (2013a). Profiles: United States of America. http://newmdb.iaea.org/profiles.aspx?ByCountry=US, last accessed 5 October 2013.

IAEA (2013b). United States of America – Policy Survey, in: http://newmdb.iaea.org/admin/Reports/Published/Policies.aspx?Pub=1&IsoCode=US&PeriodID=8&ByCountry=US&ByYear=, last access 25 August 2013.

Manaugh, G. (2009). One Million Years of Isolation: An Interview with Abraham Van Luik. BLDG BLOG. http://bldgblog.blogspot.de/2009/11/million-years-of-isolation-interview.html, last accessed 22 August 2013.

Nuclear Waste Policy Act of 1982. 96 Statutes at Large 2201, 42 U.S. Code 10101 et seq.

O'Keefe, W. (2013). Yucca Mountain and the Rule of Law. FuelFix. http://fuelfix.com/blog/2013/08/16/yucca-mountain-and-the-rule-of-law/, last accessed 15 August 2013.

Patel, P., and Ewing, R. (2013). United States launches new direction to manage nuclear waste. MRS Bulletin 38, No. 03 (2013): 206-207.

U.S. Court of Appeals District of Columbia Circuit (2013). No. 11-1271. Decided August 13, 2013. http://www.cadc.uscourts.gov/internet/opinions.nsf/bae0cf34f762ebd985257bc6004deb18/$file/11-1271-1451347.pdf, last accessed 6 October 2013.

U.S. Department of Energy (DOE) (2002). Recommendation by the Secretary of Energy Regarding the Suitability of the Yucca Mountain Site for a Repository Under the Nuclear Waste Policy Act of 1982. http://web.archive.org/web/20080917165343/http://www.ymp.gov/ym_repository/sr/sar.pdf, last accessed 23 August 2013.

U.S. Department of Energy (2010). U.S. Department of Energy's Motion to Withdraw. http://energy.gov/sites/prod/files/edg/media/DOE_Motion_to_Withdraw.pdf, last accessed 23 August 2013.

U.S. Nuclear Regulatory Commission (NRC) (2011): Fact Sheet on Nuclear Insurance and Disaster Relief Funds. http://www.nrc.gov/reading-rm/doc-collections/fact-sheets/funds-fs.html, last accessed 25 August 2013.

NRC (2012). Fact Sheet on Storage of Spent Nuclear Fuel. http://www.nrc.gov/reading-rm/doc-collections/fact-sheets/storage-spent-fuel-fs.html, last accessed 4 October 2013.

NRC (2013). Spent Fuel Storage in Pools and Dry Casks – Key Points and Questions & Answers. http://www.nrc.gov/waste/spent-fuel-storage/faqs.html, last accessed 4 October 2013.

U.S. Office of Management and Budget (OMB) (2009). A New Era of Prosperity – Renewing America's Promise. http://www.whitehouse.gov/sites/default/files/omb/assets/fy2010new era/ A_New_Era_of_Responsibility2.pdf, last accessed 25 August 2013.

U.S. Senate Committee on Environment and Public Works Majority Staff (2006). Yucca Mountain: The Most Studied Real Estate on the Planet – Report to the Chairman Senator James M. Inhofe.

Velotta, R. N. (2013, 22 August). Reid: Obamacare isn't perfect, but it's good; Yucca mountain is dead. *Las Vegas Sun*. http://www.lasvegassun.com/news/2013/aug/21/reid-obamacare-isnt-perfect-its-good-yucca-mountai/, last accessed 23 August 2013.

V. Countries with Long-Term Surface Storage for High Level Radioactive Waste

With Access to the Future
Nuclear Waste Governance in The Netherlands

Maarten J. Arentsen

1 Introduction

Nuclear waste governance in the Netherlands has been almost permanently on the national government's political agenda ever since the start of the Dutch nuclear energy program. The Dutch program was initiated by the U.S. Atoms for Peace program, which acted domestically as a guide in the development of the Dutch program. Nuclear waste became part of this nuclear program and, at certain points in time, the governance of nuclear waste itself became a hot topic, one that was fiercely disputed and much debated. The treatment of nuclear waste became controversial in Dutch society, as did nuclear energy itself (LAKA 2008). At certain times, nuclear waste became a hot issue due to local resistance to research and experimentation with underground storage. At other times, the issue was discussed regarding the Dutch position on nuclear power. Ever since the initial steps were made, the Dutch have discussed and planned to increase the country's nuclear power capacity; however as of spring 2014, the country had only one operational nuclear power plant of 450MW and one 50MW non-operational power plant, which is awaiting dismantling.

Over the years, nuclear waste became both a disputed issue and a technocratic policy issue (COVRA 2008). The policy focus is on optimal regulation in accordance with international standards and agreements, low risk, safe and strictly controlled storage, and permanent research in long-term storage options. The volume of nuclear waste is relatively small in the Netherlands and there are no signs that the nuclear landscape will be enhanced by more reactors in the coming decades. Between 2005 and 2010, two industrial consortia initiated plans for new nuclear power capacity in the Netherlands. However, these initiatives have been halted, despite the positive attitude of the country's then right-wing governing coalition. The feasibility of nuclear power in a small, densely populated country like the Netherlands is very much open to question.

This is probably one of the reasons for the transparency in nuclear waste governance in the Netherlands. The dominant approach is retrievable permanent storage to allow the reprocessing of the stored nuclear waste, should future state-of-the-art technology provide low hazard treatment and storage options. The policy is labelled "retrievable storage", with governance-by-learning as the dominant governance approach.

2 The present policy on radioactive waste storage

The current radioactive and nuclear waste policy is the result of only a few clear political decisions taken in the past.[1] It should be noted that Dutch policy distinguishes between radioactive and nuclear waste even though both types of waste are regulated under the same set of rules and regulations. The core of the present policy was decided first in 1984 and then again in 2002. In 1984, it was decided that the management and control of nuclear waste storage in the country should be centralized; the retrievable storage approach was reconfirmed in 2002.[2] The retrievable storage approach, which had already been prepared in the 1990s, states that the underground storage of highly toxic nuclear waste (the highest classification of nuclear waste) in deep geological repositories shall only be permitted if the waste remains accessible for renewed treatment and processing when state-of-the-art technology becomes available in the future. Based on research done in the 1990s, the best option for a deep repository would be underground storage in clay and salt layers in the northern part of the Netherlands. However, this kind of underground storage facility has never been realized due to massive local resistance (LAKA 2008).

Dutch nuclear waste policy is guided by and structured according to several principles. The first principle is *avoiding as much as reasonably possible the genesis of nuclear waste*. This point addresses in particular the fuel used in nuclear installations. Spent fuel can be reprocessed by enrichment, and this is indeed what occurs with the spent fuel from the Dutch nuclear power plant. The fuel is enriched in France (Borssele plant) or the United Kingdom (Dodewaard plant, which is no longer operational) and the waste from the enrichment process is sent back to the Netherlands and stored in COVRA's nuclear waste storage facility in Borssele.

If waste is produced, then the second principle states that its treatment should meet the highest standards of the radiation protection policy, providing

1 The content of this section is based on: Besluit tot wijziging van het Besluit in-, uit- en doorvoer radioactieve afvalstoffen en bestraalde splijtstoffen, het Besluit kerninstallaties, splijtstoffen en ertsen en het Besluit stralingsbescherming in verband met de implementatie van richtlijn 2011/70/Euratom.

2 Kamerstukken II, 1983/84, 18 343, nrs. 1-2 beleidsstandpunt uit 2002 Kamerstukken II, 2002/03, 28 674, nr. 1; Nota Radioactief afval van 1984.

the *highest protection of man and the environment* against the negative impact of exposure to ionizing radiation.

The third principle states that, besides the radiation protection regulation, *specific waste regulation applies* according to the standards of isolation, management, and control until the waste has lost its activity and toxicity, or until a satisfactory technology is available to reduce the environment's radiation exposure to minimally acceptable levels. This last principle implies that the nuclear waste must be stored for a period of at least one hundred years in controlled buildings at a single location. This period allows for monetary savings, as well as continued research into a safe and controlled final waste repository. Part of this regulation is that a final waste repository should allow for the future reclamation of the waste for further processing. In other words, every underground storage facility should remain accessible under all circumstances. The fourth and last principle is *the polluter pays principle*. All the costs of nuclear waste storage fall under the responsibility of those who caused the waste. Payment is required for transport, containment, and storage.

Part of the policy involves the implementation of a waste policy through centralization and public control by a single publicly owned company. This company is called Centrale Organisatie Voor Radioactief Afval (COVRA) and is responsible for the transportation and storage of nuclear waste, as well as for initiating research on longer term repository options.

The idea of retrievable storage is one of the core ideas of nuclear waste policy in the Netherlands. It means that the highest categories of waste with the longest activity should end up in a repository which will continue to be accessible for centuries. It also means that policy can keep up with state-of-the-art treatment and storage technology for nuclear waste. Dutch nuclear waste policy is therefore basically learning-oriented. An additional advantage of this policy approach is that the reclamation repository contributes to a constructive societal dialogue on the nuclear waste problem.

3 Waste types and amounts

Dutch waste classification follows the IAEA's international standards and guidelines, which distinguish between low and intermediate levels of radioactive waste, and high level radioactive waste (See Table 1 below). COVRA does not further classify the first type of waste between A-type (short lived) and B-type (long lived) radioactive waste in its communications to the media. The 2012 annual report gives some further details about quantities of radioactive waste in country. In 2012, a total of 2190 containers with low and intermediate levels of radioactive waste and 28 containers with high levels of radioactive waste were received and processed. This represented a decrease compared to previous years. Within the low and intermediate level category, the annual report also lists an intake of 425 containers of Naturally Occurring Radioactive Material. This is waste with increased levels of natural radiation stemming from industrial processes (calcinate and depleted uranium). Table 1 shows the total accumulated stored amount of radioactive waste in 2012.[3]

Table 1: Cumulative stored nuclear waste in the Netherlands (2012)

Category	Volume in m^3
A	1,490
B	5,159
C	2,686
D	1,000
Total LMRA	10,335
Total HRA	73

Source: COVRA, Annual Report (2012).

The low and intermediate level waste originates from radioactive apparatus, smoke detectors, research, and medical treatment. The amount of high level waste is limited compared to that of other countries with more nuclear facilities. The high level waste in the Netherlands comes from the sole nuclear power plant, the production of radiological isotopes, and two research reactors, one in Petten and one in Delft. The nuclear waste from the Borssele power plant is processed in France and the resulting high level waste is returned to the Netherlands for storage. Urenco, a company that enriches uranium by ultracentrifugation, has a production site in the Netherlands, but it is not known whether this company produces radioactive waste. All radioactive waste is centrally stored in buildings at one location in Borssele for at least 100 years. Thereafter, a final repository is needed, but the preferred options are still the subject of research in the Netherlands.

3 COVRA, Annual Report (2012), Vlissingen (2013).

4 Waste storage concept(s) and its costs and financing

Several concepts have guided the development of Dutch nuclear waste. First, the dominant concept of nuclear waste management is a concept of (permanent) retrievability. This states that all geological repositories should continue to be accessible to all types of radioactive waste. As indicated above, this allows state-of-the-art technology to be employed to make the waste harmless in centuries to come. This part of Dutch waste storage is a work in progress, because the final geological repository has not yet been decided on and is still the subject of research. The general expectation is that the Dutch will need to develop a geological repository to hold waste from the year 2080 on.[4] This is because of the second waste concept applied in the Netherlands, which states that all radioactive waste should be stored above ground in buildings for at least a hundred years. This period allows the highest level waste to cool down, and guarantees optimal control, isolation, and management of the waste.

The third waste concept is the application of ionizing radiation protection standards, and means that unacceptable exposure to radiation should be avoided at all times. The first step in guaranteeing this is to prevent the emergence of nuclear waste as much as possible. If waste is produced, then the principles of isolation, management, and control will apply. These principles have been effectuated by centralizing the management and regulation of radioactive waste in the Netherlands. This is reflected in the existence of a single responsible, publicly owned organization at one location, with all regulation concentrated at the national tier of government.

The fourth and final concept is that Dutch radioactive waste shall follow the international classification of radioactive waste types as discussed in the previous section. All waste types are centrally stored at a single location.

The polluter pays principle

Dutch nuclear waste management is financed according to the principle that polluters pay for the costs. These costs are divided into the costs of actual storage for at least one hundred years and the costs associated with a final geological repository, which should be covered by the capital growth fund, projected to grow to two billion euros in 2130. Two billion euros is currently the estimated cost of a final geological repository.

4 In the early 1990s, The Netherlands added a new perspective to nuclear waste disposal: a retrievable geological repository, indicating that the final geological disposal repositories should be accessible and the stored waste retrievable for treatment with future state-of-the-art technology.

This means that companies producing nuclear and radioactive waste are charged for transport and actual storage, and contribute to the capital growth fund for final disposal. The tariffs are calculated on a cost-plus basis, since COVRA is a non-profit organization.

Table 2: Costs of nuclear waste management in 2012

Category	Amount in million €
Transport and processing	13.811
Actual storage	1.109
Capital fund investment	2.379
Total income	**17.299**
Total costs	**15.925**
Cost plus	9%

Source: COVRA, Annual Report (2012).

Table 2 shows that the total 2012 budget is about €17.3m, which represents a decrease from 2011 due to a drop in the generation of nuclear and radioactive waste. Moreover, the 2012 annual report reveals COVRA's strong dependence on political and societal considerations and decisions, particularly with respect to nuclear energy. As mentioned above, the most recent initiatives for new nuclear power capacity have been cancelled, which also affects COVRA's financial position. Tariffs were increased by 3% in 2012, which resulted in a positive company result of 9%. However, in 2010 and 2011 the results were 17% and 19% respectively, indicating a poorer financial result in 2012. However, due to the polluter pays principle and COVRA's non-profit status, the customers' payments will cover the costs of storage and investment in the growth fund for a final geological repository.

5 Legal framework[5]

The first attempt at legal regulation of ionizing radiation stems from 1931, when the Dutch Parliament accepted the Ionizing Radiation Protection Act. This legislation was meant to offer protection from the application of X-ray

5 The principal reference for the content of this section is Uylenburg et al. (2007).

equipment, but never went into effect because it supposedly neglected and underestimated the professional competence of physicians. Between 1931 and the enactment of the Nuclear Energy Law in 1963, the application and management of ionizing radiation was therefore regulated under general environmental and occupational safety laws and guidelines. At that time, there were no operational nuclear sites in the Netherlands; the first nuclear power plant, the Borssele plant, went into operation in 1973.

The 1963 Nuclear Energy Law is a formally enacted law and is still in effect. It regulates all peaceful applications of nuclear and radioactive materials and waste in the Netherlands, as well as protection against ionizing radiation. The law essentially provides a legal framework, as norms and standards are all part of the rules of conduct, guidelines, and regulations enacted under the law. Safety standards, guidelines, and rules are decided internationally and the Netherlands automatically follows the international requirements in its legislation and regulation.

The regulation is based on three principles and applies one legal instrument: a conditional prohibition on the licence (permit). The first principle is the justification of the application of nuclear/radioactive materials and the production of nuclear waste. Applications should be justified and this justification is standardized through a set of permitted applications of ionizing materials with the accompanying production of ionizing waste.

The second principle is referred to as, "As Low As Reasonably Achievable" (ALARA). This principle requires the application of state-of-the-art protection measures. Protection measures are dynamic, in the sense that all results of ongoing (international) research on health and safety are constantly included as updates of the safety requirements.

The third and final principle states that if the application of materials and the production of waste is legitimate, and the best protection measures are applied, the exposure of humans to ionizing radiations may never exceed the dose limits. Similar to the safety measures, the dose limits are also the subject of constant, ongoing international research.

The dominant instrument is the conditional prohibition by licence. The licence describes in great detail the dos and don'ts relevant to the licence holder in the application of ionizing radiation. The law distinguishes between sites utilizing radioactive materials and X-ray equipment, on the one hand, and nuclear materials on the other. The governance of nuclear waste in the Netherlands is licenced under the nuclear materials regulations of Article 15b of the law, which of course also applies to the nuclear power plant in Borssele. Part of this regulation involves licencing the transport of nuclear waste. Only one company in the Netherlands, the radioactive waste company COVRA, is licenced to transport radioactive materials in the Netherlands and in the international exchange of nuclear waste.

The licencing procedure under the Nuclear Energy Law commits the licence holder to legal provisions for public participation. These provisions are part of general public law and therefore also apply to the siting and licencing of nuclear sites and nuclear waste facilities. The conditions for participation and the types of participation by the general public are prescribed in detail in the acts. An environmental impact assessment is always required when siting and licencing nuclear installations and the nuclear waste facility.

The Nuclear Energy Law is considered a complex law for two reasons. First, this is because the actual regulations are not stated in the law but are set out in general rules, codes of conduct, guidelines, and regulations enacted under the law. Second, the legal span of control is related to general environmental law in the Netherlands, but the relationship is made legally complex by the specificity of the nuclear and radioactive materials covered by the Nuclear Energy Law.

6 Institutional framework

The institutional context for Dutch nuclear waste governance has a layered structure, reaching out to the European level (Euratom) and the global level (UNSCEAR and IAEA). Within the country's borders there is the governmental layer, the science and technology layer, and the industrial layer. Recent years have witnessed significant changes in the governmental layer, as indicated in Figure 1 below. The governmental layer distinguishes between responsibility for nuclear waste policy and responsibility for nuclear waste monitoring and control. In the past both responsibilities could fall under the same ministry, but today the responsibilities belong to different ministries.

Previously six ministries were involved, each with its own responsibility and inspectorate. Apart from the economic department, all of the departments were involved in work on the themes of safety, non-proliferation, and international exchange and cooperation. On the safety front, there was a distinction between occupational safety of the workforce, the safety of the natural environment related to public safety, and food safety. Every safety focal point had its own inspectorate, including the safety of transport, which is related to environmental safety and occupational safety. Each of the ministries also had international contacts and relations with the relevant international bodies, but the ministries were always accompanied by Foreign Affairs officials, who are responsible for the country's external relations.

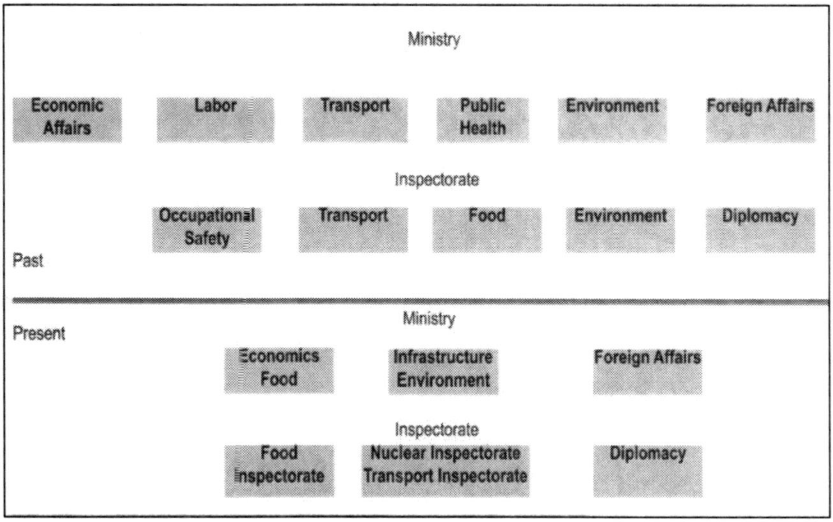

Figure 1: Past and present ministries and inspectorates

The lower half of the figure shows the governmental layer after the mergers, reflecting the situation in spring 2014. The Ministries of Labor and Public Health are no longer involved. Transport and Environment have been integrated into the new Ministry for Infrastructure and Environment, and Economic Affairs has taken over Food and the Food Inspectorate. The Nuclear Inspectorate is now an agency under the Ministry of Infrastructure and Environment and actually unites the former environmental and occupational health expertise on nuclear installations, which was previously part of two separate ministries.

With respect to the division of tasks, Economic Affairs is the licensing authority, and Infrastructure and Environment is the inspection authority. Of course, both ministries collaborate because nuclear expertise is concentrated in the Nuclear Inspectorate of Infrastructure and Environment. The transport inspectorate is part of the same Agency as the Nuclear Inspectorate.

The science technology layer consists of several types of organizations. First, the Dutch universities are involved in nuclear (waste) oriented research. Research on reactor technology is concentrated at Delft University, whereas nuclear fusion and superconductivity is concentrated at the University of Twente. The large Dutch Energy Technology Institute in Petten also hosts a reactor, but it is used essentially for the production of radiological isotopes.

The second organization is the Dutch nuclear waste company COVRA. This company is not only responsible for the transport and storage of nuclear waste of all categories, but also for initiating research on geological storage. This research is guided by research programs run by COVRA; however, the research itself is outsourced to domestic and international research organizations. COVRA additionally participates in the international research network on nuclear waste storage.

The third type of organization in the science and technology layer is the professional organization for nuclear technology and radiological safety. The size of the first organization has been quite modest since the interruption of the Dutch nuclear energy program in 1986, after the Chernobyl incident. The radiological safety profession is well organized and unites the professions responsible for radiological safety on Dutch premises and radiological workers. Both professional organizations are part of the radiological landscape in the Netherlands. They comment and advise on themes and topics that fall within their areas of expertise.

The third layer in the institutional environment is the industrial layer, which basically consists of sites in the Netherlands where X-ray equipment, radiological, and nuclear substances are used. The single operational nuclear power plant, the other two sites with running nuclear reactors, the enrichment factory Urenco, and above all, COVRA, are the major organizations in the industrial layer.

The institutional environment in the Netherlands is highly professionalized in all three layers.[6] This means that the professional standards of the various disciplines and occupations guide policy making, rule and law making, and daily activities. This is in line with the expert-based, technocratic approach adopted in the international nuclear landscape. Nuclear (waste) themes and topics, as well as radiological safety, are complex issues which require high quality, scientifically-based professional knowledge and skills. The dominant culture in the institutional environment can therefore be labeled professional and technocratic. Local and regional government bodies are, of course, involved with siting issues. However, the involvement of these bodies is very limited since all responsibility for nuclear and radioactive waste is delegated to the central government. In practice, the local and regional authorities wield zero influence because their authority can be overruled by the central government under the Dutch Nuclear Energy Law.

6 For details see Arentsen (1998).

7 Siting procedures[7]

Dutch nuclear waste storage has followed the international community in terms of ideas, research, and options. The nuclear waste siting debate nevertheless has been quite passionate and dynamic. The history of nuclear waste siting can be divided into the following periods, each with its own dominant perspective on waste storage:

Table 3: Chronology of Dutch nuclear waste management

Period	Perspective
1967-1982	Ocean dumping of LLW Temporary surface storage other waste levels Experimenting with geological storage in salt taverns
1983-1992	Temporary surface storage of all waste levels Researching geological storage in salt taverns
1992-2092	Temporary surface storage for at least one hundred years Continued research and experimentation on geological storage options
2131 onwards	Retrievable geological repository

This synopsis shows that Dutch siting history displays policy learning. In particular, in the 1970s and early 1980s, the expectation about available final geological disposal options was quite high in comparison to the expectations held by other countries in the world. The atmosphere at that time was still one of nuclear optimism. In 1973, the first Dutch nuclear reactor was connected to the grid and the then governing coalitions were headed for "more nuclear" in the coming decades.

Decisions on more nuclear capacity were centralized and became a legally grounded responsibility of the central government; they were thus no longer in the hands of the electricity industry. The nuclear ambition also made waste storage an urgent issue; at the time the salt caverns in the northern part of the country were considered suitable locations. Experimental drilling was announced and taken into consideration in three northern provinces, but these options were fiercely contested by the local population. The protests were only one of the many manifestations of the growing resistance to nuclear energy and its related problems, such as nuclear waste.

7 The principal references for this section are www.laka.org and kernenergieinnederland.nl.

The anti-nuclear coalition started organizing in those days and debated nuclear waste disposal as an issue of nuclear energy. Nuclear energy was resisted, as was experimental drilling in geological formations for nuclear waste disposal. Between 1981 and 1983 The Netherlands organized a societal debate on future energy provision in an attempt to depoliticize the anti-nuclear atmosphere in Dutch society. The attempt failed because the society-wide debate was completely dominated by the pros and cons of nuclear energy. The slogan of the final report, published in 1983, reflects the frustration of the committee in charge of the organization and content of the debate: "Those of you who think they know it all are very annoying to us, who do" (Stuurgroep MDE 1984: 2). The experts on the committee could not convince Dutch society of the need for more nuclear energy and the entire debate was considered a failure. However, the nuclear ambition was not totally diminished by the outcomes of the debate; options and policies for "more nuclear" continued until the Chernobyl nuclear incident. Only the meltdown, and not public opinion, was able to bring a halt to the country's nuclear ambitions.

However, the results of the nationwide debate on future energy initiated a change in nuclear waste policy. The 1984 White Paper on nuclear waste announced centralized surface storage for at least one hundred years as the intermediate storage option, and the continuation of research on a final geological repository. The authorities had learnt from the resistance and no longer considered any experimental "drillings" in the north of the country. The surface storage waste facility was erected in Petten, where an experimental reactor was in operation at that time. Later, the storage facility was moved to its current location in Borssele (LAKA 2008).

The COVRA plant has been completely operational since 2003 and all categories of Dutch nuclear waste are disposed in its buildings. In the meantime, research on final disposal continues, concentrating on geological formation in Boom clay and now also considering the option of international disposal. It is expected that the final geological repository will not be in operation before 2130 in the Netherlands. From that period onwards, retrievability will continue as the guiding perspective in nuclear waste storage.

COVRA's current location next to the nuclear power plant in Borssele was one of the two remaining locations recommended by the commission that advised the then-Dutch government on the location of the nuclear waste facility. The other location was Moerdijk, a large petrochemical industrial area south of Rotterdam. The municipalities of both locations only conditionally afforded their cooperation. The commission recommended Borssele because it thought that the proposed conditions could be accepted more readily than those raised by Moerdijk. In the end, the minister left the final decision on the exact location in

Borssele to the COVRA company itself. COVRA opted for the current location close to the nuclear power plant. The environment of the nuclear power plant was the best match forh COVRA's activities. An environmental impact assessment formed part of the siting procedure. At that time, COVRA opted for the most extensive procedure, allowing optimal input and participation from the neighboring institutions. The same approach was chosen in the latest procedure for licensing the planned enlargement of the buildings for HLW and LMLW. This procedure was launched in 2013. An environmental impact assessment also forms part of this procedure (see COVRA website for the assessment report).

8 Policy regarding public information and participation

Public engagement in matters of nuclear energy and nuclear waste has been quite strong in the past, particularly during the 1970s and 1980s. This participation was predominantly motivated by anti-nuclear attitudes and behavior. It was clearly manifested in the 1970s in the two Kalkar demonstrations, in which there was significant Dutch participation, and the protest against the enlargement of the Urenco site in Almelo in 1978. This latter protest was the largest anti-nuclear protest ever seen in the Netherlands, with some 40,000 participants (LAKA 2008).

The clearest manifestation of the public authorities' willingness to provide information and to participate has been the society-wide debate on the future of the Dutch energy system, which was held between 1981 and 1983. In 1985, this event was followed by a round of information events at locations nominated as possible sites for new nuclear capacity. These processes clearly show how information and participation management by the government functioned at that time. Even though the societal debate on energy had reconfirmed the anti-nuclear attitude in Dutch society, the government nevertheless decided to initiate four new nuclear reactors. It was assumed that detailed, solid information from experts would convince local populations. However, the anti-nuclear attitude was also strongly echoed in the provincial and municipal councils. Nobody wanted a nuclear power plant or a waste facility next door, and this was clearly manifested at every identified or nominated location in the country. As mentioned above, after Chernobyl the Dutch nuclear ambitions dropped to zero and stayed at zero, despite a brief revival between 2000 and 2010 when a second Dutch nuclear reactor was considered.

That period also saw the origin of research on Dutch public opinion about nuclear energy and nuclear waste (Slingeland et al. 2004). This showed that in 2002, about 21% of the population were positive about the role of nuclear in the

Dutch energy mix, as against only 10% holding that opinion in 2001. Moreover, the survey showed the close relationship between nuclear waste and nuclear energy. Most people appeared to be against nuclear because of the still unresolved waste problem. If the nuclear waste problem was comparable to problems associated with other energy wastes, then 59% of the population would have been in favor of nuclear in the energy mix. If the nuclear waste problems were "solved", the percentage of proponents of nuclear would have increased to 79%. In other words, public perception of the waste problem underlied the anti-nuclear attitude among the population.

Under the law, licensing procedures require the consultation of the public. In general, this consultation is open to those who believe they will be affected by the licensed activities. Environmental impact assessments form part of the licensing procedures and here, too, there are legally prescribed consultation rules. These rules are grounded in public law and apply to all hazardous waste licensing procedures in the Netherlands. The procedures allow every Dutch inhabitant to object to decisions taken by the public authority if he or she thinks that the decision is against his or her interest.

COVRA has learnt from the past and wants to communicate openly. All general information about nuclear waste can be found on its website and in annual reports containing all details about the types and quantities of nuclear waste. Plans on policy and strategy are also available on the website. A recent example of open communication with the neighboring municipalities and citizens is the application for a new license to enlarge the current storage buildings. COVRA initiated an environmental impact assessment procedure that legally requires the maximum communication with neighbors. In this way, COVRA wants to demonstrate its transparency to the outside world. COVRA is also open to individuals and groups who wish to visit the site and offers two documents on the website for youngsters of primary school age. In addition to COVRA, the nuclear inspectorate also provides information on nuclear energy and nuclear waste on its website.

9 Conclusions

The analysis shows that nuclear waste governance in the Netherlands is a mixture of politics and technocracy, with periodical clashes between politics, industry and society. Nuclear waste governance has always been accompanied by societal resistance. Nuclear energy was and still is a highly contested energy option, and this has left the country with only one nuclear reactor despite its nuclear ambitions in the early 1980s.

The single nuclear power plant makes the management of the high level nuclear waste relatively simple, compared to management in countries with more nuclear plants. The Dutch also decided to implement the centralized government controlled model for nuclear waste management in the 1980s, with controlled storage in buildings at a single location for at least one hundred years. This centralized coordination model was extended by integrating the concept of a retrievable geological repository for long term storage. This latter concept is the subject of continuing scientific research, both nationally and internationally. Retrievable storage helped to alleviate the controversy about nuclear energy and the accompanying nuclear waste problem. The Dutch company COVRA, which is in charge of nuclear waste treatment, also coordinates national programs on nuclear waste research under the title OPERA. This program is concerned with exploring geological repositories in Dutch salt and clay layers. International collaboration is part of the research.

The continued research indicates that learning is still an important part of nuclear waste governance in the Netherlands. Many aspects of nuclear waste treatment and storage are still unknown and require scientific research. The continued need for scientific research is perhaps one of the most important lessons learnt in Dutch nuclear waste governance. The learning process has been part of a wider international learning process aimed at discovering adequate options for nuclear waste treatment and storage. History shows that the problem was underestimated in the beginning, as nuclear waste started to emerge, or else it was simply neglected, by dumping nuclear waste at sea. The learning process also adheres to stakeholder management and the management of societal resistance. Here the Dutch experimented with different models, ranging from confrontation to organizing a society-wide dialogue. None of this succeeded in finding an ultimate way to mitigate resistance and protest. Nuclear energy and nuclear waste issues continue to split Dutch society. The latest surveys show that more Dutch inhabitants would be in favour of nuclear energy if the waste problem were solved. However, the societal debate will not blaze up again, due to a complete lack of ambition to increase the number of nuclear power plants in the Netherlands in the coming decades.

This all goes to show that the ever-present and continuing societal resistance to nuclear energy and nuclear waste cannot be mitigated only by smart communication strategy. The causes of the resistance lie deeper and go to the core of the matter. Therefore, another lesson of history is that the deeper causes of resistance extend through time, although the number of protesters at any one time might vary. During the 1980s, the protest and resistance were quite fierce, whereas now they are relatively restricted in number. However, the causes of the resistance have not changed: people are anti-nuclear because of gaps that they perceive in the management of the resource, in particular its long half-life and high radiation activity. Although the number of protesters has decreased drastically, nuclear activities are constantly and continually being monitored by critical groups. Protests, however, reveal only a handful of participants, compared to the number protesting against nuclear energy and nuclear waste in the 1980s and 1990s. The way the protests manifest themselves have also changed. Massive demonstrations no longer predominate. Protest groups actively follow policy developments and the activities of the nuclear industry, and they adopt legal means to bring the protest arguments into the policy arena. In a way, therefore, protest has become institutionalized; the form of protest has changed, but its content has not. The latest COVRA licensing procedure is a clear example of this. The procedure is being monitored by the anti-nuclear movement and is documented on their website. Members of the movement take the legal opportunities to protest the extension of the storage facilities at the COVRA site. Part of the argument against the expansion is the historical argument, which is against the uptake and application of nuclear as a technological option.

Viewed from that perspective, the context of nuclear technology and the accompanying nuclear waste has changed dramatically within the wider European context. Liberalization of European electricity markets significantly changed the prospects and the attractiveness of nuclear technology. The German decision to phase out nuclear also undermined the prospects of nuclear technology for electricity production in Europe. Both liberalization and the German decision have helped to strengthen the Dutch anti-nuclear climate in recent years. The latest initiatives to develop a second nuclear reactor were halted and there are no signs that this will change in the future. Today's electricity market in the Netherlands is focused on renewables, and energy companies face difficulties keeping their fossil-based businesses profitable. There is little room for technological adventures like investing in nuclear capacity, which is viewed as financially "unviable". History has also shown that investment in nuclear energy has societal risks for the investor, due to the overall negative perception of nuclear power in Dutch society.

Because of the societal anti-nuclear climate, it is very important that the licence to operate in the nuclear waste management field is not harmed or threatened by errors, incidents, or malfunctioning in the responsible organizations. The model of central coordination, in combination with clear regulations and internationally anchored safety standards, open communication, and transparent policy making, has provided optimal conditions for adequately functioning nuclear waste management in the Netherlands until the present day. We shall have to see whether this model will withstand future requirements for nuclear waste management in a small, crowded country like the Netherlands.

References

Arentsen, M. J. (1998). The Invisible Problem and How to Deal with it: National Policy Styles in Radiation Protection Policy in the Netherlands, England and Belgium. In Bemelmans-Videc, M. L., Rist, R. C, and Verdung, E. O. (Eds), *Carrots, Sticks and Sermons. Policy Instruments and Their Evaluation*, Transaction Publishers, London, 211-231.

Arentsen, M. J. (2006). Contested technology: Nuclear power in the Netherlands. *Energy & Environment*, 17(3), 373-382.

Centrale Organisatie Voor Radioactief Afval (COVRA) (2008). Actief voor de Toekomst. Beleidsplan COVRA 2009-2014, Borssele.

COVRA (2013). Jaarrapport 2012, Borssele.

Hagendijk, R., and Terpstra A. (2004). *Technology, risk and democracy: The Dutch nuclear energy debate (1981-1984)*, STAGE, University of Amsterdam.

LAKA (2008). Akties tegen kernenergie 1960-2007, Amsterdam.

Mulder, K. (2012). The dynamics of public opinion on nuclear power. Interpreting an experiment in the Netherlands. *Technological Forecasting & Social Change,* 79 (2012) 1513–1524.

Schepers, M. J. J., Seebregts, A. J., Lako, P., Blom, F. J., and van Gemert, F. (2007). Factfinding kernenergie, ECN, Petten.

Slingerland S., Bello, O., Davidson, M., van Loo, K., Rooijers, F., and Sevenster, M. (2004). Kernenergie op herhaling, Ratenau Instituut, Den Haag.

Stuurgroep MDE (Maatschappelijke Discussie Energiebeleid) (1984). Het Eindrapport, Leiden.

Uylenburg, R., Vogelezang-Stoute, E. M., Neerhof, R., and van der Grijp, N. M. (2007). *Evaluatie Kernenergiewet*, STEM, Maastricht.

Government documents

Ministerie van Economische Zaken (2001): Terugneembare berging een begaanbaar pad? Onderzoek naar de mogelijkheden van terugneembare berging van radioactief afval in Nederland, Den Haag.

Besluit van 27 maart 2009: houdende implementatie van richtlijn nr. 2006/117/Euratom van de Raad van de Europese Unie van 20 november 2006 betreffende toezicht en controle op overbrenging van radioactieve afvalstoffen en bestraalde splijtstof (PbEU L 337) en intrekking van het Besluit in-, uit- en doorvoer van radioactieve afvalstoffen (Besluit in-, uit- en doorvoer van radioactieve afvalstoffen en bestraalde splijtstoffen).

Tweede Kamer der Staten Generaal, 1984-1985, 18343 Nr. 6, Radioactief Afval Brief van de Minister van Economische Zaken.

Tweede Kamer der Staten Generaal, 1979-1980, 15802 Nr. 2, Energiebeleid, Nota.

Tweede Kamer der Staten Generaal, 2009-2010, 31510 Nr. 40, Energierapport 2008.

Tweede Kamer der Staten Generaal, 201-2011, 32645 Nr. 1, Energierapport 2008.

Websites

Documentation on nuclear energy and waste in The Netherlands, in: www.laka.org

Documentation history of nuclear energy and waste in The Netherlands, www.kernenergieinnederland.nl

Nuclear Inspectorate, http://www.ilent.nl/onderwerpen/leefomgeving/nucleair_en_straling/index.aspx

Breaking the Stalemate
The Challenge of Nuclear Waste Governance in Italy

Maria Rosaria Di Nucci

1 Introduction[1]

The governance of radioactive waste in Italy is burdened by the legacy of over 50 years of discontinuous nuclear research as well as by incoherent technology and industrial policies. Nuclear waste management cannot be separated from the broader political debate on nuclear power, which was and remains a highly political issue. Italy has only a very modest park of four permanently shut-down power plants. Spent fuel (SF) and radioactive waste (RW) are being temporarily stored in at least 21 sites (Nuclear Energy Agency (NEA) 2013), some of which are dedicated to fuel reprocessing and manufacturing. Most of these facilities are obsolete and decaying. Due to harsh opposition, Italy has recently abandoned its original plans for a deep geological repository and opted for a centralised surface storage facility for low and intermediate level RW, which should also temporarily house high level RW.

Italy has a rather complex decision making system, characterised in some cases by a large number of entities and by a high level of fragmentation both vertically at the government level and horizontally at the sector level. However, the jurisdiction and responsibilities for nuclear matters are centralised; the governance of radioactive waste is shaped by a limited number of actors and non-transparent top-down decisions.

The mandatory implementation at the national level of the Council Directive 2011/70/Euratom of 19 July 2011 has put pressure on decision makers. The necessary national legal framework is already in place. Nevertheless, there is still no clarity concerning the implementation of the national programme or site selection. One of the major challenges is to break the present inertia and initiate a process enabling the search for a suitable site on the basis of socio-technical and scientific criteria. These should involve inclusive and decentralised forms of

---

1 This chapter is a contribution of the Environmental Policy Research Centre (FFU) of the Freie Universität Berlin to the project, "Multi Level Governance-Perspective on Management of Nuclear Waste Disposal. A Comparative Analysis: Actors, Instruments and Institutions" that is part of the inter-disciplinary ENTRIA research platform funded by the German Federal Ministry of Education and Research (BMBF/ 02S9082B).

decision making and interactions between players and stakeholders at all levels of governance. Is there a political will to take such a path in Italy? Are there preconditions to make the necessary development understandable and transparent and to gain the political and societal acceptance necessary to move away from the present stalemate situation? In the following section the background and the major aspects of these questions and challenges will be discussed.

2 The policy for nuclear waste disposal

Italy phased out all its nuclear power plants in the aftermath of the Chernobyl disaster, but a nuclear legacy remains in form of the RW accumulated from its dismantled power plants, past reprocessing and research activities as well as from industrial and medical uses.

2.1 Historical background

Italy had a pioneering role in the early development of nuclear power in the 1950s. Nuclear energy was seen as the answer to its lack of domestic fossil fuel resources. In the 1960s, following the nationalisation of the electricity sector and the establishment of the national electricity monopolist ENEL, and due to cheap oil prices and powerful petroleum lobbies the pursuit of nuclear power was stopped (Di Nucci 2006).[2] The last order for the construction of a nuclear power plant was in 1966, for the Caorso nuclear power plant (NPP). Further plans for nuclear power development were abandoned and the country became heavily dependent on imported oil. This position changed in reaction to the oil crises of 1973 and 1979; two massive nuclear development programmes were planned, but never realised (Di Nucci 2009). Following the Chernobyl disaster in 1986, a moratorium on new nuclear plants was introduced. This was followed by a referendum in 1987, and finally a phase-out of all NPPs in 1991.[3]

In the middle of the 1990s the former electricity monopolist ENEL abandoned fuel reprocessing in its own pilot facilities and opted for reprocessing abroad and for interim dry storage of the remaining spent fuel from its nuclear

2 The influence of the state hydrocarbons company ENI was especially strong.

3 The referendum included the abrogation of a) the statutes by which the inter-ministries Committee for the Economic Programming (CIPE) could decide about the siting for NPPs in case the regions did not do it within the time stipulated by Law 393; b) the abolishment of the incentives for municipalities where NPPs were to be constructed; and c) the abrogation of the statutes allowing the electricity monopolist ENEL to take part in international activities to build and manage nuclear plants.

plants. Thanks to a contract stipulated with British Nuclear Fuel Limited (BNFL), spent fuel from the British Magnox reactor in Latina was shipped to Sellafield for reprocessing. Following the agreement between France and Italy in November 2006, a contract was signed in 2007 for the reprocessing in La Hague of 235 tonnes of used fuel from the other three Italian NPPs (NEA 2013: 12). After reprocessing, all vitrified waste will be returned to Italy by the end of 2025.

In 2008, a change of government revived the debate on the nuclear option and a feasibility study investigated the construction of three to four NPPs to be located for the most part on already existing sites. The pro-nuclear Berlusconi government introduced a package of nuclear rulings and by-laws, which included measures to simplify the licensing of siting and construction. The legislation was passed in July 2009 and envisaged six months to select sites for new NPPs. In the same year ENEL and *Electricité de France* (EdF) launched the joint venture *Sviluppo Nucleare Italia* to build at least four 1,650 MWe reactors deploying the EPWR (European Pressurised Water Reactor) technology of Areva. Following a decision by the Constitutional Court in 2011, a referendum was launched to repeal a number of laws introduced since 2008 in order to pave the way for new NPPs. The date planned for the referendum was arranged immediately after the Fukushima disaster.[4] The referendum reached a 55% turnout of eligible voters; about 94% of the voters refused plans for new reactors and discarded definitively the nuclear option.[5]

2.2 The national inventory

The Technical Guide n 26, first issued in 1985 by the nuclear safety and health protection direction (DISP) of the national energy agency (ENEA) provided the first regulatory reference on RW management. The fundamental principles to which the guide refers are the health and safety protection of the population and workers as well as environmental preservation, taking into account the impact on future generations (Ventura 2003). The guide provides waste classification as well as technical requirements. Waste is subdivided into: Category I: very low-level waste (VLLW); Category II: low or intermediate-level waste (LLW) and Category III: long-lived and/or high-level waste (ILW/HLW). The last national RW and SF inventory was the one released by ENEA in 2000. Since then the

4 The questions posed in the referendum concerned the abrogation of about 70 regulatory and legislative measures specifically established in order to enable the construction of new reactors.

5 In Italy legislative referenda require a quorum of over 50% of all eligible voters.

national inventory has been kept by Istituto Superiore per la Protezione e la Ricerca Ambientale (ISPRA).[6]

The exact RW volume stored in Italy is unknown. Available official data are mostly based on estimates of ENEA/ISPRA and *Società Gestione Impianti Nucleari* (SOGIN). The total volume of waste to be disposed is estimated to be in the magnitude of 71,000 m^3 (SOGIN 2013). Around 98% of the spent fuel has been transferred abroad for reprocessing; the remaining 2% will shortly follow (SOGIN 2013). Residual wastes estimated at 60 m^3 will be returned to Italy by 2020. The RW volume at the end of December 2011 and of the expected waste from the decommissioning of nuclear facilities (estimated at around 200-500 m^3 per year) as well as the reprocessed waste returning from the UK and France are shown in Table 1.

Table 1: Volume of radioactive waste according to category in Dec. 2011 and expected volume after 2020 (in m^3)

Waste category	**International classification**	**Source**	**Volume** (m^3) Dec. 2011	**Estimated volume** (m^3) (returned reprocessed SF & decommissioning after 2020)
Category III	ILW/HLW	Reprocessing of spent fuel, nuclear fuel cycle and R&D facilities.	1,700	
Category II	low- or intermediate-level waste (LLW)	Fuel cycle facilities; reactors; research	22,200	30,000 m^3
Category I	very low level waste (VLLW)	Industry, medical and research establishments	4,300	N/A
vitrified HLW; residual waste		Reprocessing of the spent fuel in the UK and France (235 t)		Approximately 60 m^3 of which 20 m^3 is from Sellafield

Source: adapted from NEA (2013).

6 ISPRA collects information from the operators in a data base in terms of volume, mass and physical activity, but does not elaborate data in terms of final storage (Bove et al. 2009; Capone et al. 2011). ENEA provides estimates on RW from the dismantling of facilities.

2.3 *Interim storage sites*

As is the case in other countries, Italy has no long-term repository for category III and II waste. Generally all the wastes generated by the operation and decommissioning of nuclear installations are stored in the sites of origin. In each plant, waste is treated, conditioned and stored in temporary facilities until its transport to the planned national surface repository. At the end of the decontamination operations, the temporary storage structures will be checked for residual radioactivity and dismantled. Most of this waste is stored in untreated form, and treatment and/or conditioning are pending (NEA 2013).

The Italian nuclear park consists of four NPPs of varying technologies, a number of facilities related to the fuel cycle and two experimental reprocessing plants. The removal of the spent nuclear fuel (SNF) from the four shut-down NPPs has not yet been completed. SNF from the Garigliano plant was shipped to the United Kingdom in 1987, but 63 elements are "parked" in the storage pool of the Avogadro facility at Saluggia and will be transferred to France. Whilst SNF from the power plants of Latina and Caorso was transferred respectively to Sellafield and La Hague at the beginning of the 1990s and between 2007 and 2010, SNF (39 uranium elements and eight MOX elements) is temporarily "parked" in a spent fuel pool in Trino Vercellese.

Saluggia, located in the northern Piedmont region, is the site with the highest concentration of nuclear facilities, since it also hosts the Eurex reprocessing plant. It is estimated that this site concentrates around 70% of the whole radioactivity recorded in Italy (Agnoli 2013). Spent fuel is also deposited in the pool storage of the Itrec reprocessing plant at Saluggia. Additionally, the Joint Research Centre of the European Union at Ispra, not classified as an Italian site, contains approximately 3,000 m^3 of radioactive waste.

2.4 *Evolution of the waste management strategy*

In the last decades several attempts were made to develop and implement a RW disposal concept. In 1996 ENEA established the Task Force Site tasked with undertaking conceptual and system projects, prospecting, selecting suitable sites and preparing a preliminary safety report (Luce et al. 2009).[7] In 1999, the government launched a strategy for the complete decommissioning of nuclear facilities by 2020. The underlying precondition for this was the construction of a

7 In 2002 the Task Force was renamed into "Grande Servizio Paese 3 – Sito" and conducted its activities until the ENEA´s fuel cycle activities were transferred to SOGIN in 2003 (Luce et al. 2009).

low- and intermediate-level waste repository to be used also for the temporary storage of HLWs. The Minister of Industry´s plans disclosed in December 1999 included:

- Treatment and conditioning of waste from NPPs currently in on-site storage within 10 years, with the perspective of a subsequent transport to a national waste repository.
- Site selection and construction of a national repository for LILW, within 10 years. These sites were supposed to take in the storage of ILW/HLW, especially spent fuel and waste resulting from reprocessing.
- Decommissioning of NPPs and other nuclear facilities within 20 years, with the perspective of returning all sites to unrestricted use (Di Nucci 2009).

Eventually the deadline for decommissioning was prolonged to 2024 and the possibility of reprocessing was also considered (European Commission 2007). In 1999 an agreement between the government, regions and autonomous provinces was signed to define and initiate measures to promote the safe RW management. In the framework of this agreement a working group was set up with the aim to prepare a document encompassing technical options and participatory procedures for the local population. The document was approved by the *Conferenza Stato-Regioni* in January 2002 (SOGIN 2003: 24). [8]

In 2001 the "Task Force Site" submitted to the environmental agency acting as regulatory authority ANPA (now ISPRA) a first draft of the conceptual and system projects designed for a repository for low activity RW, with the aim of starting a preliminary evaluation and of testing the acceptability of directives and methodologies for the safety analysis (SOGIN 2003: 23). In 2002 the "Task Force Site" also considered the near surface option. Following the selection criteria, around 200 areas for a surface facility were identified.

In February 2008, the Ministry of Economic Development set up a working group to investigate a centralised storage facility for the disposal of IHLW and the interim storage of HLW. They released a report in September 2008 also including a methodology and procedures for the site selection (NEA 2013). Law 75/2011 of May 2011 revoked all previous provisions; however, the Euratom Directive 2011/70 and the need for a national repository forced the government to restart the process of searching for a suitable site.

8 The so called conference of State and Regions consists of members from the central administration and the regions.

3 The current waste disposal plan, its expected costs and financing

The present disposal concept – as stated in the *Legislative Decree* 31/2010 – focuses on a waste storage solution. It specifies the construction of a central repository as a surface structure (with reversibility and retrievability options) to store approximately 80,000 m^3 of low and medium activity waste, as well as temporary storage for approximately 15,000 m^3 of high activity waste. SOGIN estimates that 60% of the cumulated waste will come from the decommissioning of nuclear installations, while the remaining 40% will come from nuclear medicine, research and industrial activities. The decision was based on the assumption that transferring the waste into a central structure "ensures maximum safety for the people and the environment and will allow for the complete restoration of environmental systems, optimizing time and costs and eliminating the need for temporary storage sites".[9] The determining requirements are: the retrievability of the wastes; long term safety; an institutional control period for at least 300 years; long-term recording of data as well as a dose limit to the population no higher than 0.01/mSv/year. The concept for a national repository should include:

- A surface repository for medium level RW.
- An interim storage facility for high level RW.
- A technology park with a centre for R&D and innovation in the field of decommissioning and RW management on location.

Technical standards for surface an near surface disposal facilities have been issued by ISPRA in . They concern: criteria for qualification of conditioned solid RW; LLW radiological characterisation for surface disposal; waste package identification procedures; packages and containers for LLW; record keeping in a near surface disposal facility; basic design criteria for an engineering LLW disposal facility; qualification criteria for the engineering barriers and monitoring system for a LLW disposal facility and siting exclusion; and investigation criteria (ISPRA 2014).

The costs for total decommissioning were estimated in 2004 at around €4bn. This estimate did not include the total liability for the final disposal of HLW and spent fuel (European Commission 2007: 53). As of June 2014 there were no official data concerning the costs, but only estimates. SOGIN speaks of "some" billions of Euros and estimates that the investments necessary to realise

9 As stated on SOGIN website, cf. http://www.sogin.it/en/about-us/the-technology-park-and-the-national-repository/the-technology-park-and-the-national-repository.html, last accessed 8 November 2013.

the technology park and the national repository would amount to €2.5bn (Energy Lab 2012). According to daily press reports, the expected investment is estimated between €1.2bn and €2.5bn (Agnoli 2014). The dismantling of all nuclear facilities and NPPs and the environmental restoration are supposed to be terminated by 2026, except for the power station located in Latina which should be completed by 2035. The estimate of the total costs for the overall environmental restoration of the nuclear sites is €4bn; for the technology park and the national repository is €2.5bn. SOGIN expects annual investments of around €250m per year. These figures have been broken down as follows:

- Dismantling of the nuclear power plants: €1.7bn
- Fuel reprocessing: €900m
- "Safe management" of plants and facilities: €1.4bn
- Realisation of the technology park and of the national repository: €1.3bn
- Research activities in the technology park: €800m
- Other supporting infrastructures: €400m

Financial resources for RW management are included in the funds allocated for decommissioning the nuclear installation. The costs for decommissioning are covered by a "decommissioning fund" created by the former electricity monopolist ENEL to cover long-term liabilities for decommissioning and spent fuel management. The accumulated fund was transferred to SOGIN at the time of its creation as an independent body, and amounted to about €750m. Between 2000 and 2004 additional €700m were channelled into the decommissioning fund (European Commission 2009).

The costs for the decontamination of nuclear sites and for waste storage are covered by a financial levy on the electricity bills paid by the final consumers. This was introduced by way of a ministerial decree in January 2000. The government agreed that additional costs caused by the decommissioning programme would be considered as stranded costs, to be recovered by a levy on the kWh price for electricity, a process established and controlled by the Authority for Electric Energy and Gas (AEEG) (Brusa et al. 2002). The levy is added to the fund and is transferred to SOGIN.

Every year SOGIN is required to submit plans about future activities and associated costs. On this basis, the energy regulator AEEG evaluates the levy on the price of electricity for the next three years.[10] It is estimated that the so called "onere nucleare" (nuclear burden) amounts to around €300/400m per year and that an average household pays €2 per year (Italpress 2013).

10 The levy becomes a part of the so called A2 component of the electricity tariff, which at present amounts to around 0.17 € cent/kWh, against an average cost of 18.9 € cent/kWh.

4 The legal framework

There are a number of laws and rulings (mostly decrees) regulating nuclear activities and RW. A milestone is Law 282/2005 promulgated for the ratification of the "Joint Convention on the Safety of Spent Fuel Management and on the Safety of Radioactive Waste Management". Further major references are the Legislative Decree 230/95 (implementation of various Euratom directives) later integrated and modified by the Legislative Decree 241/2000 as well as the Legislative Decree 314/2003 modified and converted in Law 368/2003 as well as Law 239/2004. As of today, the most important legal and normative references are Law 99/2009 and the Legislative Decree 31/2010 with their subsequent amendments. Law 99/2009 ("Directives for the development and internationalisation of undertakings as well as in energy matters") passed in July 2009 to regulate the renaissance of the nuclear programme redefined the role and mandate of SOGIN under the supervision of a new commissioner. The law included:

- Art. 25: Delegation to the government to legislate directly about siting and NPP licensing processes, nuclear fuel fabrication facilities, RW management, solutions for waste disposal and incentives for areas hosting nuclear facilities.[11]
- Art. 29: Creation of a new Nuclear Safety Agency with responsibilities for all aspects in the nuclear and radiological protection field. The agency is an independent body appointed by the Prime Minister, the Ministry of Industry and the Ministry of Environment.[12]

Legislative Decree 31/2010 of 15 February 2010, ("Discipline of the storage systems for radioactive fuel and waste as well as economic benefits [...]")[13] also as subsequently amended (Legislative Decree 41/2011), established a new licensing procedure for the construction, operation and decommissioning of the nuclear installations, including waste storage sites.[14] It regulated steps and scheduling of the siting procedures of the national site for the LLW repository and for the ILW-HLW long term disposal, including public consultations. It also

11 This was suppressed later by the legislative decree 34/2011 of 31.03.2011 later converted with modifications in Law 75/2011.

12 The press reported that the government intended to have the Agency work closely with the French *Autorité de Sûreté Nucléaire*.

13 For details, see the text of the law: http://www.normattiva.it/urires/N2Ls?urn:nir:stato: decreto.legislativo: 2010;031, last accessed 8 November 2013.

14 Details can be found in the text of the law: http://www.normattiva.it/uri-res/N2Ls?um:nir: stato: decreto.legislativo:2011;041, last accessed 8 November 2013.

built-in provisions related to the funding of the decommissioning activities and compensation measures (Art. 22) for the municipalities hosting nuclear facilities. Art. 4 states that construction and management of nuclear facilities are activities of state interest, are subject to a single authorisation upon request of the operator and are granted (in terms of Legislative Decree 66/2010) – subsequent to a consultation with the Ministry of Defence and with the respective region of the site in accordance with the unified conference – by decree of the Ministry of Economic Development in agreement with the Ministries of Environment, Infrastructures and Transportation. The response of the region is mandatory, but non binding and is to be delivered within 90 days after the request for authorisation. Should there be no reaction after this time, the Conferenza unificata will examine the matter. [15] This Legislative Decree assigned to SOGIN the task of implementing the storage concept and made it also responsible for construction and operation of the national repository.

Law 75/2011 repealed all the provisions of Law 99/2009 and of Legislative Decree 31/2010 and its amendments by the Legislative Decree 41/2011 relevant to nuclear power, turning down any future nuclear development plan in Italy. This law abrogated Articles 8 and 9 of the Legislative Decree 230/95 and changed the regulatory process by abolishing the "Technical Commission on Nuclear Safety and Radiation Protection". However, the provisions for the development of the national site for LLW disposal and ILW-HLW interim storage were kept. [16] In December 2011, following the drastic cut back in spending of the Monti government, Law 214/2011 sanctioned the abolishment of the National Safety Agency.

5 The institutional framework

Italy has a rather complex institutional setting. The most important actors are described below.

Institutional and regulatory actors

The major institution in charge of issuing the operating licence for nuclear and radioactive facilities is the Ministry of Economic Development under the

15 The "Unified Conference" is a governance body consisting of a state-region conference, state-municipalities and autonomous bodies' conference, i.e. regions, provinces, municipalities, etc. Its aims are to enhance cooperation between the state activities and other organisms and examine issues of common interest.

16 Additionally Art. 24 of Legislative Decree 1/2012 on economic development established new procedures to reduce the lead time necessary for the decommissioning licence.

technical advice of the Institute for Environmental Protection and Research (ISPRA), which is responsible for assessing and inspecting nuclear facilities and activities involving the use of radiation sources, for technical recommendations and legally binding requirements.

Radioactive waste storage and disposal activities require a concerted agreement of the Ministries of Environment, Interior, Labour and Health. Art. 5.2 of the so called Nuclear Safety Directive (Council Directive 2009/71/ Euratom of 25 June 2009) requires Member States to establish and maintain a competent regulatory authority functionally separated from any organisation associated with "the promotion or exploitation of nuclear energy or radioactive material; the production of electricity using isotopes; the management of spent fuel and radioactive waste". The fact that such activities are under the jurisdiction of the Ministry of Economic Development (an actor traditionally closely connected with the nuclear industry) is to be considered a very critical issue. In fact, in the majority of Euratom countries, the regulator is not under the supervision of the ministry in charge of energy and industrial policy.

The execution of regulatory and safety functions in Italy has a troubled history characterised by a continuing change of agencies. It started in 1964 with the creation of CNEN and later was taken over by the Nuclear Safety and Health Protection Directorate a department of the national energy agency ENEA (ENEA-DISP).[17] Under Law 99/2009 a new Nuclear Safety Agency (Agenzia per la Sicurezza Nucleare) was to be established with staff from ISPRA and ENEA-DISP. Following the cancellation of the Agency for Nuclear Safety through Legislative Decree 201/2011 its functions were temporarily assigned to ISPRA, that de facto acts as the national nuclear safety authority.

ISPRA, established in 2008 as a governmental institute with administrative and financial autonomy under the supervision of the Ministry of Environment, is in charge of the control and supervision of nuclear facilities and radiation protection. ISPRA authorisation is required for detailed designs of any structure, system and component relevant to safety in any nuclear plant. Within ISPRA, the duties of the Regulatory Body are carried out by the Nuclear, Technological and Industrial Risk Department, which has a staff of 40, including inspectors. ISPRA is also responsible for the overall national RW and SF inventory.

The creation of the short lived Nuclear Safety Agency in 2010 was much criticised for the "political" nominations of its members (Gualerzi 2010) and for the fact that the chairman, a famous oncologist and also a previous health minister, in spite of limited experience with nuclear energy safety issues took a

17 In 1994 the responsibility for safety and licensing was transferred to National Agency for Environment Protection and Technical Services (ANPA, renamed in 2002 APAT and later in 2009 ISPRA).

very strong pro-nuclear stand (Mariotti 2010). The termination of the safety agency did not generate negative reactions although it remains an objectionable anomaly that there is no dedicated, independent regulator. It is questionable whether ISPRA can live up to the criteria necessary to be a regulator, which requires not only technical competence, but also must be recognised as a trustworthy and neutral body by all stakeholders, including affected local authorities and society at large. The history, structure and organisation of ISPRA and the lack of a dedicated regulator can be considered as symptomatic for the low expectations in Italy to take on the positive (prescriptive) role of a regulator. This role should involve more than just supervising and ensuring compliance with safety, risk and technical criteria; it should also envisage participation in a wide political and societal dialogue.

Consulting Bodies
Additionally there are consulting bodies such as the *Conferenza Stato-Regioni* in charge of discussing issues where competences are shared between central and regional governments. Under the Italian constitution, the opinion of this body is non binding, but it represents a clear political message to the central government. In the past, the Italian cabinet had to consult with the *Conferenza Unificata*, the local authorities, the central government and the industry commissions of the Parliament before making a decision.

Operator/Implementer
There is a long account of industrial interdependencies leading to the birth of the operator and implementer, SOGIN, which represents a rather peculiar actor. The company can be considered at the same time a spin-off of ENEL and ENI, respectively the major electricity and hydrocarbons national companies. SOGIN started operating in 2001, but it became a group in 2004 after the acquisition from ENI of a 60% stake of Nucleco SpA, the national operator responsible for collecting and conditioning as well as with the temporary storage of RW from nuclear medicine and from R&D activities. It was also responsible for the decommissioning of the fuel cycle plants (Saluggia, Trisaia and Casaccia) and of the FN fuel manufacturing plant of Bosco Marengo. SOGIN manages and decommissions nuclear plants. In addition it is also in charge of designing, constructing and operating the national repository for LILW and the interim storage for HLW. Its strategic and operational directives are provided by the Ministry of Economic Development.

The company is financially solid and has approximately 840 employees. In 2013, revenues from decommissioning reached around €62m. [18] Due to its non-transparent management, its personnel recruiting practices and consulting services and high expenses, in the past the group has been the object of various parliamentary interrogations and has been criticised also by the Court of Auditors (*Corte dei Conti*) as well as by the energy regulator (Di Nicola 2007).

Although SOGIN is in charge of implementing a communication campaign (Art. 26e Legislative Decree 31/2010), there is almost a complete lack of information about its activities. A communication and dissemination strategy explaining rationale, location and structure of the planned national repository and the criteria upon which it is going to be selected are not known and cannot be found in any non-confidential documents. On the whole, the SOGIN presentation of domestic activities on the website is very short and vague. On the contrary, the presentation of its international activities provides sufficient details. The group is very active abroad, especially in Russia and the former Soviet Union republics.[19] Within the framework of the Global Partnership programme between Russia and Italy, SOGIN was hired to dismantle Russian nuclear submarines and to dispose of RW. A dossier on SOGIN by an investigative journalist reports that a parliamentary interrogation disclosed that the financing of the bilateral agreement between Russia and Italy to dismantle three nuclear submarines would cost €44m per year for eight years commencing in 2007 (Rovai 2009). The dossier advances the hypothesis that the Italian engagement with the nuclear submarines should be "compensated" with a RW export to Russia.

Actors involved in research and demonstration

National R&D programmes are funded by the Ministry of Economic Development and the Ministry of Education and Research; EU projects are funded within EURATOM. Amongst the other actors in charge of nuclear research, the most prominent role is played by the National Energy Agency ENEA within its Research Centres of Casaccia, Bologna and Saluggia.

---

18 In 2013, total revenues, excluding spent fuel contracts, amounted to € 184,3 (in 2012 €179.8m) and the operating result was €0.5m. The Ebitda (earnings before interest, taxes, depreciation and amortisation) amounted to €11.5m. See press release of 5 August 2014, http://www.sogin.it/SiteAssets/uploads/2014/COMUNICATI%20STAMPA%202014/com%20stampa%20SOGIN%20-%20Bilancio%202013%20e%201%5E%20semestrale%202014.pdf

19 The list of activity locations includes Russia (especially in Kola, Kalinin, Beloyarsk and Bilibino), Armenia (Metsamor), Kazakhstan (Aktau plant) and the Ukraine (Khmelnitsky and Rovno plants). Additionally SOGIN is involved in China and France (Phenix). Projects have been completed in Slovakia (at the Bohunice and Mochovce), the Czech Republic (Dukovany and Temelin), Romania (Cernavoda), Bulgaria (Kozloduy and Belene) and Lithuania (Ignalina). See www.sogin.it.

6 Siting Procedures

Over the years there have been various feasibility studies, unsuccessful attempts and government initiatives to select viable sites for a nuclear repository. Efforts to give life to voluntary mechanisms for siting have also failed. In the 1990s, the designation of a national site for waste disposal became a political priority and ENEA carried out studies on a deep geological repository and worked out a list of potential sites for LIW. Also ENEL released in 1995 a feasibility study about a centralised interim storage facility for SF and HLW. In 1996 a "Task Force Site" (TFS) was created by ENEA to develop a methodology for the selection and qualification of suitable sites. Initial activities included a feasibility study on two areas owned by the military, chosen from the locations investigated by ENEA in the period 1988-89. The study was completed in 1998 and included a performance and safety assessment (SOGIN 2003: 22).

The first phase of the TFS work was completed in 2000 and carried out a screening on the national scale based on exclusion criteria considering the indications of the IAEA and of the Technical Guide 26.[20] This screening helped to identify 8,107 areas, of which around 200 (0.6% of the national territory) had a surface larger than 300 ha. Out of these possible sites, 33 areas with favourable physical and territorial characteristics were identified in 2002 (Ventura 2003).

The list of potentially qualified national sites did not find a consensus. The Government then mandated the chairman of SOGIN acting as "Extraordinary Commissioner" to select a location for second and third category wastes. A further working group was started in April 2003 and was asked to define the criteria for the siting and to construct a final repository for low and medium activity wastes, taking into consideration both the hypotheses of a surface and a near-surface repository. In the same year, the selection criteria for the site prepared by this group was presented to the Conference of the Regions without however being discussed.[21]

After a technical evaluation, a site in the region Basilicata (Scanzano) was selected by the government, and was included in the Legislative Decree 314/2003 ("Urgent Dispositions for the collection, disposal and storage of radioactive waste").[22] Due to fierce opposition (see below), the Decree was modified by the Law 368/2003 of December 2003 including provisions to form a commission in charge of selecting a new site within 12 months. These decrees

20 See IAEA (1994): Siting of near-surface disposal facilities. Safety Series No. 111/3.1

21 See http://www.zonanucleare.com/norme/scelta_sito_rifiuti_nucleari.htm, last accessed 28 November 2013.

22 For details, see the text of the law. http://www.am bientediritto.it/Legislazione/nucleare/2003/dl%202003%20n.314.htm, last accessed 28 November 2013.

were then converted into Law 239/2004 with modifications of Art.1 (as of Official Gazette n. 215 of 13.9.2004).[23] Great difficulties were encountered with the implementation of Law 239/2004 on "restructuring of the energy sector and on the delegation to the government on the reform of the dispositions in energy in particular in relation to the site identification".

6.1 Procedures and criteria for site selection

In September 2008 the working group contracted by the Ministry of Economic Development to define a centralised storage facility for the disposal of ILLW and the interim storage of HLW released a report which included methodologies and procedures for site selection. This report was taken into consideration by Legislative Decree 31/2010, which assigned to SOGIN the task of proposing areas suitable for the siting of the surface storage facility.

Siting procedures are laid down by Legislative Decree 31/2010 and its subsequent amendments. The siting process should be accompanied by a public communication programme including a public presentation through a seminar where the central and local interested administrations would participate. It assigns to SOGIN the task to lead these proceeding. After the regulator's approval, SOGIN is expected to invite the involved regions to declare their interest within 60 days. This step is then to be followed by site investigations. Should there be no interest, SOGIN should submit to the Ministry of Economic Development a short list indicating the three most suitable sites, and within 30 days an inter-institutional committee should be created, with the participation of representatives from different ministries and regions. A request for authorisation should be submitted to the regulator within four months and the licence should be granted within 12 months.

The authorisation procedure has been simplified and requires only a so-called *autorizzazione unica* (single licence) for construction and environmental assessment. In the aftermath of the Fukushima accident, the anticipated time schedule was postponed. However, this single licence is still included in Art. 4 of the Legislative Decree 41/2011 presently in force. The same decree (Art. 8) states the technical criteria for siting, which are in line with international best practices and "ensure adequate level of safety of the population and environmental protection".

---

23 For the text of the law published in the Official Gazette, GU n.215 of 13.09.2004 which was in force until 22 August 2005, see http://www.normattiva.it/uri-res/N2Ls?urn:nir:stato:legge:2004-08-23;239, last accessed 28 November 2013.

Due to the abolishment of the Nuclear Safety Agency, in 2012 ISPRA was tasked with the development of a regulatory guide on technical siting criteria. These should serve to identify potential areas for the construction of a LLW near surface disposal facility and an interim storage for ILW and HLW. The Regulatory Guide 29 was released in June 2014 (ISPRA 2014). The siting process considers three stages in accordance with the IAEA recommendations. The first one concerns the so called "conceptual and planning stage" and the "area survey stage – regional mapping or investigation phase". In this phase, "potentially eligible" areas are to be selected and ranked. The second stage is dedicated to the "site screening phase" with the aim of selecting among the "potentially eligible" areas, the sites to be investigated in detail. The selection is performed on the basis of assessments conducted with regional scale data of possible site investigations and taking into account socio-economic aspects. The final stage focuses on the "site investigation" and "detailed site characterisation" of one or more sites with the aim of a conclusive identification of the site for the disposal facility.

The Technical Guide 29 identifies 15 exclusions criteria amongst which are: volcanic and seismic activities, geomorphological and hydraulic risk, altitude above 700 meters, distance from the coast line within 5 Km, unsuitable distance from residential areas, distance from motorways suburban roads and railway lines, nearness to industrial activities, airports and military facilities, hydrology and hydro-resources; importance of biodiversity. Among the investigation criteria are the presence of secondary volcanic and tectonic activities, presence of erosion phenomena, weather and climatic conditions, soil and groundwater conditions, hydrogeological parameters, natural habitat, availability of transport infrastructures, presence of strategically important infrastructures, etc. (ISPRA 2014).

According to unofficial estimates (Agnoli 2014) the siting process should last at least eight years (four year to obtain the licence and four years for the executive planning and construction). The map of the potential sites should be discussed in a national seminar in which the responsible ministries, regions and provinces as well as affected municipalities, research institutes and other stakeholders should take part. Action plans to enhance the safety level of waste by implementing specific treatment and conditioning projects, by refurbishing existing buildings or by constructing new storage facilities on the sites are said to be in progress.

6.2 Compensation mechanisms and socio-economic impact

Legislative Decree 314/2003 and Law 368/2003 and following modifications stated that until the disposal sites are operative, the municipalities where the nuclear installations are located will receive an annual compensation which should be commensurate to the radiological inventory of the actually stored SF and RW. Legislative Decree 31/2010 (Art. 22) specifies that the economic benefits for municipalities and regions hosting nuclear power plants would amount to €3,000/MW up to 1,600 installed MW on the site plus 20% for installed power exceeding this level per year during construction and €0.40 per MWh in the period of operation.

The economic compensation should be used for environmental restoration and sustainable development measures and is to be shared between the municipality hosting the facility (50%), the respective province (25%) and the neighbouring municipalities within a radius of 25 km from the nuclear installations (25%). For example, the compensation paid for 2011 to all hosting and neighbouring municipalities near the facilities of Eurex, Trino Vercellese, Caorso, Latina, Garigliano, ITREC, Casaccia, Ispra and Bosco Marengo amounted to €15.6m and was determined by means of the so called CIPE 41/2013 resolution published in the Official Gazette 267 of 14 November 2013.[24]

The benefits expected from the technology park and from the national repository are stated in Art. 29 of Legislative Decree 41/2011 modifying Art. 30 of Legislative Decree 31/2010. The economic compensation for the territory is subdivided according to criteria set in Art 23.4 and is to be paid by SOGIN according to criteria to be set by the Ministry of Economic Development in agreement with the Ministries of Environment and of Finance, depending on the level of radioactivity. Regional economic incentives for the potential site of the technology parks are mentioned in a press release issued by SOGIN that estimates 12,000 new jobs in connection with decommissioning and the creation of a repository (SOGIN/ Nomisma 2013).

7 Information and Participation

It was not until the middle of 1970s that public participation and organised opposition became crucial issues in connection with nuclear power. The anti-nuclear movement grew as Montalto di Castro was selected to host a BWR plant and became stronger following the Three Mile Island accident in the USA in

24 For details see Gazzetta Ufficiale, N.267 p.38-39.

1979. Following the Chernobyl disaster, opposition grew and major campaigns were initiated by 18 environmental advocacy groups and leftist political parties (Di Nucci 2009). The protest against nuclear power can be considered as the turning point of the Italian environmental movement: over one million people petitioned for a referendum and the environmental NGOs Lega Ambiente and WWF doubled their members. The newly founded Green Party received almost one million votes in the 1987 election.

Concerning nuclear waste disposal, no democratic, inclusive selection process has taken place. In 2003 a government Decree instituted an "extra-ordinary" Commissioner tasked with with the validation of a site, the approval of the economic and financial plan for it as well as with procurement and tenders for planning and constructing the national repository (Cianciullo 2003).[25] This ruling was in patent contradiction to the announced more participative forms of selection.

The strong top-down exercise of powers regarding openness and lacking participation is epitomised by the case of Scanzano Jonico. In November 2003 the southern Region Basilicata was "surprised" by the Legislative Decree 314/2003 by which the government forced a geological repository for category waste II and III in 700 meter profundity in Scanzano (La Repubblica 2003; Di Nucci 2009). Experts had identified the town's underground salt caverns as a potentially suitable repository for high-level nuclear waste. This site had been selected in spite of the criticism about its high population density and the nearness to the sea. Local residents had not been consulted (Cianciullo 2003b). To cover the initial measures on the territory and to inform the inhabitants, the decree envisaged an expenditure of €500,000 for 2003 and €2.25 m/year in the period 2004-2005 (La Repubblica 2003). The site was defined as a military defence installation of national property.[26]

For nearly two weeks, residents blocked motorways and shut down shops and businesses. Approximately 150,000 people marched in what was described as the largest demonstration held in the southern region of Basilicata (Rossano 2003). The regional council declared the area a denuclearised zone and initiated a lawsuit against the decree. As a result of the protest, the Berlusconi government was forced to withdraw from the decision to make Scanzano Jonico the site of the main nuclear waste repository, amend the decree (deleting the name of the designated location) and commission to SOGIN the search of a new site. Additionally, the government established a scientific commission and reiterated the need for a centralised procedure for the environmental impact assessment and

25 This post was covered by the then president of SOGIN, an Army general.

26 See zona nucleare. http://www.zonanucleare.com/language/english/decommissioning_italy/7_national_repository.htm, last accessed 8 November 2013.

the strategic importance of the site. The popular revolt in Scanzano set a precedent: siting procedures require an open, democratic process, where all stakeholders' interests can be discussed and where both residents' opinions and scientific arguments are considered rather than *de lege* enforcement. The government had originally drafted an agreement with regional authorities to accelerate the siting of a final repository and to ensure the necessary transparency to the general public. The Ministry of Industry set up a 'national table' with a similar purpose. SOGIN organised meetings with local authorities in order to inform them about the main strategies. The government's conduct, however, was in obvious contradiction with the allegedly adopted philosophy purporting stakeholder dialogue and public consultation (Di Nucci 2009).

In Italy there is a generally positive attitude towards NGO participation in decision making and a certain confidence in the controlling role of the Parliament. This stand is confirmed by a survey on nuclear safety carried out in 2009 by Eurobarometer (European Commission 2010: 115). Around 30% of the interviewed Italians declared that they would like to have NGOs to be consulted and to participate in the decision making process, a figure which was above average in the EU27 of 25%. The requirement for direct public participation was not so strong and 21% of the interviewees from Italy stated that they would like to be directly consulted and to participate in the decision making process (the EU27 average was 24%). Only 18% stated that they would leave the responsible authorities to decide exclusively on the matter (EU27 average 24%) whilst 21% stated that they would like the Parliament to be consulted and to participate in the decision making process (EU27 average 18%).

In 2009, a strong opposition arose especially from the ten regions with potentially suitable sites (Basilicata, Emilia-Romagna, Lazio, Liguria, Molise, Marche, Calabria, Tuscany and Umbria) that appealed against the Law 99/2009 (attributing responsibility to the government for reopening nuclear facilities) as unconstitutional. In January 2010, provisions for public consultation were stated together with compensation mechanisms for areas hosting nuclear facilities. On June 24, 2010 the Italian Constitutional Court rejected the joint appeal by the regional governments, but the government had to approve a new draft of the Legislative Decree 31/2010 on nuclear sites, in order to adjust to the decision of the Constitutional Court. Members of the new Nuclear Regulatory Agency were named by the government in November 2010, but the list needed Parliament's approval. In December 2010 a joint meeting of the Parliamentary Commissions for Environment and for Industry rejected one of the nominations, halting the Berlusconi's government plans. In the aftermath of the Fukushima disaster, the initiative "Vote Yes to stop nuclear power" started by over 60 associations promoted a referendum on June 12-13, 2011. Over 94% of respondents voted against restarting nuclear energy programmes and confirmed the results of the Chernobyl referendum showing that the majority of citizens are against nuclear power.

Following changes in the board of directors of SOGIN in 2013-2014, there have been attempts to initiate more transparent procedures for informing civil society about nuclear decommissioning and nuclear waste disposal. In March 2014 the Fondazione per lo Sviluppo Sostenibile (Foundation for Sustainable Development) launched together with SOGIN a so-called "Osservatorio per la Chiusura del Ciclo Nucleare", an organism which should also be in charge of watching technical, ecological and socio-economic aspects of nuclear sites, preparing and disseminating information. This new organisation, however, has no institutional legitimation and most grass roots organisations appear to be rather critical about its role and independence.

8 Conclusions

Since the 1990s, Italy has witnessed a move from a centralised to a decentralised model of governance granting a greater local autonomy and an administrative federalism affecting the formulation, implementation and management of public policies. Nuclear matters, however, are dealt with at a central level; there is a limited number of actors, decisions are made centrally in a non-transparent way and there has been seldom an attempt to give life to projects in accordance with regional and local development strategies. The last attempt to organise specific legislation to regulate the renaissance of the Italian nuclear programme (Law 99/2009) even included a full delegation to the government to legislate directly – without involvement of the Parliament – on NPP siting, licensing processes, radioactive waste, etc.

A fierce opposition over three decades and a contradictory industrial and nuclear policy have provoked a stalemate and rendered the mandatory search for a national site even a more difficult task than in other European countries. Thus steps to move away from the present deadlock by reinforcing competent and independent authorities considered trustworthy by the largest majority of stakeholders have not been taken. Italy has not been able to address the challenge posed by the implementation of the Council Directive 2011/70/Euratom in a consistent way and the chosen governance form appear neither efficient nor transparent.

The necessary national legislative, regulatory and organisational frameworks are already in place and, in part, even the financing schemes for spent fuel and RW management. As in other policy fields, the relevant legislation in Italy is robust, but its transposition and implementation and the related system of institutional control are not as sound. Consequently, to date there is little information on time frames and milestones or transparency policy and procedures. Moreover a serious, independent costs assessment is not available.

Actually the major governance bodies can be partly seen as a potential source for conflict: it is not acceptable that the regulator is part of the national agency for environmental protection and research and has no recognisable independence and brand as a regulator. Moreover, the role covered by the Ministry of Economic Development in the licensing process, an actor in charge of the energy and industrial policy and traditionally linked with the nuclear industry is a factor casting doubts on the neutrality of the process. The fact that the institution acting as "provisional" regulator is rather weak, and that it does not have adequate or dedicated resources to govern the nuclear waste management process, makes it vulnerable and dependent on the operator. This implies that today the operator SOGIN is an anomalous "creature" with limited accountability that for over a decade has "supervised" itself and that now influences all political and policy-making activities in the nuclear field. Its weight is strengthened by the fact that it can count on a "generous" funding system thanks to the "onere nucleare" described above.

Our analysis showed that Italy is still far away from implementing a national policy able to manage SF and RW, including long term solutions. The reinforcement of competent and independent authorities that can be considered trustworthy by the largest majority of stakeholders and the initiation of an unambiguous participation process that deserves public confidence make up the Gordian knot. The present institutional setting does not appear to offer the instruments to untie this knot. Especially the implementing body – who paradoxically is also in charge of public information and participation – has a history that does not help to make it trustworthy to the sceptical or opposing local authorities and population.

At the root of the difficult search for a national repository is unquestionably the undemocratic and non-transparent way the various governments managed the siting process in the past. Consequently, politics and policies are constrained by the past developmental paths and by events and stakeholders that have shaped this past, but also the present development. This continuity and perseverance of old policies and persistence of structures and actors are at the core of the difficult socio-technical challenges for implementing a disposal strategy. Social conflicts and opposition are deeply rooted, but are also exacerbated by the fact that technical and expert approaches have neglected socially relevant questions and have not been made transparent. Italy has failed to address these conflicts in a serious and transparent way by integrating the potentially affected local authorities and local population in the decision-making process. The governments' policies have been in obvious contradiction with the anounced stakeholder dialogue and public consultation. In spite of compensation mechanisms for the affected communities there has been no voluntary expression of interest for hosting a repository, not even by municipalities close to existing nuclear sites.

Concluding, up to now, there has been a non-transparent exercise of power regarding openness, participation, accountability, effectiveness and coherence. The adopted strategies and policies have been dictated by external events and there have been no convincing concepts for participation and dialogue. The Scanzano popular rebellion in 2003 has set a precedent: siting procedures require an open, democratic process, where all stakeholders' interests can be discussed and where both socio-economic and scientific arguments are considered and balanced against each other rather than *de lege* enforcement. It remains to be seen whether the present government will be able to learn lessons for policy from previous mistakes and deliver (in compliance with Art. 11 of the Euratom Directive) a national programme by 23 August 2015 which is based on an open process in which stakeholders and the affected local population know the "rules of the game" and consider the regulator as independent and the implementer as trustworthy.

References

Agnoli, S. (2013, 4 June). Rifiuti nucleari, tutto è pronto per il Deposito. *Corriere della Sera*.

Agenzia per la Protezione dell'Ambiente e per i servizi Tecnici (APAT) (2006). Joint Convention on the Safety of Spent Fuel Management and on the Safety of Radioactive Waste Management. First Italian National Report. http://www.isprambiente.gov.it/files/temi/italianreport.pdf. last accessed 8 November 2013.

Bove, R., Capone, M., Cherubini, N., Compagno, A., Dodaro, A., Marlia, A., and Massiet, A. (2009). Revisione critica dell'inventario nazionale dei rifiuti radioattivi ai fini dello smaltimento, Report RSE/2009/146.

Brusa, L., DeSantis, R., Nurden, P.L., Walkden. P., and Watson, B. (2002). The Decommissioning of the Trino Nuclear Power Plant, WM 02 Conference, February 24-28, Tucson, Arizona. http://www.wmsym.org/archives/2002/ Proceedings/44/421.pdf. accessed 8 November 2013.

Capone, M., Cherubini, N, Dodaro, A., Falconi, L.(2012). Revisione critica dell'inventario nazionale dei rifiuti radioattivi ai fini dello smaltimento – Anno 2011. Report RdS/2011/87.

Cianciullo, A. (2003a, 25 June). Cimiteri nucleri: slitta la decisione. *La Repubblica*.

Cianciullo, A. (2003b, 15 October). Tutti i rischi del cimitero nucleare. *La Repubblica*.

Di Nicola, P. (2007, 3 May). Favori Atomici, *L'Espresso.*

Di Nucci, M. R. (2006). The nuclear power option in the Italian energy policy, *Energy & Environment*, 17-3, 341-357.

Di Nucci, M. R. (2009). Between Myth and Reality: Development, Problems and Perspectives of Nuclear Power in Italy. In Mez, L., Schneider, M., and Thomas, S. (Eds.), *International Perspectives on Energy Policy and the Role of Nuclear Power*, Multi-Science Publishing Co, Brentwood, Essex: 279-300.

Energy Lab (2013). Le bonifiche ambientali dei siti nucleari e la gestione dei rifiuti radioattivi. Sintesi degli interventi. 13 March 2012.

European Commission (2007). Commission Staff Working Document, 'EU Decommissioning Funding Data', Second Report on the use of financial resources earmarked for the decommissioning of nuclear installations, spent fuel and radioactive waste. COM (2007) 794 final, Brussels.

European Commission (2010). Europeans and Nuclear Safety Report. Eurobarometer 324, Brussels.

European Commission (2011). Seventh Situation Report: Radioactive Waste and Spent Fuel Management in the European Union. SEC (2011) 1007 final, Brussels.

European Council (2011). Council Directive 2011/Euratom of 19 July 2011 establishing a Community framework for the responsible and safe management of spent fuel and radioactive waste. 1/199/48 *Official Journal of the European Union*, Brussels.

Giliberto, J. (2010, 24 September). Cinquanta comuni idonei per le scorie nucleari. Ecco la mappa dei siti possibili. *Il sole 24 ore.*

Gualerzi, V. (2010, 2 December). Agenzia per il nucleare. No al candidato del ministro. *La Repubblica.*

Istituto Superiore per la Protezione e la Ricerca Ambientale (ISPRA) (2014). Guida Tecnica n. 29, Criteri per la localizzazione di un impianto di smaltimento superficiale di rifiuti radioattivi a bassa e media attività. http://www.isprambiente.gov.it/files/nucleare/GuidaTecnica29.pdf

ITALPRESS (2013). Nucleare, Sogin-Nomisma. Con Decommissioning 12 Mila posti di lavoro. http://www.italpress.com/energia/33742/nucleare-sogin-nomisma-con-decom missioning-12-mila-nuovi-posti-lavoro, last accessed 8 November 2013.

La Repubblica (2003, 19 November). Scorie a Scanzano Jonico entra in vigore il decreto. *La Repubblica.*

Luce, A., Mura, A., Taglioni, A., and Zarlenga, F. (2009). Le attività condotte dall'ENEA fra il 1996 e il 2003 per il sito nazionale di deposito dei rifiuti radioattivi. RSE/2009/122.

Mariotti, A. (2010, 15 October). Nucleare: Veronesi a capo dell'Agenzia per la sicurezza, *La Stampa.*

Nuclear Energy Agency (NEA) (2013b): Radioactive waste management and decommissioning in Italy (Report), OECD Paris. https://www.oecd-nea.org/rwm/profiles/italy_report.pdf, last accessed 7 November 2013.

Rossano, A. (2003, 1 December). Scanzano, la vittoria della piazza e la forza della comunicazione, *La Gazzetta del Mezzogiorno.*

Rovai, D. (2005). Il decreto articolo per articolo e le storie di SOGIN. http://www.peacelink.it/ecologia/docs/2470.pdf, last accessed 22 October 2013.

Rovai, D. (2009). L' eredita nucleare SOGIN, un'esperienza allarmante. http://www.nonukes.it/nonukes/sogin.htm, last accessed 22 October 2013.

Società Gestione Impianti Nucleari (SOGIN) (2003). Studio per la localizzazione di un sito per il deposito nazionale centralizzato dei rifiuti radioattivi. PDN RT 002. http://www.archivio nucleare.com/files/studio-sogin-deposito-nazionale-2003.pdf, last accessed 22 October 2013.

SOGIN (2013). Company Profile, March.
http://www.sogin.it/SiteAssets/uploads/2013/SOGIN_Company%20Profile_EN.pdf

SOGIN/ Nomisma Energia (2013). Decommissioning nucleare, un nuovo business globale per creare occupazione. Press release 16 April 2013.

Ventura, G. (2003). Sistema Informativo Geografico per il sito del Deposito Nazionale dei Materiali Radioattivi. ENEA GSP3 SITO. http://www.archivionucleare.com/files/studio-gsp3-sito-deposito-nazionale-2003.pdf, last accessed 7 November 2013.

World Nuclear Association (WNA) (2007). Areva and Sogin sign reprocessing contract, *World Nuclear News*, 9 May 2007.

World Nuclear Association (2014). Nuclear Power in Italy. http://world-nuclear.org/info/Country-Profiles/Countries-G-N/Italy/#.Um5RyVMhFnw, last accessed 14 September 2014.

Subject to Political Capture?
Nuclear Waste Governance in Spain

Ana María Isidoro Losada

1 Introduction[1]

In Spain, political decision-making processes are characterised by complex, intertwined relationships and interactions between the different political-territorial levels. According to the Spanish Constitution of 1978, the central Government, has devolved or transferred power and competences to the regional authorities, which also have certain exclusive powers. Since 1983, autonomous statutes have been approved for all 17 regional governments of the country, the Autonomous Communities (ACs). Responsibilities for territorial organisation and planning, environment and economic organisation, and agriculture and public works, have been granted to the Autonomous Communities.

In general, the negotiation processes between the central government and the highly decentralized regional governments operate through party channels. At the national level, the Spanish multi-party system has been characterised at different times by the predominance either of the social-democratic centre-left Socialist Party (PSOE) or the conservative centre-right People's Party (PP) – with some minor parties participating in the coalition governments. Since these two leading national parties are also organised at the regional level, their influence and communication channels are far-reaching. However, in the last decade regional parties and new political parties grew and challenged the old-established parties. In the context of nuclear waste governance, in particular, the regional parties play a key role because of their fierce opposition to siting interim storage facilities or final deep geological repositories in their autonomous regions. The nuclear waste governance strategy in Spain should be analysed in this context. Currently, Spain belongs to the group of countries that have opted for the construction of a centralised temporary storage facility and have

1 This chapter is a contribution of the Environmental Policy Research Centre (Forschungszentrum für Umweltpolitik – FFU) of the Freie Universität Berlin to the project "Multi Level Governance-Perspective on Management of Nuclear Waste Disposal. A Comparative Analysis: Actors, Instruments and Institutions" that is part of the inter-disciplinary ENTRIA research platform funded by the German Federal Ministry of Education and Research (BMBF/ 02S9082B).

postponed the decision-making process concerning the final deep geological repository for nuclear waste. The process linked to site selection has marked the most important challenge in nuclear waste governance in Spain.

2 The status of nuclear power plants and plant life management

In Spain, there are currently five nuclear power plants, hosting seven nuclear power reactors with an installed capacity of 7,864.7 MWe.[2] Nuclear power generates approximately 20% of the nation's electricity (Foro Nuclear 2014: 8 f.).[3] The Energy Reform Bill, which passed in December 2012, reduced subsidies, negatively affecting the viability of renewable energy projects. Given that Spain currently relies on imports to cover its energy needs, political and economic reasons may again encourage the expansion of nuclear energy. Accordingly, this may lead to the generation of additional operational and decommissioning wastes (World Nuclear Association (WNA) 2013a). The afore mentioned energy reform introduced higher taxes on electricity generation, which rendered the extension of the planned lifespan of Spain's oldest nuclear power plant (NPP), Santa María de Garoña, financially unviable. Subsequently, and contrary to the operator's initial plans, the NPP was shut down in December 2012 and declared officially shut down in July 2013. A royal decree which was approved in February 2014 foresees recently-shutdown plants to apply wihtin 12 months for a renovation of their operating licences. In May 2014 Nuclenor (*Centrales Nucleares del Norte*), the owner and operator of NPP Garoña, applied for a renewal of its licence up to 2031. At the end of September Nucleor submitted a work plan which considered the requirements by the Nuclear Safety Council (*Consejo Seguridad Nuclear* (CSN)). Nuclenor plans to restart the operation of NPP Garoña by 2015 (WNA 2014).

As Spain has no domestic conversion or enrichment facilities, it relies on contracts for conversion and services abroad. Prior to 1983, certain amounts of

2 Two nuclear power reactors are located in Almaraz/Cáceres, two in Ascó/Tarragona, one in Trillo/Guadalajara, one in Vandellós II/Tarragona and one in Cofrentes/Valencia. The operation of NPP Santa María de Garoña was terminated in December 2012, its operating permit expired on July 6, 2013. The origin and development of nuclear energy in Spain dates back to the four decades of General F. Franco's military dictatorship and his ambitions to press ahead with modernisation and to strengthen Spain's international position and geopolitical influence. Nuclear development was considered both a source of energy capable of accelerating Spain's industrialisation and national development, and a means through which to turn the country into a military power (Presas I Puig 2005).

3 In 2007, energy from nuclear fission represented 17.4% – the lowest proportion – of the total electricity generated in Spain. From 2010 to 2012, nuclear power generated approximately 20% of the electricity (WNA 2013b).

spent fuel from the nuclear power plants José Cabrera and Santa María de Garoña were sent to Great Britain, and the spent fuel generated in the NPP Vandellós I was reprocessed in France.[4] As set out in a bilateral agreement with France, the waste from Vandellós I should have been returned to Spain by the end of 2010. As this did not happen, Spain has been paying a daily rate of €64,900 since January 2011 to France for temporary waste storage. The waste should return to Spain by 2017.

Spain does have a fuel element manufacturing facility, Juzbado in Salamanca, which supplies reactors in Spain and other European countries.There is also the Mestral Technology Centre, located on the grounds of Vandellós I NPP, which was set up to cover research and development needs pertaining to the dismantling of nuclear power plants. A further research and development centre is located close to the intermediate-level waste storage facility El Cabril in Cordoba. In contrast to the military ambitions of the 1940s and 1950s, there is now no clear R&D for a military nuclear programme in Spain. Today, there is little public support for nuclear energy, which is below the European Union 27 state average: According to the results of the Eurobarometer survey on nuclear safety survey of 2010, 49% of respondents think that the amount of nuclear energy in use should be reduced (European Commission (EC) 2010: 26). In June 2013, Ipsos Public Affairs evaluated at the request of Foro Nuclear the public opinion on nuclear energy in Spain. The survey revealed that 55% of the population is opposed to nuclear power (Foro Nuclear 2014: 54 ff).

In 1983, the new Socialist administration decided to stop construction on all incomplete nuclear power plants.[5] A year later, a nuclear moratorium was approved, which prohibited the construction of new nuclear power plants and paved the way to a phase-out. The subsequent application of the nuclear moratorium in 1991 implied the final cancellation of five planned power plants.

4 In 1989, just three years after the Chernobyl disaster, a serious fire accident in the nuclear power plant Valdellós I forced the plant to shut down. Currently it is the only nuclear power plant in Spain in the advanced stages of decommissioning. Beginning in 2003 and for the next 25 years, the non-removed parts of the installation will remain under the responsibility and surveillence of the National Company for Radioactive Waste Management (*Empresa Nacional de Residuos Radioactivos* (ENRESA)); the reactor box will be removed in the final dismantling stage.

5 This decision can be also regarded as an after-effect of the previous conflict phase. From 1977 to 1981, the nuclear power plant Lemóniz in the Basque Region, as well as its workers and engineers, have become target of more than 300 attacks by the armed Basque organisation *Euskadi Ta Askatasuna* (ETA). The violence escalation reached its peak in 1981, when ETA kidnapped and killed the chief engineer of the power plant. This was followed 15 months later by the assassination of the plant's general project director. The violence halted work at the plant. Consequently, the construction of the plant was never finished; to date no alternative power plants nor nuclear waste disposal projects have been launched in the Basque Region.

Similarly in 1992, the Spanish parliament decided that no further nuclear plants should come into operation in the near future. Finally, on December 30, 1994, the final shut-down of the nuclear power plants which were affected by the moratorium was enforced. As a result, the tariff burden for electricity increased and consumers had to shoulder most of the costs. In 2006, the nuclear power plant (NPP) José Cabrera was shut down and the Government approved plans for decommissioning and dismantling. In 2010 in was turned over to the National Company for Radioactive Waste Management (ENRESA). The aim is to clear the site by the end of 2017. At the same time, since the middle of 1990s, a notable policy of power plant upgrades took place and between 2010 and 2012 licence renewals for four power plants (Almaraz 1 and 2, Vandellós II, Cofrentes, Ascó 1 and 2) was approved. The licence renewal for NPP Trillo will be revised in 2014. In February 2011 the legal provision limiting nuclear plant operating lives to 40 years was removed by parliament.

3 The rise of national nuclear waste policies

In 1961, Spain was confronted for the first time with the question of nuclear waste management; at this time it became necessary to find a final destination for the waste generated by its national nuclear research reactors. In the absence of a waste management policy, the Franco regime decided to store components of the nuclear waste in a former uranium mine in Hornachuelos in the province of Córdoba (El Cabril). In 1989, a ministerial decree ruled that the national company ENRESA should develop and operate this disposal (Boletin Oficial del Estado (BOE) 1989). Since 1992, low and intermediate-level waste (LILW), including waste from decommissioning, has been centrally disposed of at El Cabril in a near-surface disposal comprising of a set of multiple barriers. The capacity of this vault-type disposal facility is considered sufficient until about 2020 (Zuloaga n.d.: 2), but forecasts indicate that after 300 years, the engineered barriers will have been fully degraded under present conditions (Zuloaga 2009: 24). In 2006, a further disposal system for very low activity waste (VLLW) was constructed on the same site and this storage went into operation in October 2008. El Cabril represents the central final repository for all national LIL- and VLL waste; by December 2012 the total accumulated volume of radioactive waste amounted to 36,613.32 m^3 (Estratos 2013: 14; ENRESA n.d.). In addition, the installation houses facilities for waste treatment and conditioning, as well as verification and characterisation of wastes (Ministry of Industry, Energy and Tourism (MINETUR) n.d. a). At the beginning of 2014, ENRESA publicly requested authorisation to double the storage capacity at El Cabril (Albert 2014).

In search of a high-level waste (HLW) disposal site, ENRESA conducted systematic geological surveys and studies from 1986 to 1996 in granite, salt, and clay formations to assess the feasibility of deep geological disposal in Spain. Around 25,000 km^2 of possible area was registered. However, due to strong public opposition, their advanced work on identifying suitable sites for further underground research was halted. In 2006, the Site Selection Plan was stopped and the government approved the establishment of an Interministerial Commission to oversee the development of a Centralised Temporary Storage Facility (CTSF, in Spanish *Almacén Temporal Centralizado* (ATC)) for high-level radioactive waste (Real Decreto 2006). As a result of the strong public opposition in the 1990s, one major requirement was that the process should comply with the principles of transparency and openness.

The national implementation of the Council Directive 2011/70/Euratom accelerated the search for and designation of the site of a ATC, which should house all of the high-level radioactive waste and spent fuel from Spanish nuclear power plants, as well as long-lived LILW that cannot be disposed of in El Cabril.

4 The nuclear waste management policy for spent fuel and high-level radioactive waste

The sixth General Radioactive Waste Plan (GRWP) is the basic reference document for national radioactive waste management strategies. Spain's installed nuclear power capacity, combined with an assumed forty year lifespan for each nuclear plant, has increased the total expected volume of conditioned low, very low, and intermediate level wastes to an estimated 176,300 m^3. At present, the expected volume of high level waste, including spent nuclear fuel, high level radioactive waste, and intermediate long-lived level waste which cannot be managed at El Cabril, is estimated to be 12,800 m^3 (MINETUR 2006).

Table 1: Radioactive waste to be managed in Spain.

Waste category	Expected volume (m^3)
low, very low and intermediate level waste (LILW)	176,300
spent fuel, high level radioactive waste and intermediate long-lived level waste	12,800

Source: 6th General Radioactive Waste Plan (2006).

With the exception of the spent fuel at Vandellós I, spent fuel is currently being stored at the sites of the nuclear power plants, both in cooling pools and in dry storage systems at the reactor sites. By 1999, construction of additional spent fuel storage capacity in the reactor pools was completed and only the nuclear power plant Cofrentes needed a second phase of re-racking, which was completed by 2009. By the end of 2012 the average usage rate of the pools of the nuclear power plants was 82% (OECD/ NEA 2013: 2, 5).

In Spain, there are currently three temporary storage facilities (in Spanish *Almacén Temporal Individualizado* (ATI)) for spent fuel and high-level radioactive waste at the sites of the nuclear power plants Trillo, which has been in operation since 2002, and José Cabrera, in operation since 2009. These facilities are both situated in the province of Guadalajara and Ascó I in Tarragona (Foro Nuclear n.d.). Recognizing that it was impossible to complete the required central interim storage facility on time, the Ministry of Industry agreed in September 2011 to build an additional temporary storage facility using dry storage on the grounds of the nuclear power plant Ascó I/II in Tarragona (ASENA n.d.). This decision was made in part because of the prolonged lifespan of the NPP Ascó; the existing cooling pools would be saturated respectively in 2013 and 2014 (Consejo Seguridad Nuclear (CSN) 2012: 174). Ultimately, in April 2013 the CSN approved the operational start of the temporary storage facility.

4.1 The centralised temporary storage facility

In 2004, the Industry Commission of the Spanish parliament recommended the development of a centralised storage facility for spent fuel and high-level radioactive waste. In the sixth Radioactive Waste General Plan, made in 2006, the construction of the Centralised Temporary Storage Facility was seen as a priority. At the same time, the development of strategies for the final disposal of high-level waste and spent fuel were neglected in these recommendations and plans. In June 2006, the CSN again favoured the option of building a centralised storage facility.

At the end of 2011, the newly elected conservative government (PP) decided to site the facility in the municipality Villar de Cañas in the province of Cuenca, a locality with 455 inhabitants that volunteered to host the site for the ATC. The construction of this facility will start in 2016, with a design which will be similar to the HABOG facility at the nuclear power plant in Borssele in the Netherlands. It will have a dry storage system with 12 vaults and natural convection air-cooling. The facility is designed to hold approximately 6,700 metric tons of used fuel and 2,600 m^3 of medium-level wastes, plus additional 12 m^3 of high-level waste – reprocessed Vandellós-I fuel – for 100 years (World Nuclear Association (WNA) 2013a). It should also store the returned HLW from reprocessing completed abroad. It has also been planned that during the next 60 years, other waste which cannot be housed at the El Cabril facility, as well as spent fuel that cannot be accommodated at the nuclear power plants where it was generated, could be stored at the Villar de Cañas facility. The adjoining construction of a research and technological centre is also foreseen; furthermore the integration of research and development initiatives should ensure the synergy effects related to technologies and provide the necessary know-how to manage the final disposal of spent fuel and HLW. The industrial park and associated investments in the development of the local and regional infrastructure should trigger further regional development. The initially calculated cost of the construction of the facility is €540m and an additional €50m for the Technology Centre (Villar de Cañas n.d.).

As a result of the high priority given to the ATC, the decision-making process for long-term disposal and related research in deep geological disposal has been postponed until 2029. This postponement is supported by the argument that more experience and knowledge at national and international levels will have been gathered by then. Consequently, a final geological disposal facility is not expected until 2050, at the earliest. According to the sixth GRWP, the focus on the decision-making process and site characterisations should be in the period between 2025 and 2040, and the construction of the final geological disposal facility should take place between 2040 and 2050.

To date, retrievability is neither an official option nor a regulatory requirement for the final disposal of high-level waste and spent fuel. However, ENRESA is considering options to facilitate easy retrieval of wastes for a period of 100 years.

4.2 Financing

The Royal Decree 1899/1984 established the system of advance payments by levy. Consequently, producers have to pay for waste to be treated and disposed of. The government sets the percentage rate which should be applied each year, where the revenue is proportional to the electricity generated by the nuclear power plants in a given period. The revenues are deposited at the National Commission of the Electric System (*Comisión Nacional del Sistema Eléctrico* (CNSE)), which transfers it to ENRESA on a monthly basis. The collected money is put into a fund specified in the sixth GRWP.

ENRESA is responsible for the financial management of the fund, and investments and follow-up are supervised by a Committee attached to the Ministry of Industry, Energy and Tourism (MINETUR). ENRESA periodically reviews the adequacy of the levy's calculation to ensure that all expenses resulting from the management of radioactive waste, as well as the dismantling and decommissioning of nuclear power plants, are covered. The money is invested in a fixed income financial assets portfolio as an interim financial investment; only possibilities with first class financial ratings are considered. The real rate of return expected from the calculation of the levy is 3.5%. Additionally, the government sets fees, in the form of taxes, on consumer electricity bills. The required income is calculated annually (Hearsey et al. n.d.: 2). In the case of small producers, separate tariffs are calculated for different types of waste. This takes into account the different treatment and disposal processes to which the waste is subjected.

According to current estimates of the MINETUR, the total costs of the Spanish nuclear waste management programme from 1985 to 2070 will be around €15bn. In comparison, the data presented in the GRWP estimated the total costs in the range of €13bn; this represents approximately 26% of the total national budget. Half of the budget was allocated for the management of spent fuel, and a fifth was allocated for the dismantling of nuclear power plants (MINETUR n.d. b: 31-32; Zuloaga n.d.).

On 28 December 2012, the *Official Gazette* published the Law No. 15/2012 on tax measures for energy sustainability. This law introduced a tax on the production of radioactive waste resulting from the generation of nuclear energy.

The applicable tax rate is:

- €2,190 per kg of heavy metal (uranium and plutonium),
- €6,000 per m^3 for production of low and intermediate level waste,
- €1,000 per m^3 for very low-level waste.

A tax on the storage of spent fuel and radioactive waste was also set. The tax rate is:

- €70 per kilogram of used spent fuel,
- €30,000 per m^3 of long-term high and intermediate radioactive waste, other than spent fuel,
- €10,000 per m^3 for low and intermediate level waste,
  €2,000 per m^3 for very low-level waste (BOE 2012: 88087-88088).

5 Main national laws and regulations in nuclear waste management

In the following section, the major legislation governing nuclear facilities and nuclear waste management in Spain is briefly described.[6]

- The Nuclear Energy Act 25/1964 and the Regulation of Nuclear and Radioactive Facilities of 1999 specify issues concerning radioactive waste management.
- The Law of 1980 stipulates the creation and regulation of the CSN.
- A 1983 Ministerial Order rules that a percentage of the consumer's electric bill shall be deducted from the electricity price in order to create a fund to cover the costs of nuclear waste management and dismantling.
- The Royal Decree (RD) 1522/1984 of July 14 established the creation of the state owned ENRESA, which operates under the supervision of the MINETUR; its responsibilities were updated by the Royal Decree 1349/2003 of October 31.
- The RD 1899/1984 established the Spanish system of advance payments by levy to cover projected future expenditures.
- The Electricity Industry Law 54/1997 regulates the electricity sector, including nuclear power, and stipulates the implementation of a fund to finance radioactive waste management, as set out in the GRWP.
- The Ministerial Order of 13 July 1998 authorizes compensation for the municipalities that provide a site for radioactive waste storage. Law 24/2005 of 18 November 2005 regulates the fees paid for ENRESA's services, which should be paid to the Fund envisaged in the GRWP.
- The RD 775/2006 stipulates the creation of an inter-ministerial commission to plan the siting of ATC facilities and its associated technological centre and the creation of a Technical Advisory Committee, consisting of six experts, mainly academic professionals. Law 11/2009 states that the management of radioactive waste, including spent fuel, and the decommissioning of nuclear facilities are assigned exclusively to the central government. The management is entrusted to ENRESA. Law 15/2012 of December 27 stipulates that fiscal measures for energy sustainability should be introduced. There are two new taxes applicable to nuclear installations and related matters: a tax on the production of spent fuel and radioactive waste from nuclear energy generation, and a tax on spent fuel and radioactive waste storage in centralized installations.

6 For additional details see Convención Conjunta (2011: 45-52).

6 The institutional framework and selected stakeholders in the nuclear waste management organization

As the main regulatory body, the Ministry of Industry, Energy and Tourism is in charge of radioactive waste management policies, i.e. enforcing nuclear legislation and granting, modifying, suspending, or revoking licences, as they are subject to a mandatory and binding report from the Nuclear Safety Council. MINETUR submits the General Radioactive Waste Plan for approval to the Government, following a report by the CSN reports. The General Direction for Energy, Policy and Mines, which is affiliated to the MINETUR, mediates between state organisms and the different stakeholders in elaborating policies concerning radioactive waste management. The Ministry of Agriculture, Food and Environment is involved in the licensing process and – in collaboration with the CSN – is responsible for the Environmental Impact Statements. In this context, Autonomous Communities and local governments are involved in the relevant geographic areas based on their competence in land planning and environment

The CSN was set up in 1980 and it is the competent organization in matters of nuclear safety and radiation protection. It authorizes all of the equipment and facilities to be built. The CSN works independently and reports directly to Parliament, which must approve radioactive waste management planning strategies and the schedule of major related processes (Convención Conjunta 2011: 55-66; OECD/ NEA 2013: 4).

The National Company for Radioactive Waste Management (ENRESA) is the Interministerial organisation in charge of spent fuel, radioactive waste management and decommissioning. It is subordinated to the MINETUR, via the Secretariat of State for Energy. ENRESA carries out research and development programmes to assess different aspects of deep geological disposal, and investigates separation and transmutation processes. Furthermore, it provides technical studies and information to the various stakeholders and draws a review of the GRWP, which then is forwarded to the MINETUR.

The *Centro de Investigaciones Energéticas, Medioambientales y Tecnológicas* (CIEMAT), is an institution under the Ministry of Economy and Competitiveness and is dedicated to nuclear research.

Other key actors at the institutional and/or political-territorial level

The governance of the current Spanish institutional framework is characterised by different political-territorial levels: the national, the regional, the provincial, and the local.

Autonomous Communities

In general, the Autonomous Communities have no specific competence in nuclear administration or management. According to Law 25/1964, Article 28, they have the right to be heard on all related subjects pertaining to the authorisation of nuclear energy installations in their territory, based on their competences in terms of development and land-use planning, as well as environmental policies. Without the consent of the ACs it is difficult to obtain a site for radioactive waste storage or disposal facility. The executive functions attributed to the Ministry of Industry, Energy and Tourism were transferred to the following Autonomous Communities and their respective regional governments: Islas Baleares, Aragón, Asturias, Basque Region, Cantabria, Castilla y León, Catalonia, Ceuta, Extremadura, Galicia, Islas Canarias, La Rioja, Madrid, Murcia, Navarra and Valencia (Convención Conjunta 2011: 56).

Municipalities

The local authorities granted municipal construction licences for the storage of radiocative waste (PGRR 2006: 80); in many cases, they tried to communicate and negotiate directly with the central government, and in doing so bypass the regional governments. Conversely, the Autonomous Communities often refused to cede power to the local level, reinforcing a certain re-centralisation of the local governments. This led to numerous conflicts and tensions between the authorities at the different political-territorial levels.

Association of Municipalities in Areas of nuclear plants (Asociación de Municipios en Áreas de Centrales nucleares (AMAC))

AMAC, the Spanish counterpart of the Group of European Municipalities with Nuclear Facilities (GMF), was created in 1990 and represents 72 municipalities located within a 10km radius around the Spanish nuclear power plants. It acts as a representative voice for the concerned municipalities, defends local interests, and supports the idea that a centralized storage facility should be planned and organized through democratic decision-making. AMAC maintains an open opposition to the enlargement of intermediate storage facilities and defends the idea of a centralized storage facility approved by consensus of the interested parties. In addition to its emphasis on the importance of information and communication, AMAC expresses a strong interest in creating an economic alternative in the nuclear regions, promoting a diversification of their economic planning through different energy industries, tourism, and agriculture. AMAC has also signed agreements with ENRESA and is one of the representative organisations within the Advisory Committee of the CSN.

7 Status of the siting procedures and eventual compensation mechanisms

Research on deep geological disposal has been on going since 1985. In 2006, the site selection plan was stopped. At this point the plan provided the information and details gathered regarding the abundance of granite and clay in the sub-soil and, to a lesser extent, salt formations suitable for housing a disposal facility. In 2006, the Parliament ratified ENRESA's plans to develop an ATC by 2010 and the CSN approved its design. The company estimated a €700m investment over 20 years and the government offered to pay up to €7.8m annually once the ATC facility would become operational. In addition, the plans included the possibility of creating an estimated 300 new jobs in the first five years.

In the same year, the Ministry of Industry carried out an information campaign targeting all municipalities potentially interested in hosting the facility. In December 2009, the government called for municipalities to volunteer to host the ATC facility and the technology centre. This call was published in the official state gazette and interested municipalities were given one month from the date of publication to submit their candidacy. The Inter-ministerial Commission, released a report classifying the different candidate municipalities after studying their locations and, in September 2010, reported the selected potential sites to the body of government responsible for the authorization of the final site (MINETUR 2010). Fourteen towns volunteered; eight of them were formally accepted, whereas three withdrew and three others were excluded because they did not fulfil the conditions (Convención Conjunta 2011: 9-10).

On the 30 December 2011, MINETUR affirmed that Villar de Cañas in the province of Cuenca/Castilla-La Mancha, a municipality with 450 inhabitants, had been selected (MINETUR 2012). According to official declarations, Villar de Cañas "meets the technical characteristics required for this type of installation and the surroundings has a high rate of unemployment, so that the project will have a positive socio-economic impact" (Estratos 2012: 10).

Despite criticisms and complaints from the various stakeholders, ENRESA bought the land (a total of 55 hectares) on 30 June 2012, paying around €10,000 per ha. The agency started its geological surveys at that time. On 13 June 2014 the anouncement of the public information process for the environmental impact assessment (EIA) and documention of the prior authorisation was published. Within a period of 30 working days the interested parties and the public have the possibility to submit statements (ENRESA 2014: 11).

8 The decision-making process and the role of political parties

Given that the bilateral contracts for conversion and services with Great Britain and France were going to expire, respectively in 2010 and 2011, at the end of 2009, the Zapatero-led socialist government raised the urgent need to create an ATC in Spain. In 2010, after receiving the report from the Inter-ministerial Commission which presented the municipalities that volunteered for the subsequent selection process, the government opened a negotiation with the respective Autonomous Communities. In all cases, the ACs with candidate municipalities, which were willing to host the ATC, were opposed to siting the facility. The Autonomous Communities complained that they were being ignored and excluded from the initial siting process. This lack of participation and inclusion of the regional authorities caused considerable dissent. It also provoked an apparent tension between the municipalities, which wanted the installation of the facility for economic reasons, and the ACs. This led to delays in the decision-making and project implementation processes.

In Castilla-León the regional government initially agreed, provided that the prolonged existence of the nuclear power plant in Garoña would be considered. Afterwards, it changed its position and was against the storage facility. In Extremadura, the conservative party (PP) in the regional government rejected the storage location, arguing that they already hosted the nuclear power plant Almaraz and the former uranium mine La Haba in Badajoz, where probably thousands of tons of unclassified (radioactive) waste from the national research reactor JEN were buried during Franco's dictatorship (El Escarabajo Verde 2011). The speaker of the PP complained that the Zapatero government had taken advantage of the financial crisis in the municipalities and criticized that the Autonomous Communities were set aside while more protagonism was given directly to the municipalities (La información 2010). In particular, the regional governments of Catalonia and Castilla-La Mancha publicly opposed the siting of the facility, despite the willingness of some municipalities in their regions to host the ATC. In 2010, following the initiative of the socialist president of Castilla-La Mancha, the majority of the regional government voted against the ATC. In many cases, the controversy appeared to be dictated by a Not In My Back Yard (NIMBY) attitude. In some cases, regional politicians even stated that they would agree on siting the facility outside of their own region.Since 2011, the president of Castilla-La Mancha has been the secretary-general of the conservative party (PP) and a faithful supporter of nuclear power (Hernanz 2010). At the end of 2011, when the Rajoy conservative Government (PP) was only a few days in office, MINETUR surprisingly announced that Villar de Cañas, in the AC Castilla-La Mancha, had been selected. In the report by the

Interministerial Commission in 2010, this municipality was fourth on the list of the potential sites (MINETUR 2010: 51). In January 2012, AMAC contested the decision to construct the ATC at Villar de Cañas and pointed out that in the decision-making process, only the positive decision of the political party in the Autonomous Community and the municipality of Villar de Cañas had been considered; technical criteria was not. It is worth pointing out that at this time, all governments of the different political-territorial levels were in the hands of the conservative party. Additionally, a local NGO criticized that the construction of the ATC was awarded to a subsidiary of Iberdrola, where the husband of the president of the AC Castilla-La Mancha is member of the Board of Directors. This conflict of interest might also explain the bias towards the selection of Villar de Cañas (Eco Quijote 2013).

Considering Spain's persisting economic crisis, it is not surprising that municipalities located in depressed regions are attracted by the promise of jobs and local investment tied to the ATC. Considering these circumstances, public participation can readily translate into popular approval for project siting. In the case of Villar de Cañas, local organisations criticised that an ENRESA foundation sponsored several cultural events in the municipality and tried to influence public opinion in its favor (Cuenca news 2013).

9 Information policy and participation

In 1961, the Franco regime decided to store nuclear waste in El Cabril/ Hornachuelos. This storage took place without formal authorisation and without informing the involved municipalities and the public. Then, in 1975, the Ministry of Industry officially approved the use of El Cabril to accommodate nuclear waste. However, this decision was again made without informing or allowing for the participation of the affected municipalities. The lack of local community support and their claim to be informed and involved in the process became evident when, in 1988, the mayor of Hornachuelos ordered the immediate stop of work and the closure of the disposal facility. This however, produced no real results.

In the 1970s and 1980s, conflicts related to the construction of the NPP Lemóniz in the Basque Region represented not only one of the major anti-nuclear issues, but also strongly influenced policy change. In July 1977, approximately 200,000 people took part in one of the most important anti-nuclear demonstrations of all time, in Bilbao.

In 1996, due to strong public opposition, ENRESA's advanced work on identifying suitable sites for further underground research and general policymaking was halted. Local actors and environmental groups accused ENRESA of not having informed the public of these processes. Consequently, large demonstrations were organized; the first one took place in 1996 in Belalcázar with 10,000 participants, followed by a demonstration in 1997 in Villanueva with 15,000 supporters, and finally a demonstration in 1998 with 20,000 participants in Torrecampo and the involvement of all municipalities in the province of Córdoba/Andalucía.

Given the strong public opposition during the 1980s and 1990s, calls for increased public information and transparency emerged, making the implementation of more participatory and dialogue-orientated policy-making in the selection of the ATC site of pivotal importance. Wanting to push the agenda of radioactive waste management to the national level and hoping to count on broader public support, the regulatory body CSN, as well as ENRESA and the Spanish Association of Municipalities in Nuclear Areas, launched the Communities Waste Management Spain initiative (COWAM) (2004-2006). They concentrated on a communication and information strategy, based on the experiences of the COWAM projects funded within the Euratom Framework Programme. The COWAM Spain programme specifies a methodology to steer the search for solutions for the identification of suitable sites. Following the parliament's approval to develop an ATC in 2006, the AMAC announced that the members of the Association were committed to supporting the government in the siting process. COWAM organised a number of informational meetings, seminars, and discussions in the seven nuclear areas. The meetings were open to the various stakeholders involved in the decision-making process: the mayors of nuclear areas, councillors, local governments, associations, professionals, and enterprises (Kopetz and Martell 2009).

On 29 December 2009, the candidacies were published in the State Official Gazette. At the beginning of March 2010, the Ministry of Industry notified the candidates to submit their applications within 20 days. In addition, 44 institutions and organisations were specifically notified about the formal procedure. At the end of this process, a total of 14,420 affirmations made by municipalities were registered.

The local council of Ascó, a municipality which has hosted a nuclear power plant for decades, was one of the volunteering municipalities. It was supported by various stakeholders from academia and trade unions, as well as some local mayors. AMAC, together with the Ascó local council, supported the creation of a consortium that would jointly manage the technology park, which was planned alongside the storage facility. In addition, different political, social, and economic actors debated on the perspectives of, and visions for local development in their territory. Despite this positive attitude the Government by means of an

Agreement of the Cabinet of Ministers deciced in favour of the candidature of Villar de Cañas. AMAC considered the decision to site the ATC at Villar de Cañas to be "hasty and insufficiently thought out". In October 2013, the party *Los Verdes*, publicly disagreed with the siting because of geological studies which stated that the suitability of the site was questionable, due to the serious geological risks connected to the existing karst formation in the subsoil and the related water filtration (Los Verdes 2013). *Greenpeace Spain* and *Ecologistas en Acción* criticise that the Aarhus Convention on Access to Information, Public Participation in Decision-Making and Access to Justice in Environmental Matters had not been applied. Neither the necessary transparency nor public participation mechanisms to consult and involve the public in the decision-making process regarding the storage facility were considered. In January 2014, *Ecologistas en Acción* denounced responsible authorities for refusing to provide the geological reports related to the site at Villar de Cañas (dclm.es 2014).

Despite the significant share of nuclear energy in Spain's energy mix, public opinion data regarding nuclear safety related topics finds 83% of the public respondents felt uninformed (EC 2010: 89). According to the results of the Eurobarometer on nuclear safety survey of 2010, on the whole, Spain shows a high non-response rate to the statements that "nuclear energy helps to decrease dependence on imported fuels" and "nuclear energy helps to limit climate change". This result surprises as these attitudes are typical in the group of countries where domestic energy sources do not include nuclear power. The majority of respondents (73%) consider nuclear power plants to represent a risk, which ranks Spain second in the category, "fear of nuclear power" (EC 2010: 49). Spain, like countries with no nuclear power plants and those which record a high level of opposition to nuclear energy, tends to prefer a location for a new NPP in "a country outside the EU, under the supervision and guidance of the responsible authorities and legislation of that country" (36%) (EC 2010: 38). A majority of Spaniards think that the risks of nuclear power as an energy source are greater than its benefits (64%) (EC 2010: 72). There is a strong lack of confidence (44%) concerning the safe management of radioactive waste (EC 2010: 63).

10 Conclusions

Since the beginning of nuclear waste storage in the 1960s, there has been a notable lack of transparency related to the nuclear waste management policies in Spain. Decisions were generally made by politicians in consultation with a limited group of involved stakeholders behind closed doors, and the relation to the public is characterised by misinformation. Based on the historic reasons mentioned above and because of recent experiences, there is considerable public mistrust regarding nuclear energy and nuclear waste management policies. The low level of transparency displayed in the decision-making procedures in the placement of the ATC has not stimulated major societal support; on the contrary, these procedures are constantly being questioned. At a first glance, it seems that in the context of the ATC, the democratic decision-making and the legitimacy of the results have been undermined by processes of appropriation through politics ("political capture"), where nuclear waste governance issues have not only been used for party-politics, but also reflect procedures based on clientelism and patronage. In addition, the non-deliberative policy and decision-making style and the reluctance of national authorities to actively cooperate with the Autonomous Communities have created a complex situation, with highly diverging interests. The criticism expressed by AMAC, the anti-nuclear organisations and the local groups shows that even though the government and "the Establishment" have made a decision, the lack of consensus will provoke further challenges for the implementation of the project in general. Furthermore, future governments will be confronted with major challenges related to the siting of a deep geological disposal facility.

Thus, two main problems can be underscored: the lack of trust among the different actors involved in the project and the highly non-transparent administrative practices. Therefore, it is necessary to ensure the involvement of all key political-territorial tiers and authorities, and to establish a constant communication and information system among the stakeholders. This may lead to facilitating a fairer involvement of actors in search of solutions and to a democratic political decision-making process, thereby ensuring better trust and guaranteeing procedures which would open the door to a successful socio-political consensus.

References

Albert, M. J. (2014, January 27). Enresa pide permiso para duplicar el almacén radioactivo de El Cabril. *El País*. http://www.elpais.com/ccaa/2014/01/27/andalucia/13908 20694_552928.html, last accessed 7 February 2014.

ASENA (n.d). Luz verde Asco II para construir almacen temporal de residuos individualizado. http://www.asena-consulting.es/noticias/luz-verde-asco-ii-para-construir-almacen-temporal-de-residuos-individualizado/, last accessed 7 February 2014.

Boletin Oficial del Estado (BOE) (1989, 2 November). Orden de 31 de octubre de 1989 por la que se otorga a la «Empresa Nacional de Residuos Radioactivos, Sociedad Anónima», autorización para la construcción de la ampliación de la Instalación

Nuclear de Almacenamiento de Residuos Radiactivos Sólidos de Sierra Albarrana, BOE-A-1989-25837.

BOE (2012, 28 December). Ley 15/2012, de 27 de diciembre, de medidas fiscales para la sostenibilidad energética, No. 312. http://www.boe.es/boe/dias/2012/12/28/pdfs/BOE-A-2012-15649.pdf, last accessed 19 January 2014.

Convención Conjunta (2011). Convención Conjunta sobre seguridad en la gestión de combustibles gastado y sobre seguridad en la gestión de residuos radioactivos. 4to informe nacional, octubre 2011.

Consejo de Seguridad Nuclear – CSN (2012). *Informe del Consejo de Seguridad Nucler al Congreso de los Diputados a al Senado.* http://www.csn.es/images/stories/ publicaciones/periodicas/ informeanual2012.pdf, last accessed 7 February 2014.

Cuenca news (2013, 30 March). Protesta musical contra clientelismo en torno al cementerio nuclear. http://www.cuencanews.es/index.php?option=com_content&view=article&id=22980:protesta-musical-contra-clientelismo-en-torno-al-cementeri onuclear&catid=1:local &Itemid=596, last accessed 8 February 2014.

dclm.es (2014, January 5). *Los responsables del cementerio nuclear en Villar de Cañas se niegan facilitar los informes geologicos*, http://www.dclm.es/noticia.php?id=22012, last accessed 8 February 2014.

European Commission – EC (2010). *Eurobarometer on nuclear safety*. http://www.ec. europa.eu/energy/nuclear/safety/doc/2010_eurobarometer_safety.pdf, last accessed 8 February 2014.

Eco Quijote (2013). La Plataforma contra el cementerio nuclear pide la dimisión de Cospedal. http://www.ecoquijote.blogspot.com.es/2013/04/el-atc-cospedal-y-lopez-del-hierro-ante.htnml, last accessed 8 February 2014.

El Escarabajo Verde (2011, January 12). 'El escarabajo verde' descubre los secretos de la mina La Haba, este viernes en La 2. http://www.rtv.es/rtve/20110112/escarabajo-verde-descubre-secretos-mina-haba-2/394543.shtml, last accessed 8 February 2014.

Empresa Nacional de Residuos Radiactivos (ENRESA) (n.d.): Datos operativos y radiológicos de El Cabril. http://www.enresa.es/actividades_y_proyectos/rbma/datos_operativos_radiologicos_cabril, last accessed 7 February 2014.

ENRESA (2014). *Dossier de Prensa. Almacén Temporal Centralizado*. Retrieved from http://www.enresa.es/files/multimedios/dossieratcjulio14.pdf, last accessed 11 October 2014.

Estratos (2013). El almacén centralizado de El Cabril recibió 2.892 metros cúbicos de residuos radioactivos en 2012. http://www.enresa.es/files/multimedios/estratos, last accessed 7 February 2014.

Estratos (2012). *ATC – Villar de Cañas en Cuenca, albergará el ATC*, invierno 2012, N° 101.

Foro Nuclear (n.d.). La energía nuclear en Castilla – La Mancha. http://www.foronuclear. org/ recorrido_energia/fornuc/5/5_4_3/index.html, last access 8 February 2014.

Foro Nuclear (2014). Resultados y perspectivas nucleares para el año 2014. http://www.foronuclear.org/resultados-y-perspectivas-nucleares-2014/, last accessed 1 September 2014.

Hearsey, C.J.; Emmery, D.; Kunsch, P.; Bollingerfehr, W., and Webster, S. (n.d.). *The Financing of Radioactive Waste Storage and Disposal*. http://ec.europa.eu/energy/nuclear/studies/doc/other/ euradwaste99fs.pdf, last access 07.02.2014.

Hernanz, M. (2010, February 1). El parlamento manchego rechaza la construcción del ATC con el voto en contra del PP. *Rtve.es*. http://www.rtve.es/noticias/20100201/parlamento-manchego-rechaza-construccion-del-atc-voto-contra-del-pp/315545.shtml, last accessed 8 February 2014.

Kopetz, I., and Martell, M. (2009). *COWAM in Practice*. Spanish Prospective Case Study, September 2009. http://www.cowam.com/IMG/pdf_D1-11_FINAL_SPAIN_ENG.p pdf, last accessed 7 February 2014.

La información (2010, February 1). El PP extremeño rechaza el ATC en la región, que 'ya ha contribuido más que solidariamente al sistema energético'. http://noticias.lainformacion.com/economia-negocios-y-finanzas/energia-nuclear/el-pp-extremeno-rechaza-el-atc-en-la-region-que-ya-ha-contribuido-mas-que-solidariamente-al-sistemaenergetico_wSxwYLkpKnO25K fagxFrJ5/, last accessed 8 February 2014.

Los Verdes (2013). Los Verdes piden al gobierno de Rajoy la paralización del proyecto de almacén de residuos nucleares en Villar de Cañas por los riesgos geológicos de la zona. http://www.verdes.info/los-verdes-piden-a-rajoy-paralizar-proyecto-de-almacen-de-residuos-nucleares-en-villar-de-canas-por-riesgos-geologicos, last accessed 8 February 2014.

Ministry of Industry, Energy and Tourism – MINETUR (2006). 6th General Plan of *Radioactive Waste*.

MINETUR (2010). Comisión Interministerial Emplazamiento ATC: Informe de propuestas de emplazamientos candidatos, 20 September 2010. http://www.emplazamientoatces/Candidatos/proyectos/Informe_propuesta_emplazamientos_candidatos.pdf, last accessed 7 February 2014.

MINETUR (2012). *Acuerdo designación Emplazamiento ATC*, 20 January 2012. http://www.emplaza mientoatc.es/Documents/Acuerdo_ATC.pdf, last accessed 7 February 2014.

Presas I Puig, A. (2005). Science on the Periphery. The Spanish Reception of Nuclear Energy: an Attempt at Modernity? *Minerva: A Review of Science, Learning & Policy*, 43(2), 197-218.

Real Decreto (2006, 23 June). REAL Decreto 775/2006, de 23 de junio, por el que se crea la Comisión interministerial para el establecimiento de los criterios que deberá cumplir el emplazamiento del almacén temporal centralizado de combustible nuclear gastado y residuos de alta actividad, y de su centro tecnológico asociado. http://legislacion.derecho.com/real-decreto-775-2006-por-el-que-se-crea-la-comision-interminnisterial-para-el-establecimiento-de-los-criterios-que-debera-cumplir-el-emplazamiennto-del-almacen-temporal-centralizado-de-combustible-nuclear-gastado-y-residuos-de-alta-actividad-y-de-su-centro-tecnologico-asociado, last accessed 7 February 2014.

Villar de Cañas (n.d.). Plataforma logística del ATC. http://www.villardecanas.es/ATC/proyecto.pdf, last accessed 7 February 2014.

World Nuclear Association – WNA (2013a). *Nuclear Power in Spain.* http:www.world-nuclear.org/ info/Country-Profiles/Countries-O-S/Spain/, last accessed 7 February 2014.

WNA (2013b). *Nuclear Share Figures, 2002-2012.* http://www.world-nucler.org/info /Facts-and-Figures/Nuclear-generation-by-country/, last accessed 7 February 2014.

WNA (2014). *Nuclear Power in Spain.* http://www.world-nuclear.org/info/Country-Profiles/ Countries-O-S/Spain/, last accessed 11 October 2014.

Zuloaga, P. (2009). *El Cabril Low and Intermediate Level Radioactive Waste Disposal Facility – Long-term performance assessment.* http://www.cresp.org/PACOP/Presentations/Day1July132009/11_Zuloaga_Near_Surface_Disposal.pdf, last accessed 7 February 2014.

Zuloaga, P. (n.d.). *New developments in low level radioactive waste management in Spain.* http://www.euronuclear.org/events/topseal/transactions/Paper-Session-III-Zuloaga.pdf, last accessed 7 February 2014.

Contributors

Maarten J. Arentsen is an associate professor of Energy Innovation at the Department of Management and Governance of the University of Twente. He is one of the co-founders of CSTM, the University's institute for sustainable development research. Since 1991, he has chaired the energy section of CSTM. In this role, Arentsen develops, conducts, supervises, and coordinates research projects on energy policy, energy market reform, and (green) energy innovation, with a special focus on technological and institutional change. Between 1995 and 2004, he was the vice-director of CSTM. Arentsen publishes in national and international journals and books and teaches at the undergraduate and postgraduate levels. Email: m.j.arentsen@utwente.nl

Burkhard Auffermann is a senior research fellow at the Finland Futures Research Centre (FFRC) at the University of Turku. He holds a PhD in Political Science from the Freie Universität Berlin (1994). He was a professor of International Relations at the Universities of Tampere and Helsinki, Jean-Monnet Chair for European Integration, and a research fellow in the Ministry for Foreign Affairs, Helsinki. Auffermann is an expert in Futures Studies and Security research, with specific research interests in energy and climate policies. Email: burkhard.auffermann@utu.fi

Anne Bergmans holds a PhD in social sciences and a master's in international politics and maritime economics. She works at the University of Antwerp as senior research fellow in the Faculty of Political and Social Sciences, and as a lecturer and programme director in the Master of Safety Sciences at the Faculty of Law. Anne's work is situated in the fields of science, technology governance, and sociology of the environment, with a particular focus on public and stakeholder engagement in radioactive waste management. She has been involved in a number of international research projects in this field: CARL (2004-2007), MoDeRn (2009-2013), and InSOTEC (2011-2014).
Email: anne.bergmans@uantwerpen.be

Achim Brunnengräber is an adjunct professor (Privat-Dozent) in the Department of Political and Social Sciences, Freie Universität Berlin. He is the coordinator of the "Multi Level Governance-Perspective on Management of

Nuclear Waste Disposal" research project at the Environmental Policy Research Centre (FFU). Key teaching and research areas include: global and multi-level governance; international political economy; and international environmental, energy and climate policy. Email: achim.brunnengraeber@fu-berlin.de

Martin Bursik graduated in environmental studies at the Faculty of Natural Sciences at Charles University, Prague. He has served as a consultant for energy and environment. Bursik is a former Czech minister of the environment and deputy prime minister. During the Czech presidency of the European Council in 2009, he represented the EU at the international climate negotiations with major world economies. He implemented the Environmental Tax Reform and the Green Investment Scheme programme. Bursik founded and chaired the Czech Renewable Energy Chamber. He is a lecturer on environmental politics at New York University, Prague and on renewable energy sources of energy at the Environmental Faculty of the Agricultural University, Prague.
Email: martin.bursik@me.com

Maria Rosaria Di Nucci is a senior researcher at the Environmental Policy Research Centre (FFU) of the Freie Universität Berlin and advises – as an independent consultant – the European Commission and European research organisations. Di Nucci holds a master's degree in Economics from the University of Rome and a PhD from the University of Sussex. Formerly, she held an executive position at the Ministry of Environment of the Land Berlin and was a lecturer in Industrial Economics at the Berlin Technical University. Her research interests include: energy policy, in particular smart grid and smart metering, and comparative nuclear policy. She is a member of the "Multi-Level-Governance-Perspective on Management of Nuclear Waste Disposal" research project at the FFU. Email: dinucci@zedat.fu-berlin.de

Richard A. Forrest is a doctoral researcher at the Environmental Policy Research Centre of the Freie Universität Berlin, where he is preparing his dissertation on the role of NGOs in climate change advocacy. He received his bachelor's and master's in Asian Studies from the University of Michigan and a master's degree in Public Policy (environmental policy specialization) from the University of Maryland. Forrest has also had a long career in development, working primarily for NGOs in the United States and Japan to advance knowledge-sharing in support of sustainable development. Email: raforrest@gmail.com

Peter Hocke is a political scientist and a senior fellow at the Institute for Technology Assessment and Systems Analysis (ITAS) at the Karlsruhe Institute

for Technology (KIT). He studied political science, philosophy, and sociology at the University of Regensburg and the Freie Universität Berlin. In the 1990s, he was a research fellow at the Berlin Social Science Center (WZB). Since 2001, Hocke has been working at ITAS and since 2006 has been a member of the "Expert Group on the Swiss deep disposal" (ESchT) of the German Federal Ministry for the Environment. Areas of work include: technology assessment, final disposal as a technology conflict, nanotechnology, and mass media.
Email: peter.hocke@kit.edu

Ana María Isidoro Losada is a research fellow at the Environmental Policy Research Centre of the Freie Universität Berlin and cooperates in the "Multi-Level-Governance-Perspective on Management of Nuclear Waste Disposal" research project. Presently, she is preparing her doctoral thesis on democracy and power questions in the context of Venezuela. Her main research interests include: democracy, social inequalities, participation, spatial politics, Latin America, and multi-level-governance. Isidoro Losada holds a Diploma in Landscape Architecture and Planning; she studied Political Science, Sociology, and History at the Universities Hannover and Paris.
Email: isidoro.losada@fu-berlin.de

Tomas Kåberger is professor of Industrial Energy Policy at Chalmers University of Technology and holds visiting positions at Zhejiang University in China, Technische Universität München, and Lund University. He serves as the executive board chairman of the Japan Renewable Energy Foundation and on the board of investment foundations and energy technology companies in Sweden. He has served on several government commissions in Sweden and has been director general of the Swedish Energy Agency.
Email: tomas.kaberger@chalmers.se

Jari Kaivo-oja is the research director of the Finland Futures Research Centre (FFRC) at the Turku School of Economics at the University of Turku. He holds a Master of Social Sciences (international economics) and a doctorate of Administrative Sciences (regional planning) from the University of Tampere. He is an adjunct professor at the Faculty of Science at the University of Helsinki and at the Faculty of Social Sciences at the University of Lapland. His main research interests include: foresight research, energy and climate policy, low-carbon energy transitions, innovation research, the service economy, management and planning sciences (especially transition management), sustainable development, ubiquitous r/evolution, robotics, conflict management, and global development issues. Email: jari.kaivo-oja@utu.fi

Beate Kallenbach-Herbert is head of the Nuclear Engineering & Facility Safety division of the Oeko-Institute in Darmstadt, Germany. Her main research interests include the technological and governance-related aspects of nuclear waste management in the national and international context. She has served as a member of the German delegation in conferences on the "Joint Convention" on Nuclear Waste Management at the IAEA and is a member of the Nuclear Waste Management Commission (ESK) that advises the German Federal Ministry for the Environment, Nature Conservation, Building, and Nuclear Safety (BMUB) on nuclear waste matters. Email: b.kallenbach@oeko.de

Sophie Kuppler, holds a Master of Science and is a junior fellow and doctoral student at the Institute for Technology Assessment and Systems Analysis (ITAS) at the Karlsruhe Institute for Technology (KIT). She studied environmental sciences, as well as technical and socio-economic planning at the Technical University of Cottbus (Germany), the University of Roskilde (Denmark), and the Yarmouk University (Jordan). Kuppler has been at ITAS since 2009. Areas of work encompass: the effects of deliberative policy approaches, nuclear waste governance, governance of technology conflicts, public participation in decision-making, environmental movements, and policy instruments.
Email: sophie.kuppler@kit.edu

Erik Laes is a chemical engineer and holds additional degrees in philosophy and environmental science. In 2006, he obtained a PhD in engineering sciences. The topic of his doctoral study was the role of nuclear energy in a sustainable energy supply, with an emphasis on governance and the use of participative technology assessment methodologies as a decision support. In July 2009, he joined VITO in the 'Transition, Energy and Environment' unit, where he is currently involved in projects concerning energy policy analysis, sustainability evaluation of energy technologies, transition management, and smart grids/smart energy users.
Email: erik.laes@vito.be

Markku Lehtonen is a research fellow at the University of Sussex (SPRU – Science and Technology Policy Research, The Sussex Energy Group) and a visiting researcher at the Ecole des Hautes Etudes en Sciences Sociales (EHESS), Paris. He holds a Master of Science in environmental protection from the University of Helsinki (1994) and a PhD in environmental economics from the University of Versailles (2005). In 2012-2014, he conducted research at Ifris, Université Paris-Est Marne-la-Vallée on the role of evaluation of the socio-economic aspects of radioactive waste management in France. His other areas of

specialisation include governance and controversies in the areas of nuclear energy and biofuels, as well as the role of expertise in policymaking.
Email: m.lehtonen@sussex.ac.uk

Jyrki Luukkanen is the research director at the Finland Futures Research Centre (FFRC) at the University of Turku. He earned a Doctorate of Technology in systems theory in Tampere University of Technology (1994). Luukkanen is an adjunct professor of Futures Studies in University of Turku. He has directed more than 50 international and domestic research projects and has served as a consultant for international organisations. His research interests include: energy and climate policy, sustainable development indicators, modelling, material flow analysis, and development studies. Luukkanen has authored and co-authored a large number of publications on these issues. Email: jyrki.luukkanen@utu.fi

Gordon MacKerron was the director of SPRU from 2008 until the end of 2013 and is now a professor of Science and Technology Policy at the University of Sussex. He was previously the director of the Sussex Energy Group at SPRU from April 2005 to November 2008. He is an economist specialising in energy and environmental economics, with degrees in economics from the Universities of Cambridge and Sussex. MacKerron's academic career has focused on the economics and policy issues of electricity and especially nuclear power. He has published and broadcast widely in these fields. Between 2003 and 2007, MacKerron was the chair of the Committee on Radioactive Waste Management, an independent body charged with recommending the best approach to long-term radioactive waste management to the UK Government.
Email: gordon.mackerron@sussex.ac.uk

Lutz Mez is a senior adjunct professor (Privat-Dozent) of Political Science in the Department of Political and Social Sciences, Freie Universität Berlin. In April 1986, he co-founded the Environmental Policy Research Centre (FFU) and served as its executive director until April 2010. Since November 2009, he has been the coordinator of the Berlin Centre for Caspian Region Studies (BC CARE). In 1993/94, Mez was a visiting professor at the Department of Environment, Technology, and Social Studies at Roskilde University, Denmark. He holds a doctoral degree and a diploma in political science from Freie Universität Berlin. His major research area is comparative environmental and energy policy, with particular reference to nuclear and electricity policy.
Email: lutz.mez@fu-berlin.de

Miranda Schreurs is a professor of comparative politics and the director of the Environmental Policy Research Centre. She was previously an associate professor in the Department of Government and Politics, University of Maryland, College Park. Schreurs is member of the German Environment Advisory Council, the Berlin Climate Council, and the Enquete Commission New Energy for Berlin, and chair of the European Environment and Sustainable Development Councils. She was also a member of Ethics Commission for a Safe Energy Supply appointed by Chancellor Merkel after the Fukushima nuclear accident. Schreurs has a PhD from the University of Michigan and a MA and BA from the University of Washington. Her areas of specialty include comparative and international environment and energy policy and politics.
Email: miranda.schreurs@fu-berlin.de

Jantine Schröder holds a Master in Philosophy and additional degrees in International Relations & Conflict Prevention and Sustainable Development & Human Ecology. Since 2008, she has worked at the Belgian Nuclear Research Centre on the Programme of Integration of Social Aspects into Nuclear Research. From 2011 to 2014 she also worked within the Society and Environment research team at the University of Antwerp; the team coordinated InSOTEC, a European project on the socio-technical challenges related to geological disposal of radioactive waste. Her research is thematically focused on radioactive waste management and energy governance; it is theoretically based in science and technology studies, philosophy of science, and technology and sustainable development studies. Schröder's work is methodologically driven by approaches that enable interaction between disciplines, interests, perspectives, and fundamental and applied research.
Email: jantine.schroder@sckcen.be

Pertti Suomela is a researcher at the Finland Futures Research Centre (FFRC) at the University of Turku. He holds a Bachelor's degree in Administrative Sciences (Politics of the Environment and the Regions) from the University of Tampere and is currently working on his Master's thesis. Previously, Suomela worked in the ICT and mobile phone industries. His research interests are focused on climate change and other global environmental problems.
Email: pertti.suomela@gmail.com

Johan Swahn is the director of the Swedish NGO Office for Nuclear Waste Review, MKG. He leads the organisation's work in reviewing the Swedish nuclear industry's licence application for a permit for a Swedish final repository for spent nuclear fuel. Swahn has a Master of Science in Engineering Physics and a doctorate in Science, Technology, and Global Security from the Chalmers University of Technology, Göteborg Sweden. He is a member of the International Panel on Fissile Material, IPFM. Email: johan.swahn@mkg.se

Jarmo Vehmas holds a PhD in administrative sciences and Bachelor of Science in Power Engineering. Currently, he works as Regional Manager at the Finland Futures Research Centre (FFRC), the Turku School of Economics, and the University of Turku. He is also an Adjunct Professor (Docent) of Environmental Policy at the Faculty of Biological and Environmental Sciences, University of Helsinki. Vehmas has coordinated the EU FP7 research project "Synergies research project in Multi-scale Inter-linkages of Eco-social Systems" (2008-2011) and participated in many international and national research projects, on topics such as energy and climate policies and scenarios, the assessment of different dimensions of sustainable development, and eco-innovations.
Email: jarmo.vehmas@utu.fi

CPSIA information can be obtained at www.ICGtesting.com
Printed in the USA
LVOW11s1748070415

433633LV00001B/149/P

9 783658 089610